Introduction to Logistics Systems Management

Wiley Series in

Operations Research and Management Science

Operations Research and Management Science (ORMS) is a broad, interdisciplinary branch of applied mathematics concerned with improving the quality of decisions and processes and is a major component of the global modern movement towards the use of advanced analytics in industry and scientific research. *The Wiley Series in Operations Research and Management Science* features a broad collection of books that meet the varied needs of researchers, practitioners, policy makers, and students who use or need to improve their use of analytics. Reflecting the wide range of current research within the ORMS community, the Series encompasses application, methodology, and theory and provides coverage of both classical and cutting edge ORMS concepts and developments. Written by recognized international experts in the field, this collection is appropriate for students as well as professionals from private and public sectors including industry, government, and nonprofit organization who are interested in ORMS at a technical level. The Series is comprised of three sections: Decision and Risk Analysis; Optimization Models; and Stochastic Models.

Decision and Risk Analysis

Forthcoming Titles

Barron • *Game Theory: An Introduction*, Second Edition

Introduction to Logistics Systems Management

Second Edition

Gianpaolo Ghiani

Department of Innovation Engineering, University of Salento, Italy

Gilbert Laporte

HEC Montréal, Canada

Roberto Musmanno

Department of Electronics, Informatics and Systems
University of Calabria, Italy

A John Wiley & Sons, Ltd., Publication

This edition first published 2013

Registered office
John Wiley & Sons Ltd, The Atrium, Southern Gate, Chichester, West Sussex, PO19 8SQ, United Kingdom

For details of our global editorial offices, for customer services and for information about how to apply for permission to reuse the copyright material in this book please see our website at www.wiley.com.

Library of Congress Cataloging-in-Publication Data

Ghiani, Gianpaolo.
[Introduction to logistics systems planning and control]
Introduction to logistics systems management / Gianpaolo Ghiani, Gilbert Laporte,
Roberto Musmanno. – Second edition.
pages cm
Includes index.
ISBN 978-1-119-94338-9 (hardback)
1. Materials management. 2. Materials handling. 3. Business logistics. I. Laporte, Gilbert. II. Musmanno, Roberto. III. Title.
TS161.G47 2013
658.5–dc23

2012033914

A catalogue record for this book is available from the British Library.

ISBN: 978-1-119-94338-9

Set in 10/12pt Times-Roman by Laserwords Private Limited, Chennai, India

To Laura, Allegra and Vittoria
To Ann and Cathy
To Maria Carmela, Francesco and Andrea

Contents

Foreword

Logistics is concerned with the organization, movement and storage of material and people. The term 'logistics' was first used by the military to describe the activities associated with maintaining a fighting force in the field and, in its narrowest sense, describes the housing of troops. Over the years the meaning of the term has gradually generalized to cover business and service activities. The domain of logistics activities is providing a system's customers with the right product, in the right place, at the right time. This ranges from providing the necessary subcomponents for manufacturing, to having inventory on the shelf of a retailer, to having the right amount and type of blood available for hospital surgeries. A fundamental characteristic of logistics is its holistic, integrated view of all the activities that it encompasses. So, while procurement, warehouse management and distribution are all important components, logistics is concerned with the integration of these and other activities to provide the time and space value to the system or corporation. Excess global capacity in most types of industry has generated intense competition. At the same time, the availability of alternative products has created a very demanding type of customer, who insists on the instantaneous availability of a continuous stream of new models. So the providers of logistics activities are asked to do more transactions, in smaller quantities, in less time, for less cost and with greater accuracy. New trends such as mass customization will only intensify these demands. The accelerated pace and greater scope of logistics operations have made planning-as-usual impossible. Even with the increased number and speed of activities, the annual expenses associated with logistics activities in the United States have held constant for the past several years, around 10% of the Gross Domestic Product. Given the significant amounts of money involved and the increased operational requirements, the management of logistics systems has gained widespread attention from practitioners and academic researchers alike. To maximize the value in a logistics system, a large variety of planning decisions has to be made, ranging from the simple warehouse-floor choice of which item to pick next to fulfil a customer order to the corporate-level decision to build a new manufacturing plant. Logistics management supports the full range of those decisions related to the design and operation of logistics systems.

There exists a vast amount of literature, software packages, decision support tools and design algorithms that focus on isolated components of the logistics system or isolated planning in the logistics systems. In the last two decades, several companies have developed Enterprise Resource Planning (ERP) systems in response to the need of global corporations to plan their entire supply chain. In their initial implementations, the ERP systems were primarily used for the recording of transactions rather than for the planning of resources on an enterprise-wide scale. Their main advantage was to provide consistent, up-to-date and accessible data to the enterprise. In recent years, the original ERP systems have been extended with Advanced Planning Systems (APSs). The main function of APSs is, for the first time, the planning of enterprise-wide resources and actions. This implies a coordination of the plans among several organizations and geographically dispersed locations.

So, while logistics management requires an integrated, holistic approach, its treatment in courses and textbooks tends to be either integrated and qualitative or mathematical and very specific. This book bridges the gap between those two approaches. It provides a comprehensive and modelling-based treatment of the logistics processes. The major components of logistics systems (storage and distribution) are each examined in detail. For each topic the problem is defined, models and solution algorithms are presented that support computer-assisted decision making and numerous application examples are provided. Each chapter concludes with case studies that illustrate the application of the models and algorithms in practice. Because of its rigorous mathematical treatment of real-world management problems in logistics, this book will provide a valuable resource to graduate and senior undergraduate students and practitioners who are trying to improve logistics operations and satisfy their customers.

Marc Goetschalckx
Georgia Institute of Technology
Atlanta, United States

Preface

Logistics lies at the heart of modern economies. From the steel factories of Pennsylvania to the port of Singapore, from Nicaraguan banana fields to the postal delivery and solid waste collection companies in any region around the world, almost every organization faces the common problem of getting the right materials to the right place at the right time. Indeed, the fierce competition in today's markets has made it imperative to manage logistics systems more and more efficiently. In this context, quantitative methods have proved able to achieve remarkable savings.

This textbook has grown from a number of undergraduate and graduate courses on logistics that we have taught to engineering, computer science and management science students. The goal of these courses is to give students a solid understanding of the analytical tools necessary to reduce costs and improve the service level in logistics systems. The lack of a suitable textbook had forced us in the past to make use of a number of monographs and scientific papers which tend to be beyond the level of most students. We therefore committed ourselves to developing a quantitative textbook written at a level accessible to most students.

In 2004 we published with Wiley a book entitled *Introduction to Logistics Systems Planning and Control*, which was widely used in several universities throughout the world. The 2004 edition of the book received the Roger-Charbonneau award from HEC Montréal as the best pedagogical textbook of the year. In view of this success, we proceeded to prepare a substantially revised edition now entitled *Introduction to Logistics Systems Management*.

This new edition puts more emphasis on the organizational context in which logistics systems operate. It also covers several new results and techniques that have arisen in the field of logistics over the past decade. This book targets an academic as well a practitioner audience. On the academic side, it should be appropriate for advanced undergraduate and graduate courses in logistics and supply chain management. It should also serve as a methodological reference for practitioners in consulting as well as in industry. We make the assumption that the reader is familiar with the basics of operations research and statistics, and we provide a balanced treatment of forecasting, logistics system design, warehouse management and freight transport management. In our text, every topic is illustrated by a numerical example so that the reader can check his or her understanding of each concept before moving on to the next one. We provide at the end of each chapter case studies taken from the scientific literature, which illustrate the use of quantitative methods for solving complex logistics decision problems. Finally, every chapter ends with an exhaustive set of exercises.

Gianpaolo Ghiani (gianpaolo.ghiani@unisalento.it)
Gilbert Laporte (gilbert.laporte@cirrelt.ca)
Roberto Musmanno (musmanno@unical.it)

Exercises and Website

This textbook contains questions and problems at the end of every chapter. Some are discussion questions, while others focus on modelling or algorithmic issues. The answers to these problems are available at the book's website (`http://www.wiley.com/go/logistics_systems_management`), which also contains additional material (FAQs, a list of references, software, further modelling exercises, links to other websites etc.).

Acknowledgements

We wish to acknowledge all the individuals who have helped in one way or another to produce this text. First, we are grateful to the reviewers whose comments were invaluable in improving the organization and presentation of the book. Similarly, we are indebted to Valentina Caputi, Antonio Igor Cosma, Lucie-Nathalie Cournoyer, Emanuela Guerriero, Rosita Guido, Demetrio Laganà, Emanuele Manni, Francesco Mari, Marco Pina, Ornella Pisacane, Claudia Rotella, Francesco Santoro and Antonio Violi for their scientific and technical support.

About the Authors

Gianpaolo Ghiani is Professor of Operations Research at the University of Salento, Italy, where he teaches Logistics courses at the Faculty of Engineering.

Gilbert Laporte is Professor of Operations Research at HEC Montréal, Canada Research Chair in Distribution Management, adjunct Professor at Molde University College, Norway; the University of Bilkent, Turkey; and the University of Alberta, Canada; and visiting professor at the University of Southampton, United Kingdom.

Roberto Musmanno is professor of Operations Research at the University of Calabria, Italy, where he teaches Logistics courses at the Faculty of Engineering.

List of Abbreviations

1-BP	One Bin Packing
2-BP	Two Bin Packing
3-BP	Three Bin Packing
3PL	Third Party Logistics
AGVS	Automated Guided Vehicle System
AHP	Analytical Hierarchic Process
AIRT	Inter-Regional Association of Transplantation
ANN	Artificial Neural Network
APS	Advanced Planning System
AP	Assignment Problem
ARP	Arc Routing Problem
AS	Air-Stop
AS/RS	Automated Storage/Retrieval System
ASM	Absolute Scoring Method
ATSP	Asymmetric Travelling Salesman Problem
B2B	Business-to-Business
BF	Best Fit
BFD	Best Fit Decreasing
BL	Bottom Left
BTT	Baxter Transfusion Therapy
CDC	Central Distribution Centre
CEO	Chief Executive Officer
CIR	Inter-Regional Centre
CL	Carload
COT	Cut-Off Time
CPFR	Collaborative Forecasting and Replenishment Program
CPL	Capacitated Plant Location
CPP	Chinese Postman Problem
CRP	Continuous Replenishment Program
DC	Distribution Centre
DDAP	Dynamic Driver Assignment Problem
EAN	European Article Number
EDI	Electronic Data Interchange
ELC	European Logistics Centre

EOQ	Economic Order Quantity
ERP	Enterprise Resource Planning
EU	European Union
EUPP	European Union Pfizer Plant
FBF	Finite Best Fit
FCNDP	Fixed Charge Network Design Problem
FDA	Food and Drugs Administration
FF	First Fit
FFD	First Fit Decreasing
FFF	Finite First Fit
FIFO	First-In, First-Out
GIS	Geographic Information System
GOD	Great Organised Distribution
GPS	Global Positioning System
ICT	Information and Communication Technology
IP	Intger Programming
IRP	Inventory-Routing Problem
ISO	International Organization for Standardisation
KPI	Key Performance Indicator
LB	Lower Bound
LFCNDP	Linear Fixed Charge Network Design Problem
LIFO	Last-In, First-Out
LMCFP	Linear Minimum Cost Flow Problem
LMMCFP	Linear Multicommodity Minimum Cost Flow Problem
LP	Linear Programming
LTL	Less-Than-Truckload
MAD	Mean Absolute Deviation
MAPD	Mean Absolute Percentage Deviation
MCTE	Multiple-Commodity Two-Echelon
MIP	Mixed Integer Programming
MMCFP	Multicommodity Minimum Cost Flow Problem
MRP	Material Requirement Planning
MS	Multiple Source
MSE	Mean Squared Error
MSrTP	Minimum spanning r-tree problem
MTO	Make to Order
MTS	Make to Stock
NAPM	National Association of Purchasing Managers
NCC	National Classification Committee
NF	Network Flow
NITp	Nort-Italy Transplantation
NMFC	National Motor Freight Classification
NP	Non-Polynomial
NRP	Node Routing Problem

NRPCL	Node Routing Problem with Capacity and Length Constraints
NRPSC	Node Routing Problem – Set Covering
NRPSP	Node Routing Problem – Set Partitioning
NRSPTW	Node Routing and Scheduling Problem with Time Windows
NTC	National Transplant Centre
OCST	Centre-South Organization of Transplantation
PCB	Printed Circuit Board
PZ	Pick Zone
RDC	Regional Distribution Centre
RFId	Radio Frequency Identification
R-LMMCFP	Relaxed Linear Multicommodity Minimum Cost Flow Problem
RPP	Rural Postman Problem
RSM	Relative Scoring Method
RTSP	Road Travelling Salesman Problem
RZ	Reserve Zone
SC	Shipment Centre (Section 6.12)
SC	Set Covering
SCSE	Single-Commodity Single-Echelon
SCTE	Single-Commodity Two-Echelon
SNDP	Service Network Design Problem
SPL	Simple Plant Location
SQI	Supplier Quality Index
SS	Single Source
SSCC	Serial Shipping Container Code
STSP	Symmetric Travelling Salesman Problem
TAP	Traffic Assignment Problem
TL	Truckload
TS	Tabu Search
TSP	Travelling Salesman Problem
UB	Upper Bound
VAP	Vehicle Allocation Problem
VRDP	Vehicle Routing and Dispatching Problem
VRP	Vehicle Routing Problem
VRPMT	Vehicle Routing Problem with Multiple Trips
VRSP	Vehicle Routing and Scheduling Problems
ZIO	Zero Inventory Ordering

1

Introducing logistics

1.1 Definition of logistics

According to a widespread definition, logistics (from the Greek term *lógos*, which means 'order', or from the French *loger*, which means 'allocate') is the discipline that studies the functional activities determining the flow of materials (and of the relative information) in a company, from their origin at the suppliers up to delivery of the finished products to the customers and to the post-sales service. The origins of logistics are of a strictly military nature. In fact, this discipline arose as the study of the methodologies employed to guarantee the correct supply of troops with victuals, ammunitions and fuel and, in general, to ensure armies the possibility of moving and fighting in the most efficient conditions. Indeed it was the Babylonians, in the distant 20th century BC, who first created a military corps specialized in the supply, storage, transport and distribution of soldiers' equipment. Logistics was applied exclusively in a military context until the end of Second World War. Subsequently, it was extended to manufacturing companies in order to determine all the activities aimed at ensuring the correct purchasing, moving and managing of materials. Logistics problems are also increasingly present in the service sector, for example in the distribution of some services such as water and gas, in postal services, in urban solid waste collection, in the maintenance of road and electricity networks and in the post-sales activities of manufacturing companies (*service logistics*).

1.2 Logistics systems

From the point of view of companies, logistics is seen as a system (the logistics system), which includes not only all the functional activities determining the flow

---

Introduction to Logistics Systems Management, Second Edition. Gianpaolo Ghiani, Gilbert Laporte and Roberto Musmanno.

of materials and information, but also the infrastructures, means, equipment and resources that are indispensable to the execution of these activities.

A logistics system is made up of facilities, where one or more functional activities are carried out (e.g. storage and distribution). Figure 1.1 shows a schematic representation of a logistics system in which the manufacturing process of the finished goods is divided into a transformation phase and an assembly phase, performed in different centres. At the start are the suppliers of materials and components which feed the final manufacturing process. The end part represents a typical two-level distribution system with a tree structure. The Central Distribution Centres (CDCs) are directly supplied by the production plants, while each Regional Distribution Centre (RDC) is connected to a single CDC which has the task of serving the customer, who can also be dealers or retailers.

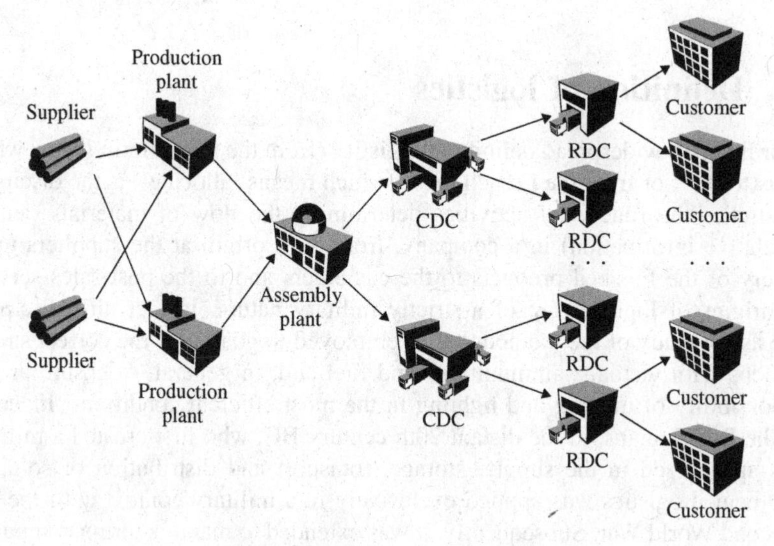

Figure 1.1 Example of a logistics system.

At each facility the flow of materials is temporarily interrupted, generally in order to change their physical-chemical composition, ownership or appearance. In all cases, each logistic activity carried out involves costs which affect the value of the product, constantly adding to it as it draws nearer the facilities closest to the final customer. This added value can be spatial (following e.g. distribution activities) or temporal (owing to storage activities).

Galbani is the Italian leader in the milk and dairy products sector and one of the main actors in the pressed pork market. The Galbani group is currently made up of three independent operational societies, one of them called biG Logistics. This company has the task of managing the logistics activities of the whole group. The logistics system is organized in such a way as to guarantee an efficient synchronization of the internal production and distribution processes of the products, both for the Great Organized Distribution (GOD) and for the channel represented by the traditional retail shops. The distribution network of the company is organized on two levels: there are, between the production plants and the destination markets, a central warehouse and 11 distribution platforms. This solution allows the minimization of the transport times and of the storage times of the goods in the warehouses. As a result, it favours a rapid delivery of the products, strictly respecting the refrigeration chain (deliveries within 12 hours for national distribution and 24 to 36 hours for abroad). The daily products are dispatched directly by the production plants to the central warehouse, located in the area of Ospedaletto Lodigiano, considered a barycentral position with respect to the national markets of the Galbani group. The central warehouse serves, in turn, the second-level platforms with the orders mixed according to their destination (see Figure 1.2). The platforms receive the entering flow of goods from the central warehouse and supply both the Distribution Centres of the GOD and the so-called satellites. The satellites are small-sized stores without stockpiles, with vans used for distribution to retailers. The van operates as a truly travelling store. There are 111 satellites distributed throughout the national territory, with a coverage radius on the provincial scale.

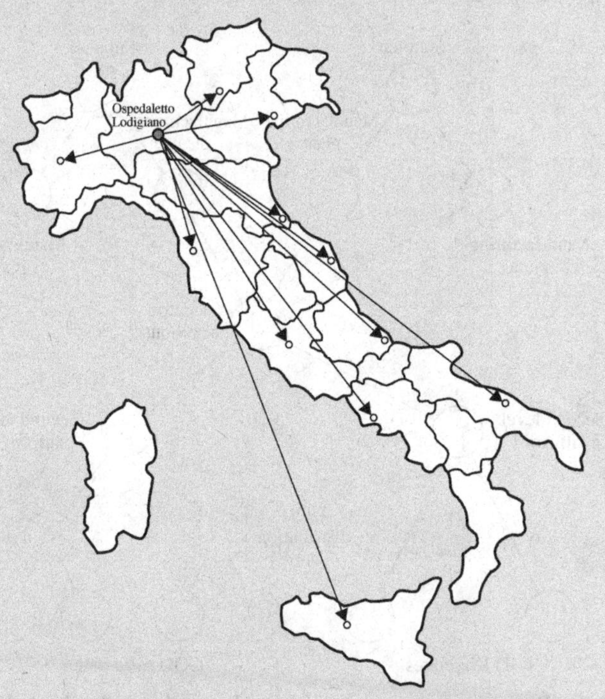

Figure 1.2 Geographical position of the central warehouse (in grey) and the 11 distribution platforms of the logistics system of Galbani.

A logistics system can be represented by means of a directed (multi)graph $G = (V, A)$, where V is the set of facilities, and A is the set of links existing among the facilities used for the flow of materials (see Figure 1.3). There can be several arcs between a pair of facilities, representative of alternative forms of transport services, different routes, or different products.

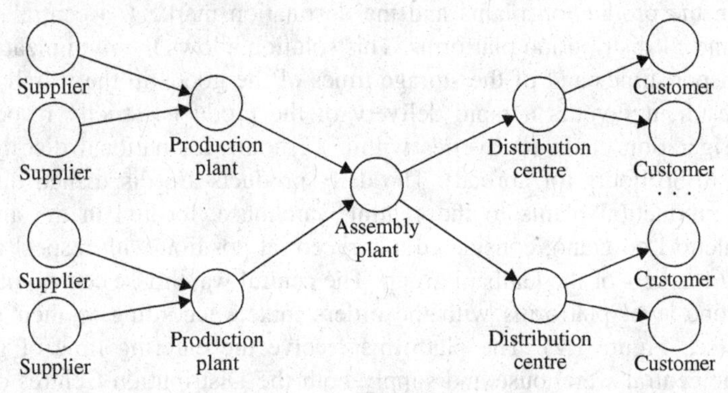

Figure 1.3 Representation of a logistics system system by a directed graph.

The logistics system of Galbani can be represented by the directed graph of Figure 1.4.

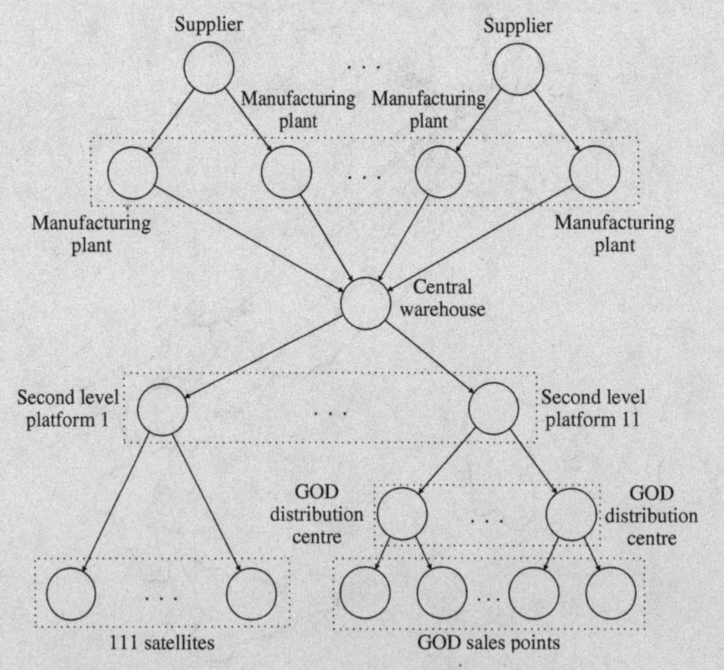

Figure 1.4 Representation by a directed graph of the logistics system of Galbani.

1.2.1 Logistics activities

Logistics activities are traditionally classified depending on their location with respect to the production and distribution process. In particular, *supply logistics* is carried out before the production plants and consists in the management of raw materials, materials and component parts supply as a function of the company's production plan. *Internal logistics* is carried out in the production plants and consists in receiving and storing materials, in picking them up from the warehouse to feed the production lines and in successively moving the semi-finished goods up to packaging and storing the finished product. Finally, the *distribution logistics* activities are carried out after the production plants and before the market. They supply the sales points or the customers. In this schematization, the supply logistics and the distribution logistics are collectively called *external logistics*.

Storage and *distribution* of the finished products are the primary logistics activities, and particular attention will be paid to them in this text. Logistics activities can be conducted by the company itself or can be entrusted to a third party (3PL, or *Third Party Logistics*). These choices are made by the company according to the same logic on which 'make or buy'-type decisions are based. They assume an in-depth knowledge of the nature of the costs that the company bears (fixed costs, variable costs, direct costs, and indirect costs).

When the multinational company Gillette decided to reorganize its logistics system in Turkey in 1999, it entrusted the Exel company with the execution of a series of logistics activities on its behalf, including distribution (both at national and international levels), customs issues, storage of finished products, and repackaging and labelling of the products.

1.2.2 Information flows and logistics networks

Within a logistics system, with the exception of the cases where recycling of product wrapping is provided or where defective components or products are returned, the flow of materials typically moves from the suppliers to the processing and assembly plants, thence to the sales points and finally to the customers. The flow of materials is integrated with an information flow which follows the opposite direction: in the logistics systems of the MTO-type (*Make to Order*), for example, customers' orders influence the production plan and the latter determines the demand for materials and components of the processing and assembly plants. Analogously, in MTS-type (*Make to Stock*) logistics systems, market information (demand recorded in the past, results of possible market surveys etc.) is used to forecast the sales and therefore affects the mode of distribution, as well as the production and supply plans.

The production centres of the Galbani group determine the daily production plan on the basis of recorded predictive data, among others, from the volumes distributed the previous working day. In this case, the 11 logistics platforms gather the sales data recorded both at the GOD and at the satellites, and this information is transmitted to the central warehouse and from there to the production plants.

Hellena is a Dutch company that produces biscuits. The ordering of raw wheat flour is done by fax. Once an order is received and the goods prepared, the supplier provides for the dispatch of the product which must be accompanied by the delivery note. This document must therefore be issued before delivery or dispatch of the goods with specifications of the main elements of the operation (serial number, date, quantity and description of the goods transported etc.) and issued with a minimum of two copies (one must be retained and filed by the supplier and the other must be consigned to the customer together with the transported goods). In this context, two information flows are activated. The first, relative to the sending of the fax, travels in the opposite direction from the transport of the order (from the customer to the supplier) and also uses a different channel. The second, the delivery note, accompanies the consignment, using the same channel as the goods and travelling in the same direction.

An existing information flow (created through fax, email etc.) between a pair of facilities can be represented by an arc. This means that the information network, that is, the set of information flows, can also be represented by a directed (multi)graph, analogously to the network of materials flow.

Networks of materials flow and information networks give rise to *logistics networks* (see Figure 1.5).

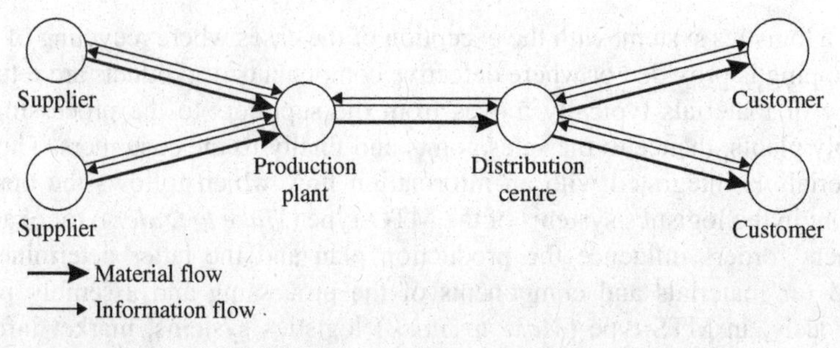

Figure 1.5 Representation of a logistics network.

1.2.3 Case of more products

When a company has to simultaneously handle several products, the logistics system inevitably becomes more complex. It is convenient to group the products into *classes* of different importance, so that the logistics activities can be organized for each class of product and not for each single product.

The *ABC* classification, which is the most widespread method for this purpose, allows the goods to be subdivided into three classes, called *A*, *B* and *C*, on the basis of the value of the products. The value of the products is typically measurable by means of the revenue generated by them in a reference time span (e.g. a year). In this way, class *A* is made up of the set of products achieving a corresponding high percentage (e.g. 80%) of the overall annual revenue. Class *B* is made up of the set of articles associated with the following 15% of the revenue, whereas class *C* is made up of the remaining articles. The classification is achieved by ordering the list of products in non-increasing values, and successively selecting the articles in the resulting order, up to a predetermined cumulated value.

On the basis of the so-called 80–20 principle, or Pareto principle, founded on the observation that, in 19th-century Italy, 20% of the population possessed 80% of the wealth, class *A* will account for a modest fraction of the products. In contrast, class *C*, despite affecting overall revenue only slightly, could be made up of numerous products.

This observation suggests different operating modes with regard to logistics. For example, it is convenient to adopt a distribution over large geographical areas for products of class *A*, using more CDCs and RDCs, with high stock levels of the products. On the other hand, the distribution of class *C* products can be made by using a single CDC and reducing the stocked quantity of products to a minimum.

Blucker is the owner of an Irish plant manufacturing products for the building industry. The Cork warehouse is for the storage and distribution to wholesalers of products in the water-based dispersion adhesives category. There are a total of 15 products. Information about the revenue and the annual amounts sold to wholesalers is provided in Table 1.1. The *ABC* (80–15–5) classification of the products by annual revenue can be deduced from Table 1.2, obtained by ordering the products on the basis of non-increasing revenue value. Class *A* products make up 78.42% of the annual revenue, whereas they represent only 40.66% of the overall weight of products sold. Class *B* products represent 15.15% of the annual revenue and have a relative weight of 32.22%, whereas class *C* products make up only 6.43% of annual revenue; the weight of these products is equal to 27.12% of the overall weight of all products sold in the year. The cumulative percentages of the annual

Table 1.1 Annual revenue and amounts sold of Blucker products.

ID	Article	Revenue [€]	Quantity [kg]
1	FIL12	324 764	38 614
2	BG1	109 000	33 452
3	BG2	959 800	24 522
4	BG3	86 540	25 545
5	P	341 280	24 767
6	TX	156 984	19 768
7	K0	762 250	32 234
8	K1	119 150	17 669
9	K2	51 206	22 600
10	K3	80 596	32 574
11	P-L1	144 625	30 578
12	P-L2	553 600	31 400
13	P-L3	35 608	33 560
14	P-L4	123 720	18 768
15	P-L5	102 300	35 287

Table 1.2 *ABC* classification of Blucker products.

ID	Article	Annual revenue (€)	Annual amounts sold (kg)	Cumulated annual amounts sold (%)	Cumulated annual revenue (%)	Class
3	BG2	959 800	24 522	5.82	24.29	*A*
7	K0	762 250	32 234	13.47	43.58	*A*
12	P-L2	553 600	31 400	20.92	57.59	*A*
5	P	341 280	24 767	26.80	66.23	*A*
1	FIL12	324 764	38 614	35.97	74.45	*A*
6	TX	156 984	19 768	40.66	78.42	*A*
11	P-L1	144 625	30 578	47.91	82.08	*B*
14	P-L4	123 720	18 768	52.37	85.21	*B*
8	K1	119 150	17 669	56.56	88.23	*B*
2	BG1	109 000	33 452	64.50	90.98	*B*
15	P-L5	102 300	35 287	72.88	93.57	*B*
4	BG3	86 540	25 545	78.94	95.76	*C*
10	K3	80 596	32 574	86.67	97.80	*C*
9	K2	51 206	22 600	92.03	99.10	*C*
13	P-L3	35 608	33 560	100.00	100.00	*C*
	Total	3 951 423	421 338			

amounts sold and of the annual revenue for each of the 15 products are plotted in Figure 1.6. The same figure exhibits the 80–20 curve of equation $y = [(1 + \alpha)x] / (\alpha + x)$, which best fits the plotted values (y is the cumulative percentage of the annual revenue, x is the cumulative percentage of the amounts sold and $\alpha = 0.238$ is obtained by using the least squares method, see Exercise 1.5).

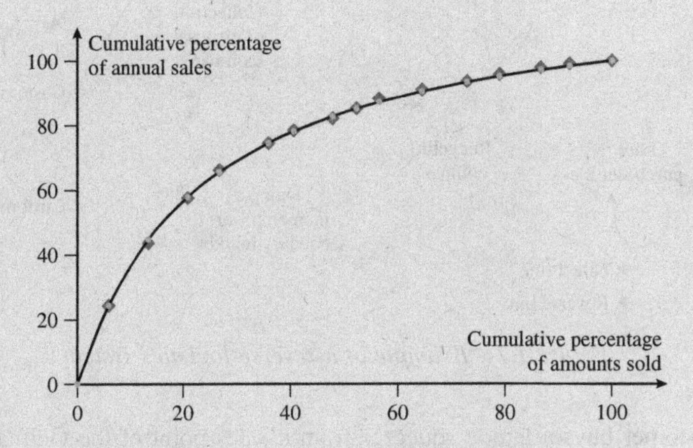

Figure 1.6 80–20 curve for Blucker products.

1.3 Reverse logistics

The life cycle of a product does not finish with its delivery to the end consumer. In fact, it is possible that the product can become obsolete, damaged or otherwise nonfunctional, and must therefore be discarded or sent back to its origin for possible repair.

Reverse logistics is the sector of logistics dealing with product flows (unsold items or returns) from their final destination to the initial producer, or to a facility dedicated to their treatment. Examples of reverse logistics' functional activities include control in the facilities to avoid the unjustified return of products which are only apparently nonfunctional, recovering and collecting unsold items, transporting returns in dumps or disposal centres, or operating in secondary markets. A possible schematization of the flows of materials in a logistics system both direct and reverse is shown in Figure 1.7.

As can be seen from the figure, to effectively and efficiently manage the flows of materials and related information from the point of consumption to the point of origin, reverse logistics may require connections of the original network and the use of specific reverse links. This kind of approach is oriented toward the possibility of regaining value from products that have exhausted their life cycle.

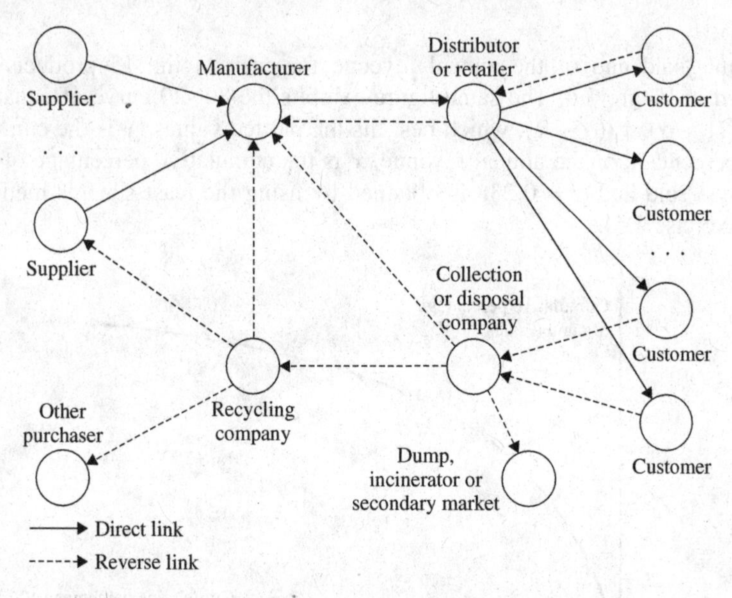

Figure 1.7 Example of a reverse logistics system.

A customer buys a lemon squeezer from a sales point of the German chain MediaMarkt which he subsequently finds to be defective. The customer takes it back to the sales point, which verifies the real defect of the lemon squeezer and then substitutes it with a new, functioning one. The retailer returns the defective lemon squeezer to the appropriate collection centre. This centre credits the sales point with a lemon squeezer, and therefore a debit to the manufacturer is created. The lemon squeezer is sent to the manufacturer who repairs it and sends it to a secondary market. In this way, the manufacturer obtains an added value on the defective product.

1.4 Integrated logistics

Until now logistics has been discussed as an operational tool within a company. It is indeed in this direction that many companies have operated until the 1980s. At that time, however, the increase in competition at all levels (raw materials, finished products, high consumption goods, capital equipment etc.) and a marked reduction in product lifespan have translated into a greater flexibility for the companies, that is, into an increased capacity to adapt more rapidly to the needs of a market in evolution. It is in this economic context that *integrated logistics* arose, that is, the coordinated management, according to a systemic vision, of the logistics activities of different companies involved in the management of the materials and information flows, with the aim of maximizing the overall profitability.

The management of an integrated logistics system yields an increase of the relationships not just among the various functions of a company (marketing, production and logistics) but also in the interactions among different partner companies, which yields an overall competitive advantage.

Integrated logistics can be realized in two different alternative forms.

The first case (*efficiency approach*) relies on so-called *intelligent relations*, that is, on the stipulation of contracts of a strictly operative nature that do not modify the company's own strategies but tend to speed up exchanges with the partners and lead to a reduction of waste and of activities that do not provide an added value.

In the second case (*differentiation approach*), the company tries to forge exclusive alliances with the partners, thus generating unique and privileged relationships that are not replicable by competitors and generate an added value with respect to the competition.

A case of intelligent relations is that of Calzedonia, which uses an ERP software. The system is available on a web portal for the person in charge of purchasing, and provides all the information about orders in real time. When the company decided to adopt this system, it notified its suppliers that orders would no longer be placed by fax, but rather by email. All the confirmations or changes to orders made by email are integrated within the ERP software of the company and flow together into the portal. Thanks to this software, the users of the purchasing office can dedicate their time to activities that generate added value (identify the best prices, create a climate of mutual interests with the suppliers, coordinate themselves etc.) and concentrate on problems of delays to supplies. The suppliers feel more controlled (the delivery date of materials is a precise date) and also more responsible with regard to respecting deadlines. All this leads to a better planning of production by the company and, in general, to a greater competitiveness of the integrated logistics system.

An example of a differentiation strategy is the alliance between Nokia, a Finnish company specializing in the manufacture of cellular telephones, and Yahoo!, an American company supplying broadband services, electronic mail etc. The agreement is not only about mutual support and market share increase but also a challenge to rival companies, such as Apple and Google, which offer wireless services. By making available the integration of its OVI Maps, Nokia has become an exclusive global supplier for the navigation services and maps of Yahoo! On the other hand, Yahoo! made its own messaging technology available to Nokia exclusively. Yahoo! has therefore become the official supplier of Ovi Mail and Ovi Chat. To guarantee a maximum quality service for its own customers, the two companies benefit from advantages relative to their respective global distribution structures and the joint strength of their own brands.

1.5 Objectives of logistics

In this chapter, logistics has been introduced without mention of the objectives to be pursued. These can be characterized by three variables: costs, profits and service level.

Costs. The costs of logistics activities are the financial resources consumed by the company when carrying out these activities. They are divided into fixed and variable costs. The main cost categories of a logistics system are summarized in Table 1.3.

Table 1.3 Main cost categories of a logistics system.

Main cost categories	Fixed costs	Variable costs
Storage costs	Administrative costs Running costs of storage centres	Insurance policies Financial burdens and opportunities costs Deterioration costs Obsolescence costs
Operational management costs	Administrative costs of issuing and computing orders	Loading and unloading goods costs Movement costs Stock control and management costs Packaging costs Deferment of takings Forfeits
Stockout costs		Lost sale Loss of customer Loss of image
Transport costs	Devaluation of means of transport Rental of means of transport	Insurance costs Variable transport costs
Plant and equipment costs	Rates of plant devaluation	Rental fees (variable according to volume)

Profits. Logistics activities affect company profits, even though, contrary to costs, the impact of logistics operations on sales is difficult to quantify. For this reason, the sole objective of profit maximization is not very practical from the point of view of logistics.

Service level. The service level encompasses the overall degree of customer satisfaction and depends on numerous factors (indicated collectively as *marketing* *mix*), connected to the product characteristics, price, promotional offers and mode of distribution. It is possible to quantify the service level, as will be shown in this chapter, by using suitable indicators.

Company profits are directly connected to the service level offered to customers. For example, it has been experimentally tested that an effective and efficient organization of the distribution service yields a direct increase in market share.

One possible objective of logistics is to minimize costs in a reference time horizon (e.g. a year), while keeping the service level unchanged. Alternatively, the objective could be to determine the optimal service level for maximum profit (difference between revenues and costs) in a reference period. In general, the maximization of profit is obtained for high (but less than maximum) values of service level.

Ecopaper is a Turkish company producing various kinds of paper (glossy, chemical, adhesive etc.) as well as derivative products (e.g. gift-wrapping paper). It distributes them to various kinds of customers (specialized shops, GOD, advertising agencies etc.). The market in which it operates is highly competitive, and delivery time is seen to be a key factor for the company's success.

Ecopaper can act on the distribution system, in particular on the number and location of the distribution centres, guaranteeing different delivery times. The logistician has made available the following estimates about the annual costs of distribution and the relative sales, as a function of different levels of logistic service (see Table 1.4).

Table 1.4 Annual estimate of sales, costs and profits (in M€) of Ecopaper.

	Orders dispatched within 3 days [%]				
	60	70	80	90	95
Sales	4.00	5.00	7.00	9.00	10.50
Costs	1.80	3.00	3.50	6.00	7.10
Profits	2.20	2.00	3.50	3.00	3.40

As can be seen from the table, organizing the distribution system so as to guarantee 80% of deliveries within three days of issuing an order leads to a maximization of profits. This is due to the savings that can be generated by optimizing the number of distribution centres.

1.5.1 Measures of the service level

The most widespread way of measuring the service level of a logistics system is through the quantification of the *order-cycle time*, defined as the time interval from the issuing of an order (or request of service) to the delivery of the product (or completion of the service). The main logistical components of the order-cycle time are the order processing time (checking for errors in the order, preparation of the shipping documents, updating of the store inventory etc.), the availability of the products in the warehouse, the assembly time of the products making up the order (withdrawal from a storage point or creation of packaging for transport) and the shipping time (movement of products from the storage point to delivery point, including loading and unloading of the goods).

In general, it is not possible to know a priori with certainty the duration for each of these operations, given that several internal and external random factors affect the company's logistics system. For this reason, each component of the order-cycle time can be formally represented only by a continuous random variable, which has an unknown probability distribution but can be estimated on the basis of historical data gathered within the company. The two most significant quantities used to characterize a random variable are the mean and the standard deviation. These can be estimated by the sample mean and the sample standard deviation.

Consequently, the order-cycle time can also be represented by a continuous random variable, whose probability distribution is obtained from the probability distributions of the individual time components.

The English company MobilTrust has an order-cycle time for its products which essentially depends on two elements: the assembly time and the transport time. Five hundred observations on assembly time and 252 recorded transport times are available. The minimum observed value of assembly time is 2.3 days and the maximum is 15.9 days, whereas the minimum transport value is 6.9 days and the maximum is 13.2 days. For simplicity, it is convenient to express the available historical values in integer numbers of days. Table 1.5 shows the number of observations and the related discretized time values.

Let X and Y be the independent continuous random variables associated, respectively, with the assembly time and with transport time. The set of values observed for the assembly time is indicated by Ω_X ($\Omega_X = \{2, 3, \ldots, 16\}$). Let h_i be the number of observations recorded for every value $x_i \in \Omega_X$ (e.g. $h_3 = 4$). Analogously, let Ω_Y be the set of values observed for the transport time and k_j, the number of observations recorded for every value $y_j \in \Omega_Y$.

Table 1.5 Historical data of assembly time and transport time (in days) of MobilTrust.

Assembly		Transport	
Number of observations	Time	Number of observations	Time
1	2	19	7
4	3	27	8
4	4	54	9
18	5	65	10
38	6	48	11
56	7	25	12
69	8	14	13
96	9		
72	10		
68	11		
41	12		
18	13		
12	14		
2	15		
1	16		

The sample mean $\bar{X}$ and the sample standard deviation S_X of X are obtained in the following way:

$$\bar{X} = \frac{\sum_{i=1}^{|\Omega_X|} h_i x_i}{\sum_{i=1}^{|\Omega_X|} h_i} = 9.13 \text{ days;}$$

$$S_X = \sqrt{\frac{\sum_{i=1}^{|\Omega_X|} h_i (x_i - \bar{X})^2}{\sum_{i=1}^{|\Omega_X|} h_i - 1}} = 2.3 \text{ days.}$$

Similarly, the sample mean and the sample standard deviation of the transport time ($\bar{Y}$ and S_Y) can be calculated:

$$\bar{Y} = 9.9 \text{ days;}$$

$$S_Y = 1.55 \text{ days.}$$

Using a simple hypothesis test, it can be verified that the two random variables can be assumed to be normally distributed with a mean and standard deviation equal to the sample values.

The random variable Z associated with order-cycle time is, therefore, the sum of the independent random variables X and Y. Since X and Y are normally distributed, Z is also a normal random variable, with mean and standard deviation estimated by the sample values:

$$\bar{Z} = \bar{X} + \bar{Y} = 19.03 \text{ days};$$

$$S_Z = \sqrt{S_X^2 + S_Y^2} = 2.77 \text{ days}.$$

Figure 1.8 shows the probability density functions corresponding to the observed values of the random variables associated with assembly time, transport time and overall supply time.

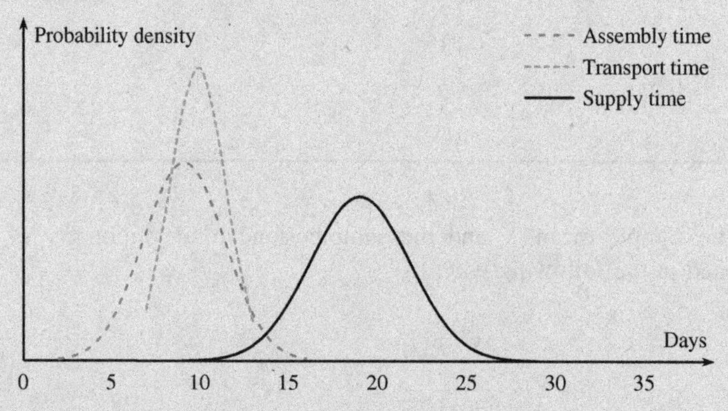

Figure 1.8 Density probability functions of assembly time, transport time and supply time for the problem of MobilTrust.

The coefficient of variation, which is defined as the ratio of the standard deviation and mean of the order-cycle time, can be used as a measure of the *reliability* of the service offered to the customers. The lower the value of this index, the greater the reliability.

Lugan is a Chilean chain of supermarkets which entrusts the task of supplying all its sales points to a transport company. Since the key factor for being competitive in the reference market is punctuality with the deliveries, the choice must be in favour of the most reliable operator. Two transport companies (A and B) were tested, to carry out the deliveries starting from the same distribution centre to the same sales point. The sales point was chosen among

Table 1.6 Historical data of service times (in hours) of operator A in the problem of Lugan.

Number of observations	Loading time	Journey time	Unloading time	Overall time
7	0.20	2.08	0.11	2.39
8	0.22	2.09	0.10	2.41
11	0.22	2.13	0.07	2.42
19	0.22	2.18	0.13	2.53
32	0.19	2.38	0.10	2.67
19	0.24	2.40	0.07	2.71
10	0.17	2.69	0.12	2.98
8	0.17	2.78	0.09	3.04

Table 1.7 Historical data of service times (in hours) of operator B in the problem of Lugan.

Number of observations	Loading time	Journey time	Unloading time	Overall time
32	0.23	1.50	0.60	2.33
7	0.25	1.50	0.10	1.85
10	0.19	1.30	0.50	1.99
8	0.20	1.50	0.78	2.48
8	0.13	1.55	0.05	1.73
7	0.36	2.30	0.07	2.73
32	0.25	1.35	0.75	2.35
10	0.17	1.20	0.09	1.46

those whose product demand, in value, has maintained unchanged levels over time. The data shown in Tables 1.6 and 1.7 were obtained from observations in the field.

The overall service times of the two suppliers can be represented by means of normally distributed independent random variables T_A and T_B normally distributed, with a mean and a standard deviation estimated by the sample values:

$$\bar{T}_A = 2.65\,\text{hours};$$

$$S_{T_A} = 0.19\,\text{hours};$$

$$\bar{T}_B = 2.19\,\text{hours};$$

$$S_{T_B} = 0.33\,\text{hours}.$$

On the basis of the mean value of the overall service time of the two operators, the company would tend to choose operator B ($\bar{T}_B < \bar{T}_A$). However, the values of the coefficients of variation of the two variables, equal respectively to $V_{T_A} = S_{T_A}/\bar{T}_A = 0.073$ and $V_{T_B} = S_{T_B}/\bar{T}_B = 0.151$, suggest, in contrast, choosing operator A, who proves to be the most reliable of the two.

Other possible measures of service level can be ascribed to the level of transport efficiency (e.g. the percentage of dispatches completed within a prescribed time) and to the integrity, precision and completeness of the orders (e.g. the percentage of orders dispatched within a prescribed time).

1.6 Management of the logistics system

Within a company, the management of the logistics system is the transverse process of planning, organizing and controlling the logistics system.

Planning means taking the best decisions possible, according to the predetermined objectives of the logistics system.

Organizing refers to organizing the human resources directly involved in logistics activities within a company organizational chart, so as to attain the company's objectives, effectively and efficiently.

Control means measuring the performances of the logistics system according to the qualitative and quantitative standards requested by the company management, and possibly initiate corrective actions when the results are not in line with the objectives.

The planning, organization and control phases of the logistics system arise in sequence: planning human resources involved in logistics can occur only after the planning phase of the logistics system. In the same way, it is not thinkable to measure to which extent the objectives of the logistics activities have been attained if the company organizational chart is not first defined in terms of human resources assigned to these activities.

1.6.1 Planning phase

The planning of the logistics system prevalently considers the following *decision-making areas*: forecasting, location, supply, storage and distribution.

Forecasting is the process of estimating the uncertain parameters that characterize the logistics system. The forecasts serve to correctly size the logistics system, to define the production capacity and the correct stock level, to work out plans and production programmes, to organize transport and so on. Chapter 2 will be dedicated to forecasting methods in logistics.

Location is the activity by which the optimal location of the facilities is determined, both in the planning phase of a logistics system and in the reorganization of an existing one. The main problems of location in the logistics context will be illustrated in Chapter 3.

Supply is the decision-making area that concerns all the logistics activities relative to the purchasing of raw materials, semi-finished goods or supply services. The decisions regarding this area largely depend on the specific company context. Chapter 4 will deal with the main decision-making problems regarding management of the suppliers.

Storage and distribution are decision-making areas for which logistics activities are of primary importance.

In these contexts, but not exclusively, logistics planning can be organized at three different decision-making levels: strategic, tactical and operational.

Strategic decisions, or long-term choices, have a long-term effect (more than a year) on the logistics system and typically involve major financial investments; therefore, these decisions are unlikely to be reversible in the short term. They are generally based on forecasts relative to aggregated data (e.g. regional demand for families of similar products). *Tactical decisions*, or medium-term choices, refer to the use of available resources and are usually based on forecasts. They are carried out with an annual, seasonal or monthly frequency. *Operational decisions*, or short-term choices, concern the definition of weekly or daily work plans for the staff and for the material resources. They use data from the surrounding environment (orders issued by customers, information on the state of the warehouses and on the availability of vehicles, news of strikes in the transport sector etc.) and the results of forecasts. Logistics planning relative to the areas of storage and distribution will be dealt with specifically in Chapters 5 and 6. Table 1.8 classifies logistics decisions depending on the strategic, tactical or operational level, which will be discussed in the following chapters.

The organizational and control activities of the logistics system will be the object of a deeper examination in Sections 1.6.2 and 1.6.3.

1.6.2 Organizational phase

The various organizational decisions made by a company determine its distribution of responsibilities and tasks. These are influenced by different factors, such as the sector in which it operates, the culture of the company (beliefs, values and expectations shared by the members of the organization), the technology employed (the greater the use of advanced technology, the slimmer the organizational structure) and the company size (small companies typically have a sole decision maker, whereas in medium-large ones power must be delegated and functions must be created to establish the relationships of authority).

Logistics in a company is a primary activity which can be illustrated by adopting a traditional mode of presentation of organizational structures: *functional*, *divisional* and *matricial*.

Table 1.8 Examples of strategic, tactical and operational decisions within the decision-making area of storage and distribution.

Decision-making area	Planning level		
	Strategic	Tactical	Operational
Storage	Warehouse planning Selection of warehouse equipment Choice of warehouse layout	Allocation of the products at storage points Choice of inventory policies for the products in stock in the warehouse	Pickup of products from the storage area Consolidation of products in the loading unit
Distribution	Choice of transport mode Fleet sizing and composition	Freight assignment on the transport network Transport service network design Vehicle assignment Crew rostering Determination of the vehicles to be rented	Vehicle routing Repositioning of vehicles and empty containers Consolidation of the shipping orders

1.6.2.1 Logistics in companies organized according to the functional model

The functional structure is based on principles of work subdivision and specialization. Similar activities, requiring analogous skills and the same kind of resources, are grouped within *functions*. A function is associated to a department, headed by a director.

In a functional structure, the management level deals with strategic decisions, whereas the tactical and operational decisions are delegated to functional areas.

The functional structure is suitable to small- or medium-sized organization. It is directed towards objectives of efficiency which generally offer products or services with a low degree of diversification, and use a stable technology in an economic environment of little uncertainty.

The advantages provided by such a structure can be found in the efficient use of the resources available and in the simplicity of the hierarchical control and communication within the functional areas.

In contrast, the interdependence among work flows is not sufficiently taken into consideration, and it is possible that a lack of coordination occurs among the functions because of divergences of interests and objectives. A functional-type structure (which by nature is not very flexible) is, in fact, often incapable of tackling production diversification or increased turbulence in the outside environment. Moreover, an increase in the company size often comes with a loss of control (slowing of the decision-making process) because each department becomes overloaded with responsibilities.

The functional structure cannot explicitly contain a logistics function (see Figure 1.9). This is possible only if there exists a real climate of support among the various company functions. Coordination among the areas of responsibility for key logistics activities, such as inventory management and transport, can be obtained in an informal way by means of incentives systems, for example.

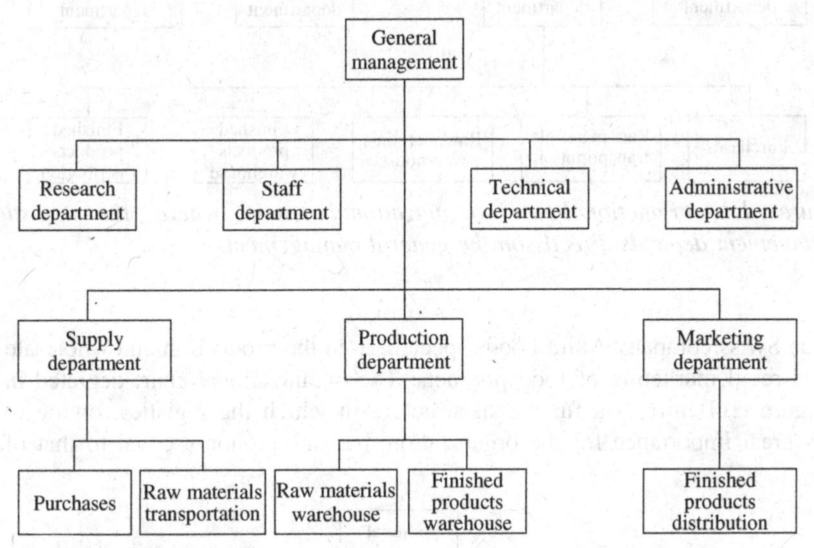

Figure 1.9 Functional-type organizational chart where the logistics activities are separate.

A similar structure is suitable, for example, for organizations achieving a great production output for many customers, widely distributed over a territory. These are generally food, chemical and clothing companies, which do not have a particular supply problem (a limited amount of raw materials) and whose production plants often run on a continuous cycle. One of the most difficult problems is the distribution of finished products since it is necessary to contain the distribution costs and, at the same time, be competitive in a typically mature market.

Alternatively, the logistics function can depend directly on the general management (see Figure 1.10).

The number of underlying activities is the correct compromise among the objectives of efficient coordination (which drives towards their reduction) and technical effectiveness (which drives towards their subdivision). Each of these is under the control of a manager who answers directly to the logistics department to whom the coordination activities are entrusted.

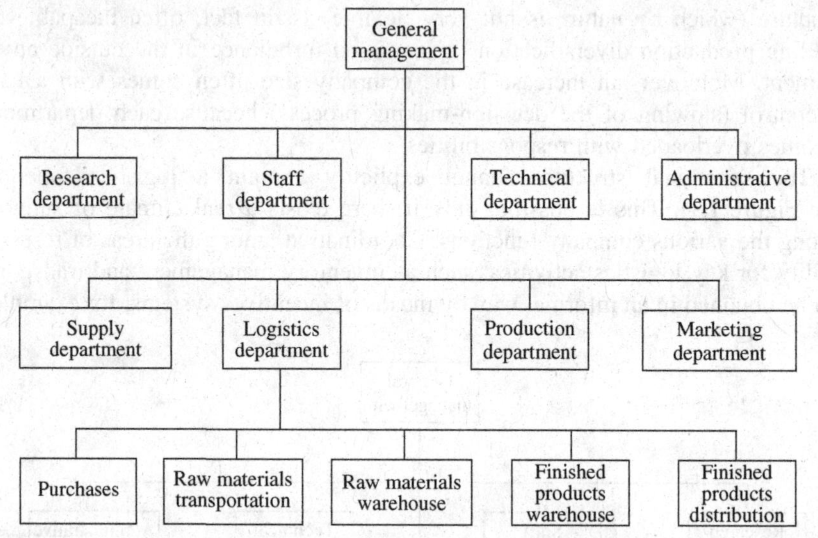

Figure 1.10 Functional-type organizational chart where the logistics management depends directly on the general management.

The Swiss company Akira Foods specializes in the production and wholesale and retail marketing of food products. The organizational chart depicted in Figure 1.11 mirrors a functional structure in which the logistics, owing to its great importance for the organization, has an autonomy equal to that of

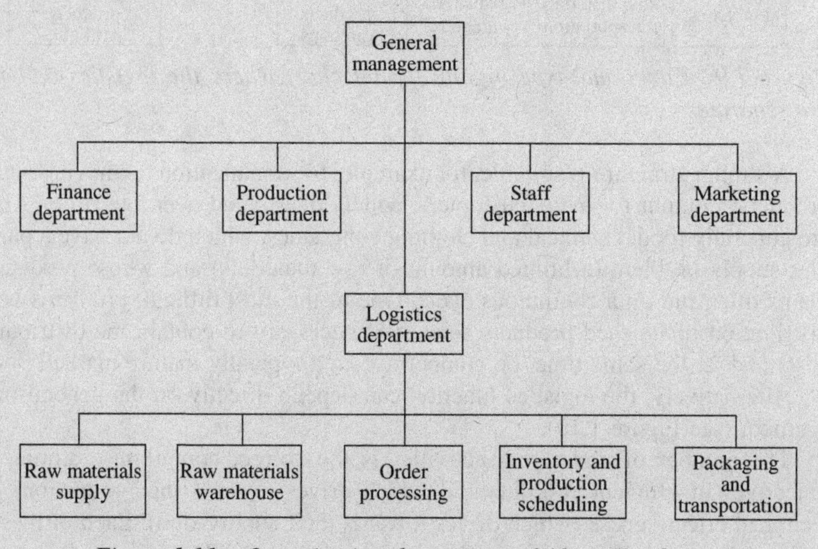

Figure 1.11 Organizational structure of Akira Foods.

the other functions and coordinates numerous activities. Through the use of this organizational structure, the company is able to guarantee high levels of service to its customers represented by numerous retailers and supermarkets operating in the Swiss market.

An organizational model of this type is also suitable for companies with after-sales logistics problems, such as those arising in the electronic, mechanical and automobile sectors. In these cases, a company should prefer a structure in which the logistics function is highly developed and depends directly on the general management.

To optimize the company's commercial policies and its distribution process, it may be preferable to subordinate the logistics functions to the marketing department (see Figure 1.12).

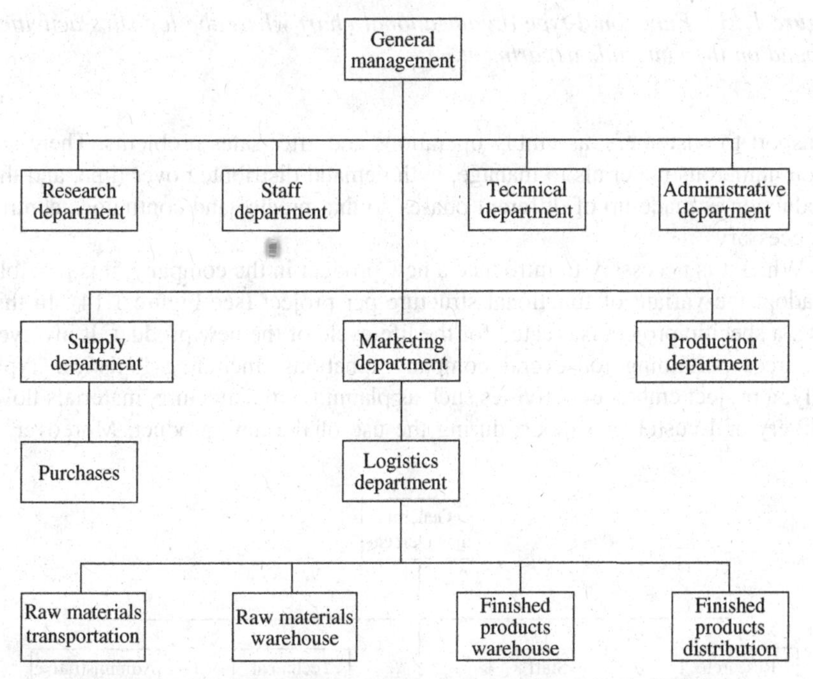

Figure 1.12 Functional-type organizational chart where the logistics department depends on the marketing department.

Another functional-type structure variant is one in which the materials department controls the logistics activities (see Figure 1.13). This choice is adopted by companies that manufacture complex products (in general to order). After the production phase, the problems of distribution and marketing are limited to

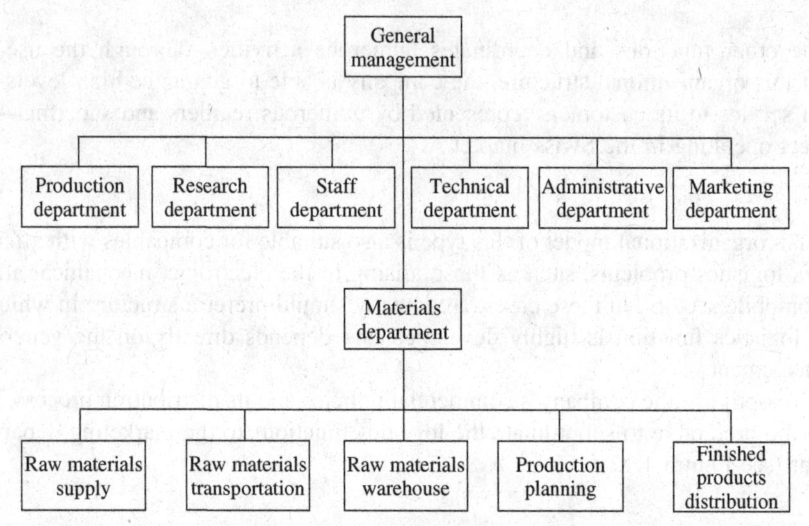

Figure 1.13 Functional-type organizational chart where the logistics activities depend on the materials department.

transport to customers, assembly operations and after-sales problems. There are often numerous materials to manage, with demand distributed over time, and the production is made up of different phases so that precise and continuous control is necessary.

When it is necessary to introduce a new product in the company, it is possible to adopt the variant of functional structure per project (see Figure 1.14). In this case, a specific project is created for the life cycle of the new product. It involves resources pertaining to several company functions, including logistics. Typically, a project embraces activities such as planning, manufacture, materials flow, delivery and customer support during the use of the new product. Moreover, a

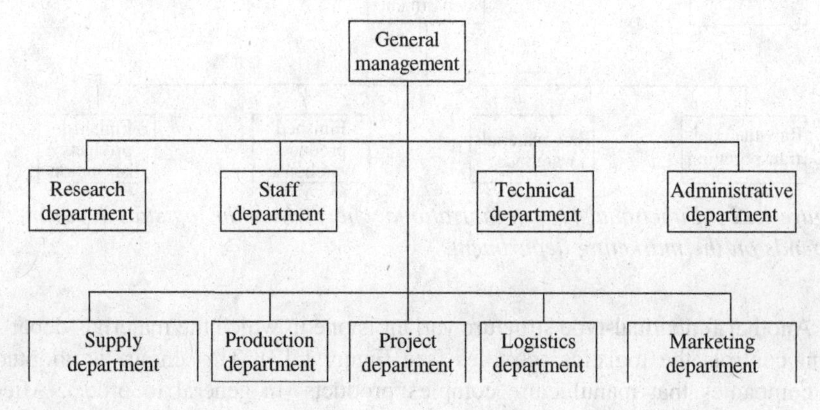

Figure 1.14 Functional-type organizational chart with a project manager.

temporary staff person becomes the project manager, whose task is to coordinate all the resources involved so as to avoid conflicts and to optimize collaboration.

1.6.2.2 Logistics in companies organized according to the divisional model

Divisions are completely autonomous operational structures that behave like independent businesses. They plan, create and market products or services within their own areas of competence. Each division, entrusted to a manager who has total responsibility for it, is organized entirely according to a functional-type logic.

The subdivision of the company organizational chart into divisions can be operated on the basis of products or services, geographical areas or markets. The general management is responsible for strategic decisions. It deals with the allocation of resources to the different divisions, chooses the portfolio products, checks on the divisional managers and coordinates the latter by means of appropriate staff entities.

The single divisions are responsible for product or service strategies and operational decisions. This structure is also adapted to large companies (it allows the creation, at least ideally, of an unlimited number of divisions), even operating in turbulent competitive markets (divisions allow more flexibility in the organization, which is needed to quickly react to outside changes).

By adopting a divisional logic, a company is in a better position to face the proliferation of products or services (possibly within a diversification strategy) and the continuous technological development of the market to which it belongs.

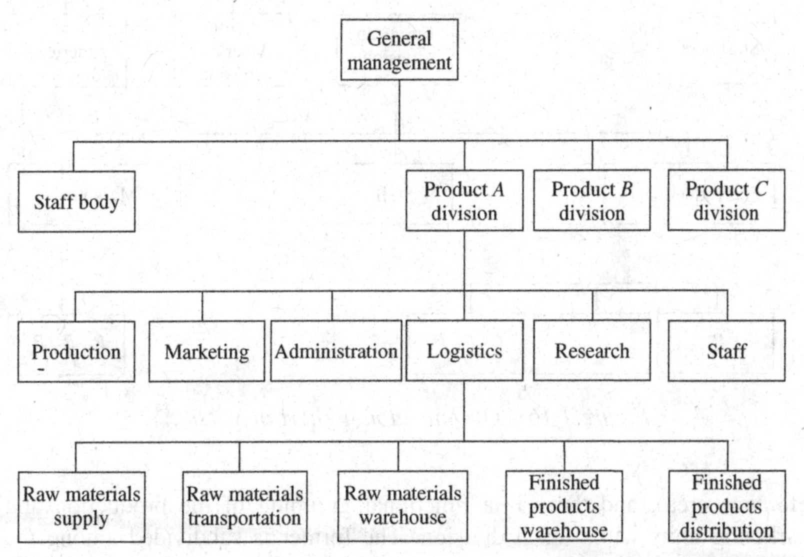

Figure 1.15 Divisional-type organizational chart (subdivision into divisions on the basis of products).

However, if the power of the divisional managers is favoured, the opportunities for exchange of knowledge among specialists are reduced. The duplication of jobs not only generates an increase in structural costs but also can create rivalries between the different divisions. Moreover, there is a risk that short-term objectives, in particular immediate divisional results, will be favoured, and that the general business objective of the company will be lost.

The logistics function (as illustrated in Figure 1.15) is placed like the others within each division for which, as already mentioned, a functional logic is used. This means that each division can adopt its own organization, as discussed in this chapter.

Varsth is a multinational company registered in Denmark. The production plants of the company are distributed in three geographic areas: Europe, North America and South America. The company produces and markets three lines of product: washing machines, television sets and small electrical appliances, respectively P_1, P_2 and P_3. The production of the three lines occurs independently, whereas the commercial channels used for the sales are common to all products. The organizational structure of Varsth (represented in Figure 1.16) is the divisional type (on the basis of the specific geographic

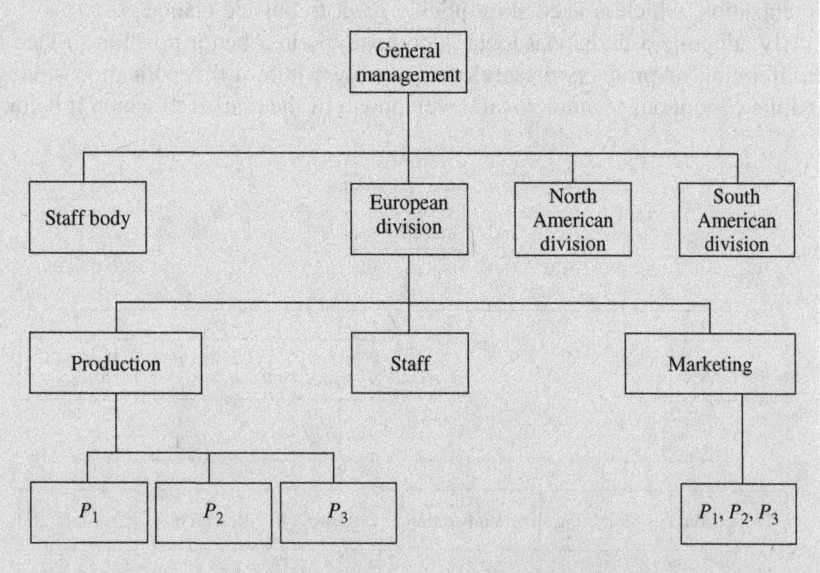

Figure 1.16 Organizational chart of Varsth.

reference area), and there is a functional grouping of the production and marketing units within each division. The former is subdivided among the three product areas, whereas the latter is identical for all three lines.

1.6.2.3 Logistics in companies organized according to the matrix model

The matricial organizational structure is adopted by large-size companies that create complex products with a short, medium or long high-technology life cycle. These companies conduct multiple projects simultaneously, relative to several products or business areas, each of which must satisfy the specific requirements of their customers or users (also by means of high differentiation strategies).

The matrix structure couples the organization's functional elements with its divisional elements. Technicians and specialists belonging to different functions are in fact assigned to one or more project groups, coordinated by a project manager. There are therefore two simultaneous lines of authority (function and project) even if usually, depending on the specific case, one dimension proves dominant.

Such an organizational structure is well suited to turbulent economic markets and favours elements like motivation and staff development. The resources are in fact specialized and coordinated so as to attain specific results effectively and efficiently.

The dual hierarchical dependence can, however, generate confusion and conflict owing to the presence of different commands. Therefore, one needs to establish various coordination mechanisms to favour cooperation among managers, with resulting increased operating costs. Moreover, it is possible that the fragmentation of objectives will generate a lack of overall vision in the business.

The activities of function logistics are assigned to the different projects that involve them.

The functional manager is responsible for the logistics system overall but does not have direct authority over the component activities where he or she shares authority with different project managers.

A simplified example of a matricial organizational chart is provided in Figure 1.17.

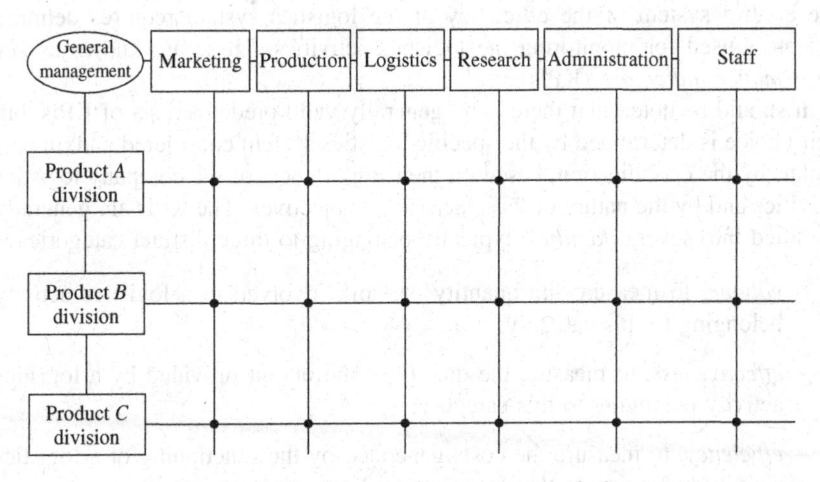

Figure 1.17 Matricial organizational chart.

Fashiondream is a company operating in the Swiss market. It makes and markets three different products: clothes, underwear and shoes. A specific project is coordinated by a manager for each product. In order to use the staff of the different functions adequately, the project groups share resources so that the employees develop the in-depth knowledge that is necessary to serve the three product lines effectively. Figure 1.18 depicts the organizational structure of Fashiondream.

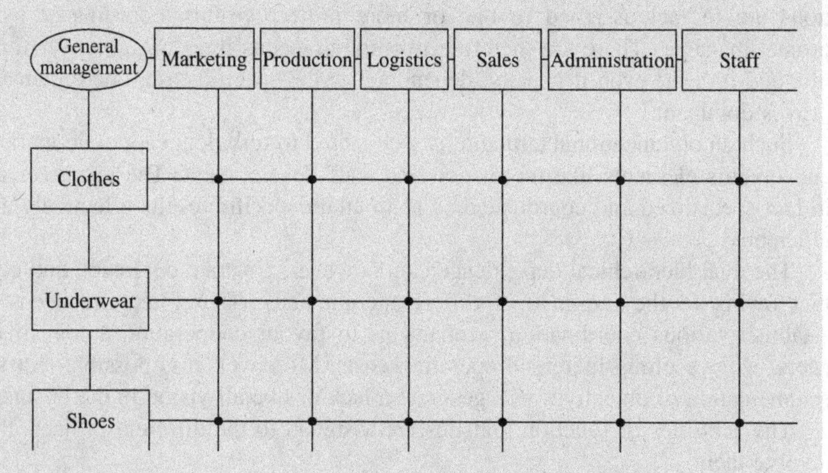

Figure 1.18 Fashiondream matricial organizational chart.

1.6.3 Control phase

The control system of the efficiency of the logistics system requires defining indicators used for monitoring the logistics activities. These are known as *key performance indicators* (KPIs).

It should be noted that there is no generally valid predefined set of KPIs, but their choice is determined by the specific logistics system considered and, in particular, by the classification, based on their importance, of the company logistics activities and by the nature of these activities' objectives. The KPIs are generally classified into several *families*, typically belonging to three distinct categories:

- *volume*, to measure the quantity of work involved in a logistics activity belonging to this category;
- *effectiveness*, to measure the quality of the output provided by a logistics activity belonging to this category;
- *efficiency*, to measure the costs generated by the functioning of a logistics activity belonging to this category.

Each KPI is measured according to a suitable metric that can vary from case to case. To make all the measures homogeneous and reach a synthetic and coherent 'control panel', a normalization operation for each single indicator is needed, transforming it from the initial scale of measure (strictly correlated to the individual KPI) into a single and homogeneous scale.

To summarize, the phases that make up the control system of the logistics system by means of a KPI control panel are the following:

1. classification of the logistics activities and definition of the objectives to pursue;
2. identification of the KPIs subdivided by families;
3. creation of the control panel.

Borg is a Canadian company producing wooden objects. Owing to many management problems resulting from a recent organizational restructuring, the new logistician has been charged with the monthly monitoring of the most critical logistics activities.

When designing a logistics control system by means of a KPI control panel, it was found that the most significant problems were connected to the high number of complaints received about errors in dispatched orders, frequent delivery delays, an incorrect policy of stock management, a disproportionate overstaffing in the warehouse and its effect on costs, the inefficiency of the transport system and its impact on distribution.

The logistician therefore identified 19 KPIs, subdivided into eight families and calculated with a monthly frequency. These KPIs are described in Table 1.9, in which the computing method used for each of them is also indicated.

The KPIs were calculated in March of the present year and were normalized in the interval [1, 10], using the min-max normalization procedure (which will be described in detail in Section 2.4.7) on the historic series composed of the last 12 dates available for each KPI (from April of the previous year to March of the present year). The values obtained are shown in Table 1.10.

The logic of the normalization procedure is that for each KPI, independently from its meaning, the normalized value equal to 1 corresponds to the worse value (e.g. the minimum number of orders or the maximum number of complaints per month), whereas 10 is the best one.

Every KPI within its own family was weighed in a suitable way (with the sum of the weights of the KPIs of every family equal to 1), so as to find a single efficiency value for every family of KPI. Table 1.11 shows these values, together with the weights chosen for each KPI in the month considered.

Table 1.9 KPIs for the Borg logistics system.

Family	KPI	Computing method
Dispatch of orders	Number of orders	Total orders received in the month/days in the month
	Complaints	Number of complaints per month
	Extent of completeness	Order lines dispatched overall in the month/order lines inserted overall in the month
	Errors on order lines	Order lines with errors found by complaints in the month/order lines inserted and not cancelled in the month
	Errors on orders	Number of orders with delivery in the month/total orders received in the month
Punctuality	Deliveries dispatched within delivery time window	Number of lines dispatched in the month within the delivery time window/total number of lines dispatched in the month
	Value of deliveries dispatched within delivery time window	Value of quantity dispatched in the month within the delivery time window/total value of quantity dispatched in the month
Staff	Warehouse employees	Monthly average of number of employees daily employed in the warehouse
	Effective employees	Monthly average of effective daily employees/monthly average of nominal daily employees
	Productivity of warehouse employees	Monthly total average of daily movements/monthly average number of warehouse employees
Warehouse	Warehouse movements	Monthly average of daily movements
	Pick-up operations	Monthly number of movements/monthly number of order lines dispatched
Inventory	Recoveries	Monthly cumulated value of economic-financial savings deriving from recoveries
	Warehouse value	Overall monthly value of inventory

Table 1.9 (*continued*)

Family	KPI	Computing method
Transport	Deliveries per journey	Monthly average number of customers served by a route
	Trip saturation	Monthly average amount of goods dispatched per vehicle/vehicle capacity
	Trip forecast	Trips planned in the month/ effective trips in the month
Costs	Budget	Monthly costs/monthly budget
Reliability	Reliable deliveries	Monthly order lines delivered correctly on time/total monthly order lines

Table 1.10 Normalized KPI values for the Borg logistics system in the considered month.

Family	KPI	Value
Dispatch of orders	Number of orders	4.72
	Complaints	6.21
	Extent of completeness	5.77
	Errors on order lines	7.83
	Errors on orders	6.34
Punctuality	Deliveries dispatched within delivery time window	6.33
	Value of deliveries dispatched within delivery time window	4.21
	Warehouse employees	5.40
Staff	Effective employees	7.23
	Productivity of warehouse employees	6.42
Warehouse	Warehouse movements	6.27
	Pick-up operations	5.96
Inventory	Recoveries	4.46
	Warehouse value	6.24
	Deliveries per journey	7.28
Transport	Trip saturation	6.34
	Trip forecast	7.22
Costs	Budget	7.25
Reliability	Reliable deliveries	4.88

Table 1.11 Performance of the KPI families for the Borg logistics system in the considered month.

Family	KPI	Value	Weight	Family performance
Dispatch of orders	Number of orders	4.72	0.20	
	Complaints	6.21	0.25	
	Extent of completeness	5.77	0.18	6.13
	Errors on order lines	7.83	0.17	
	Errors on orders	6.34	0.20	
Punctuality	Deliveries dispatched within delivery time window	6.33	0.60	5.48
	Value of deliveries dispatched within delivery time window	4.21	0.40	
Staff	Warehouse employees	5.40	0.35	6.31
	Effective employees	7.23	0.30	
	Productivity of warehouse employees	6.42	0.35	
Warehouse	Warehouse movements	6.27	0.60	6.15
	Pick-up operations	5.96	0.40	
Inventory	Recoveries	4.46	0.30	5.71
	Warehouse value	6.24	0.70	
	Deliveries per journey	7.28	0.30	
Transport	Trip saturation	6.34	0.50	6.80
	Trip forecast	7.22	0.20	
Costs	Budget	7.25	1.00	7.25
Reliability	Reliable deliveries	4.88	1.00	4.88

The logistician has determined that the minimum value to be achieved for each KPI should be 6, and the objective should be 10.

The control panel was constructed using a radar graph (see Figure 1.19), with greater visual and information impact.

The control panel allowed the logistician to identify the areas needing priority action (i.e. those with a KPI value lower than 6), at the organizational and planning levels.

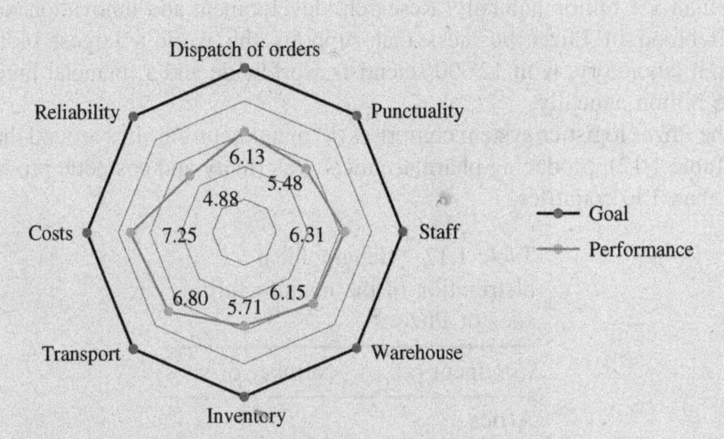

Figure 1.19 Control panel of the Borg logistics control system for the considered month.

1.7 Case study: The Pfizer logistics system

The Pfizer Pharmaceuticals Group is the largest pharmaceutical corporation in the world. Its mission is 'to discover, develop, manufacture and market innovative, value-added products that improve the quality of life of people around the world and help them enjoy longer, healthier, and more productive lives'. The Pfizer range of products also includes self-care and well-being products and health products for livestock and pets.

Founded in 1849 by Charles Pfizer, the company was first located in a modest red-brick building in the Williamsburg section of Brooklyn, New York, United States, that served as office, laboratory, factory and warehouse. The firm's first product was Santonin, a palatable antiparasitic which was an immediate success. In 1942 Pfizer responded to an appeal from the US government to expedite the manufacture of penicillin, the first real defense against bacterial infection, to treat Allied soldiers fighting in World War II. Of the companies pursuing mass production of penicillin, Pfizer alone used the innovative fermentation technology. Pfizer manufactures some of the most effective and innovative active ingredients including atorvastatin, whose medicine is the most prescribed cholesterol-lowering one in the United States; amlodipine, which belongs to the calcium channel blocker dihydropyridine class, and is used as an anti-hypertensive; azithromycin, the most-prescribed brand name oral antibiotic in the United States, and sildenafil citrate, a breakthrough treatment for erectile dysfunction.

With a portfolio that includes five of the world's 20 top-selling medicines, Pfizer sets the standard for the pharmaceutical industry. Ten of its medicines are ranked first in their therapeutic class in the US market, and eight earn a revenue of more than $ 1 billion annually. Research, development and innovation represent the lifeblood of Pfizer business that supports the world's largest biomedical research laboratory, with 12 000 scientists worldwide and a financial investment of $ 6 billion annually.

The Pfizer logistics system comprises 58 manufacturing sites around the world (see Table 1.12), producing pharmaceutical, veterinary and cosmetic products for more than 150 countries.

Table 1.12 Geographical distribution of the manufacturing sites of Pfizer.

Continent (s)	Number of sites
Africa	7
Asia	13
Australia	2
Europe	16
Americas	20

Because manufacturing pharmaceutical products requires highly specialized, developed and costly machines, each Pfizer plant produces a large amount of a limited number of pharmaceutical products for the whole international market of the company (see Table 1.13).

Table 1.13 Features of some of Pfizer's European plants.

Country	Number of plants	Number of products	Productivity rate (in millions of items per year)
Belgium	1	29	6.5
France	1	14	2.4
Germany	1	3	11.4
Italy	3	182	87.1
United Kingdom	1	8	5.0

In the following, the attention will be focused on the logistics system of Pfizer with respect to a cardiovascular product, named Alpha10. The product is packaged in blisters, each containing 20 tables of 5 or 10 milligrams. Alpha10 is produced in a unique European plant (EUPF plant) for an international market including 90 countries (see Figure 1.20). Every year the plant produces over 117 million blisters. The product expires 60 months after its production and must be stored at a temperature varying between 8° and 25 °C.

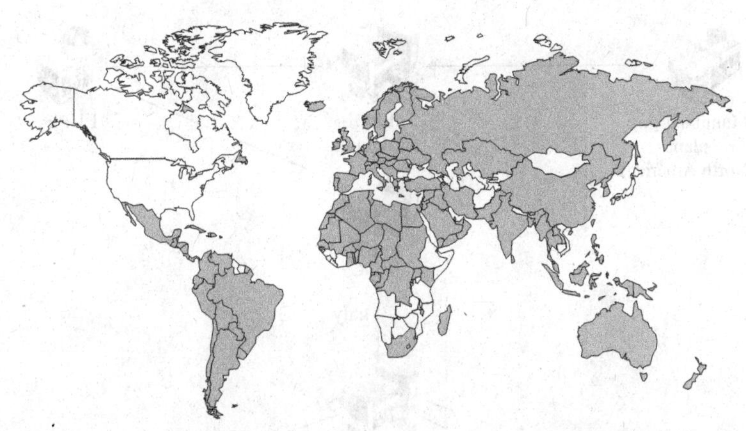

Figure 1.20 Geographical selling areas (in grey) for the product Alpha10 of Pfizer.

The main component of Alpha10 is a particular active pharmaceutical ingredient, based on a Pfizer property patent, manufactured in a North American plant. Its packages are transferred by air to the European Logistics Centre (ELC), located in Belgium, which in turn replenishes the EUPF on a monthly basis (see Figure 1.20). Freight transportation between the ELC and the EUPF is performed by overland transport providers (e.g. Danzas). The EUPF plant manufactures ALFA10 tablets that are subsequently packaged into 120 blister boxes and sent weekly to a third-party CDC.

In Italy, Alpha10 is distributed, together with other products of Pfizer, to both hospitals and pharmacies, using two different channels (see Figure 1.21).

Hospitals (about 2 000) are supplied directly by the company, throughout a CDC and seven RDCs. Hospitals may be supplied by more than one warehouse, depending on stock levels. Transportation is performed by specialized haulers in refrigerated vans.

Pharmacies (about 16 000) are supplied through wholesalers (about 200, managing about 300 distribution centres of medicines at wholesale). Wholesaler orders are collected directly by Pfizer and shipped weekly by the CDC. The CDC is able to deliver the product in any Italian location within at most 60 hours. Wholesalers receive orders from pharmacies very frequently (up to four times a day). Pharmacies expect the wholesalers to deliver medicines within 4 to 12 hours (it is worth noting that, in Italy, pharmacies have a high contractual power over the wholesalers). Therefore, the average revenue of the wholesalers is low, due to the high logistics costs that have to be paid for guaranteeing a high service level to the pharmacies.

The Pfizer logistic network, also extended to the wholesalers, uses a specific information system, named Manugistics.

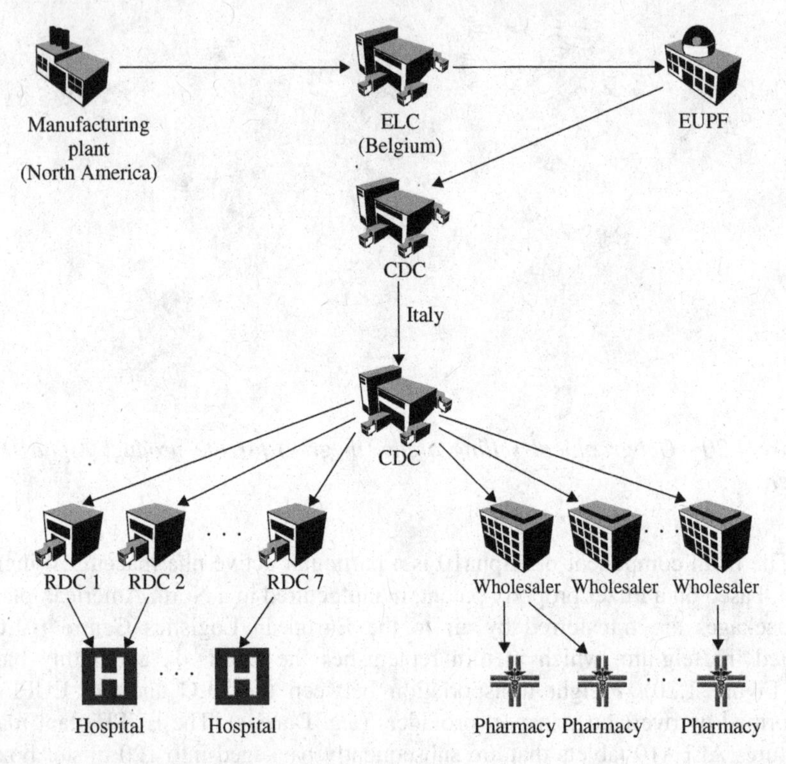

Figure 1.21 Pfizer logistics system representation for the product Alpha10.

1.8 Questions and problems

1.1 Adama is a French manufacturer of photovoltaic panels. The company has a production plant in Rennes, which supplies four warehouses located in Angers, Bourges, Clermont-Ferrand and Montauban. The warehouses directly supply the installers, grouped, for this purpose, into 10 operative districts spread across the country. The installers belonging to the same operative district are served from a single warehouse. Each installer returns defective photovoltaic panels to the corresponding warehouse; these panels are then sent to a repairing centre located in Poitiers. Model the logistics system of Adama through a directed graph and give a graphical representation of it (hint: the installers of the same operative district are assumed to be concentrated in a unique point).

1.2 Explore the websites of companies producing beverages, and represent the logistics system of one of them.

1.3 Discuss the main differences in managing a logistics system of a service company (in particular, a bank) and a production company (in particular, a chemical industry).

1.4 What could represent a direct link between a CDC and the assembly plant of the logistics system depicted in Figure 1.1?

1.5 Florim is an Albanian company specialized in manufacturing ceramic products. The company realizes its revenue by selling six products, as reported in Table 1.14.

Table 1.14 Annual revenue (in €) and amounts sold of the products of Florim.

Product	Sales	Quantity
1	350 000	2 700
2	160 000	2 200
3	920 000	2 500
4	125 000	1 500
5	360 000	4 200
6	160 000	1 900

Let $y = [(1 + \alpha_1)x] / (\alpha_1 + x)$ be the equation of the 80–20 curve C_1 defined such that the first 21% of products sold correspond to 68% of the annual sales; similarly, let $y = [(1 + \alpha_2)x] / (\alpha_2 + x)$ be the 80–20 curve C_2, obtained by assuming that the first 21% of products correspond to 62% of the annual sales. Check which of the curves C_1 and C_2 is a better approximation of the actual trend of the cumulative percentage of the annual sales with respect to the cumulative percentage of the amounts sold.

1.6 Zuick is a German import-export company of household appliances. The company, whose headquarters is located in Hannover, distributes 15 products whose weekly amounts sold and corresponding sales are reported in Table 1.15.
The company is investing additional financial resources in two products, K-505 and K-506, for which the logistician proposes an intensive distribution strategy, involving more CDCs and increasing the stocking levels. By using an *ABC* classification (20–30–50) of the products by the weekly amounts sold, verify whether the distribution strategy proposed by the logistician is correct and, if not, modify it accordingly.

1.7 El.Ma is an American distributor of electrical equipment. The warehouse in Columbus, Ohio, has 18 products to store and sell. Monthly sales and average monthly stock values are reported in Table 1.16. Make an *ABC* classification (80–10–10) of the products by their monthly sales and by monthly average stock values, respectively. Which inventory policy should El.Ma adopt for the product named ‘locking release 24V’?

Table 1.15 Weekly amounts sold and corresponding sales (in €) of the products of Zuick.

Product	Quantity	Sales
K-501	155	119 806
K-502	64	31 448
K-503	70	25 607
K-504	66	24 406
K-505	61	15 196
K-506	58	13 112
K-507	60	10 106
K-508	197	11 395
K-509	154	9 489
K-510	56	8 664
K-511	74	13 955
K-512	208	16 283
K-513	164	14 085
K-514	71	12 984
K-515	163	123 935

Table 1.16 Monthly sales and average monthly stock values (in €) of the El.Ma products.

ID product	Description	Sales	Stock values
1	Digital starter	22 356	980
2	Differential block 4P	147 800	3 667
3	Land trolley	10 450	1 174
4	HCS cable	65 980	2 030
5	Engine control unit 380V CA	18 654	652
6	Contacter 24–60V CC	27 580	1 721
7	Control builder	19 768	558
8	Universal dimmer	46 225	1 015
9	BRI interface	8 766	775
10	Circuit breaker 10KA	80 350	3 159
11	Motoadaptor	13 746	1 100
12	OPC server	57 558	3 111
13	Spring relay	7 852	733
14	Electronic delayer	9 785	724
15	Sectioner	32 400	894
16	Locking release 24V	12 328	1 020
17	TMA360	15 980	1 058
18	Control unit for release	11 900	1 062

1.8 Barilla is an Italian multinational food company which 'has significantly believed in B2B e-commerce and, in particular, in EDI (*Electronic Data Interchange*), especially in terms of Web-EDI, which supports the order-delivery-invoice cycle and, also, collaborative processes, such as CRP (*Continuous Replenishment Program*) and CPFR (*Collaborative Forecasting and Replenishment Program*)' (Mauro Viacava, CEO of Barilla). From the point of view of integrated logistics, examine the meaning of EDI, CRP and CPFR, and explain how these are presumably used in the logistics system of Barilla.

1.9 Assume that, for a certain company, the estimated annual sales to service level curve $r(l)$ has been determined by means of a simulation method. The resulting equation is $r(l) = 950\,000\,l - 328\,000\,l^2$, where l denotes the percentage of customers served within 24 hours (e.g. if $l = 0.7$, 70% of customers are served within 24 hours); $r(l)$ is expressed in \$. The annual logistics costs (in \$) are estimated as 280 000, 320 000, 380 000, 410 000, 460 000 and 510 000, with respect to the following values of service level offered to customers: 50%, 60%, 70%, 80%, 90% and 100%, respectively. Determine the service level at which the maximum estimated annual profit is achieved.

1.10 There exist five design alternatives for a logistics system, whose costs and service levels are plotted in Figure 1.22. Determine which alternatives should be taken into account as a possible design solution and which ones should be discarded.

Figure 1.22 Costs and service levels of five alternatives for the design of the logistics system of Problem 1.10.

1.11 Electrolux is a Dutch company that has recently decided to start production and sale of a new energy-efficient light bulb. During the phase of logistics system design, two alternatives are taken into account:

- Use the foreign manufacturing plant located in Tartu (Estonia), where the unit production cost is €0.97 (the cost of raw materials purchase is included). The transport cost to the CDC of Groningen is €16 per box, where a box contains 100 units of the product. For simplicity, it is assumed that this cost includes also inventory costs at the CDC. The CDC of Groningen supplies two RDCs, situated in Delft and Eindhoven; their annual demands are 28 000 and 35 000 boxes, and the transport costs per box are €9 and €10, respectively;
- use the national manufacturing plant of Dordrecht, where the unit production cost is €1.38. This facility supplies the RDC of Delft and Eindhoven and the unit transport costs are, in this case, €2.5 and €3.0 per box, respectively.

Determine which alternative to prefer, considering the minimization of production and transport costs as the logistics objective to pursue.

1.12 Tranexpress is an international freight forwarder. Its service time consists of two components: the time required for preparing paperwork for customs, coordinating customs inspections, warehousing and consolidation, container loading (time for additional services) and transport time. The recorded data related to some past observations are reported in Table 1.17. Characterize the total service time from a statistical point of view, by computing the sample mean and the standard deviation of the corresponding random variable. Compute a quantitative index for measuring service reliability.

Table 1.17 Observed data related to additional services and transport times (in days) of Tranexpress.

Additional services		Transport	
Number of recorded data	Time	Number of recorded data	Time
1	21	20	120
3	26	29	132
4	33	56	148
9	41	66	162
18	52	52	178
35	60	26	197
42	98	16	209
22	107		
13	114		
2	122		
1	129		

1.13 Norsk is a Danish company that specializes in food product for daily consumption. It has five associated subsidiaries in the European Union and a network of distributors in North America. Recently, the company has decided to redesign its distribution network in Scandinavia where 140 distribution centres have been transformed into simple stores without administrative functions. The administrative functions have been concentrated in 14 logistics centres with regional character, and forecasting activities based on data analysis have been centralized in the company headquarters. List and classify the decisions taken at the business management level during the reorganization phase of the logistics system.

1.14 Explain why stocks of products are usually increased in one or more facilities in the absence of logistics coordination.

1.15 An ineffective management of a logistics system can produce instability. One of the effects due to mismanagement is known as the *bullwhip effect*: a small fluctuation in customer demand can cause amplified fluctuations in material flows orders over time via the upstream supply chain itself. This particular phenomenon was first recognized by the managers of Procter & Gamble who noticed significant variability of sellers' orders despite the fact that consumer demand remained constant; this is because the orders were issued on the basis of limited information regarding the final demand of the product (in this case, Pampers disposal diapers). In general, distorted information along the logistics system gives rise to a growing imbalance between demand and supply; a distorting effect creates the same problems throughout the supply chain. For example, decisions about the management of the CDC in a company are, in general, based on the level of stocks and on the orders issued in a logistics node immediately downstream of the CDC, such as an RDC, without knowing the end-user demand. The orders issued by the RDC are used to estimate the mean and standard deviation of demand. These estimates are the basis of decisions on reorganization issues. For example, in the (s, S) method (see Section 5.3.2.2), an order is issued whenever an inventory level falls below a given reorder point s and the inventory level is then increased to an order-up-to-level S. As the perceived demand varies, the parameters S and s are updated and order quantities are also changed. In light of this, show that a typical bullwhip effect in a logistics system, made up of a production plant, a CDC, a RDC and a retailer, is like the one reported in Figure 1.23 (where it is assumed that a sudden 10% increase in end-user demand occurs). Can this effect be reduced through more and better information sharing?

1.16 Define the functional organizational structure of a petrochemical company showing the position of the logistics activities. Compare advantages and disadvantages when a divisional organization is adopted.

1.17 Ajt Solar is an Indian company of photovoltaic products. To control the logistics system, some critical logistics activities are monthly monitored. Three families of KPI are considered: order processing, inventory management and transport. The KPIs, their families and the computing method used for each of them are shown in Table 1.18.

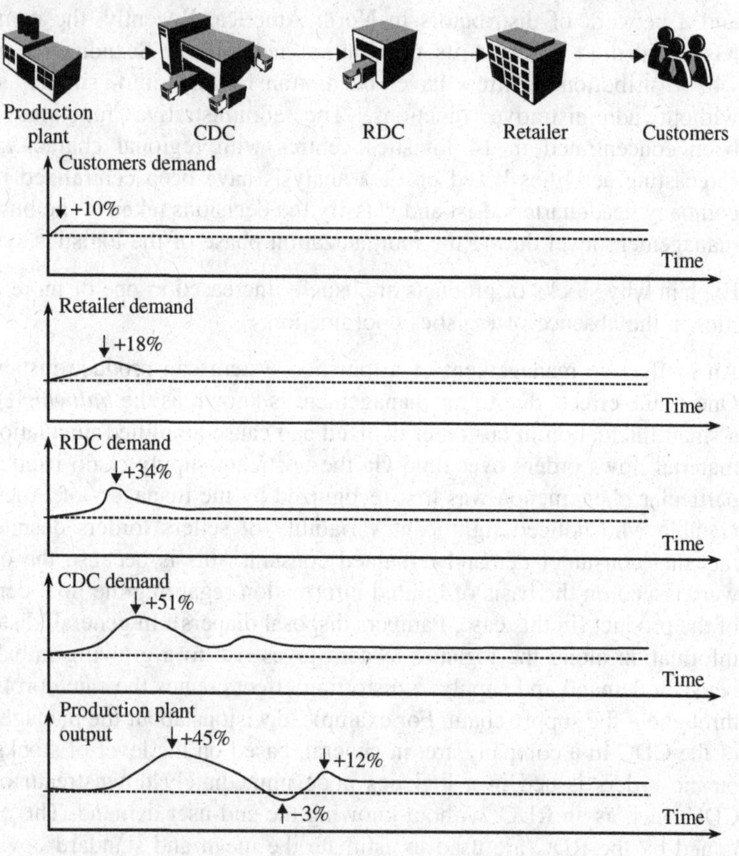

Figure 1.23 The bullwhip effect for the logistics system of Problem 1.15.

Table 1.18 KPIs and computing method used of the Ajt Solar company.

Family	KPI	Computing method
Order processing	Orders	Number of orders in a month/ number of weekdays in a month
	Complaints	Number of complaints a month
Inventory	Stocks values	Monthly inventory value
Transports	Deliveries per journey	Average number of customers monthly served with a journey
	Planned journeys	Monthly number of planned journeys/ monthly number of journeys carried out

Table 1.19 KPI values for the last 15 months for the Ajt Solar problem.

Period	Orders	Complaints	Stocks values ($)	Deliveries per journey	Planned journeys
1	154.26	21	470 800	15.48	1.07
2	151.04	12	500 800	36.76	0.97
3	161.23	16	533 000	18.94	1.13
4	145.33	24	565 900	33.07	1.14
5	158.66	14	567 700	31.15	1.13
6	171.25	16	471 900	40.37	1.10
7	98.66	31	522 200	23.35	0.83
8	102.45	8	531 000	14.33	1.00
9	134.74	12	509 800	39.80	0.93
10	147.24	16	579 700	18.37	1.20
11	133.54	21	548 300	26.04	1.15
12	154.81	18	458 700	30.10	1.00
13	148.82	20	542 100	36.60	0.95
14	124.31	13	524 500	26.52	1.11
15	164.03	11	567 400	33.46	1.00

The values of each computed KPI in the past 15 months are reported in Table 1.19. In the current month (with 23 working days), the company has recorded the following data: 3 568 orders, 15 complaints, $ 560 400 in stock value, 24.25 customers on average served on a journey, 4 950 planned journeys and 4 902 journeys made. Build a control panel for monitoring the three identified KPI families, according to the methodology illustrated in Section 1.6.3. Show how the actions on the logistics system can change depending on the weights assigned to each KPI belonging to the first and third families.

2

Forecasting logistics requirements

2.1 Introduction

Forecasting is an attempt to determine in advance the most likely outcome of an uncertain variable. Planning and controlling logistics systems need predictions for the level of future economic activities because of the time lag in matching supply to demand. Typical decisions that must be made before some data are known include the main phases of the network planning process (e.g. facility location and capacity purchasing) as well as production scheduling, inventory management and transportation planning. Logistics requirements to be predicted include customer demand, raw material prices, labour costs and lead times. See Table 2.1 for a list of the main forecasts required by the various planning and control decisions.

Forecasting methods are equally relevant to every kind of logistics system, although they are crucial for MTS systems (see Section 1.2.2), where inventory levels have to be set in every facility.

Forecasting is based on some hypotheses. No forecasting method can be deemed to be superior to others in every respect. As a matter of fact, in order to generate a forecast the historical data must show some degree of regularity. For instance, the historical data pattern must remain nearly the same in the future or the data entries must depend to some extent on the past values of a set of variables.

Long-term, medium-term and short-term forecasts. Forecasts can be classified on the basis of the time horizon they refer to. *Long-term* forecasts span a time horizon from one to five years. Predictions for longer time periods are very

---

Introduction to Logistics Systems Management, Second Edition. Gianpaolo Ghiani, Gilbert Laporte and Roberto Musmanno.

Table 2.1 Main forecasts required by logistics systems planning and control.

Decision area	Forecast
Purchasing	Price of raw materials, components and semi-finished goods Availability of raw materials, components and semi-finished goods
Facility location	Fixed location costs Variable location costs Demand of the logistic nodes to be served
Inventory	Picking rate of the products in the warehouse Inventory costs
Distribution	Travel times Customer demand (both location and time pattern)

unreliable, since political and technological issues come into play. Long-term forecasts are used for deciding whether a new item should be put on the market, or whether an old one should be withdrawn, as well as in designing a logistics network. Such forecasts are often generated for a whole group of commodities (or services) rather than for a single item (or service). Moreover, in the long term, sector forecasts are more common than corporate ones. *Medium-term* forecasts extend over a time period ranging from a few months to one year. They are used for tactical logistical decisions, such as setting annual production and distribution plans, inventory management and slot allocation in warehouses. *Short-term* forecasts cover a time interval ranging from a few days to several weeks. They are employed to schedule and reschedule resources in order to meet medium-term production and distribution targets. As service requests are received, there is less need for forecasts. Consequently, forecasts for a shorter time interval (a few hours or a single day) are quite uncommon, except for highly dynamic environments such as urban express parcel delivery.

The role of the logistician in generating forecasts. Medium- and long-term forecasts are hardly ever left to the logistician. More frequently, this task is assigned to marketing managers who try to influence forecasts, for example by launching an advertising campaign for those items whose sales are in decline. On the other hand, the logistician often produces short-term forecasts.

Arves, an American company headquartered in Kansas City, produces electronic games. In 2001, in order to counter the decline in sales expected for the subsequent year, which would have a serious economic-financial consequence, sales management has considered the possibility of increasing the company's market share by 3%, by means of the following actions:

- reducing the retail price by 15%;
- Internet sales promotions through the 'pay-per-click' option.

Forecasting approaches can be classified in two main categories: qualitative and quantitative methods.

2.2 Qualitative methods

Qualitative methods are mainly based on expert judgement or on experimental approaches, although they can also make use of simple mathematical tools to combine different forecasts. Qualitative methods are usually employed for long- and medium-term forecasts when there are not enough data to use a quantitative approach. This is the case, for example, when a new product or service is launched on the market, when a product packaging is changed or when the forecasts are expected to be affected by political changeovers or technological advances. The most widely used qualitative methods are *management judgement*, the *Delphi method* and *market research*.

In management judgement, a forecast is developed by the workforce, for example the company's management or sales force. As a rule, the workforce can provide accurate estimates since it knows a lot about the company's business, including shifts in customers' behaviour and the profile of prospective customers.

China currently has a share of about 30% of world sales of luxury articles. Experts of McKinsey estimate an annual growth of the Chinese market for these articles equal to about $ 15 billion for the next five years.

In the Delphi method, a series of questionnaires is submitted to a panel of experts. Every time a group of questions is answered, new sets of information become available. Then a new questionnaire is prepared by a coordinator in such a way that every expert is faced with the new findings. This procedure eliminates the bandwagon effect of majority opinion. The Delphi method terminates as soon as all experts share the same viewpoint. This technique is mainly used to estimate the influence of political or macro-economical changes on data patterns.

The Delphi method has been used recently to estimate the tourist flows in the Lazio coastal region in Italy. The group of experts involved was made up of 800 hotel managers and tour operators, coordinated by a team of 10 employees of the Lazio regional authority.

Market research is based on interviews with potential consumers or users. It is time consuming and requires a deep knowledge of sampling theory. For these reasons, it is used only occasionally, for example when deciding whether a new product should be launched.

Tienda is a Spanish company manufacturing and marketing aromatic oils. Its lemon-scented oil, obtained from citrus fruits and olives, has recently been introduced. To forecast its demand, the company has asked a market research company to carry out a survey. For this purpose, a sample of 1 455 customers were selected in 32 Spanish supermarkets in the area of Seville. Each questionnaire allowed researchers to estimate the probability that the customer would buy the new product. On the basis of these data, a forecast of the new product's sales was generated.

Table 2.2 summarizes the features of the main qualitative forecasting methods.

Table 2.2 Features of the main qualitative methods.

Feature	Forecasting method		
	Management judgement	Delphi method	Market survey
Time horizon	Medium or short	Medium or long	Medium
Effort	Moderate, if based on the company's experts	Large	Large
Costs	Low, if based on the company's experts	High	High
Data required	No	No	No
Accuracy	Low	High	Moderate

2.3 Quantitative methods

Quantitative methods can be used every time there are enough data. Let y_t, $t = 1, \ldots, T$, be the sequence of the T past observations of the variable to be forecasted, arranged according to the time of their outcome (*time series* or *historical data*). For the sake of simplicity, it is assumed that all the periods are equally spaced in time. The choice of most suitable technique depends on the nature of the variable to be predicted, as well as on the amount and quality of the available data.

2.3.1 Graphical representation of time series

The simplest and most intuitive way to study a time series is to use a graphical representation (a simple Cartesian diagram (t, y_t)) which allows one to interpret the data more easily than in a table. The visual analysis of the diagram can be used as a support to most complex methods (e.g. to identify a linear trend in the data or to detect possible outliers) or even to visually extrapolate the data. In order to emphasize the main features of the time series (which is discrete in nature), the T dots corresponding to the past observations are often connected through a continuous line.

Figure 2.1 provides a graphical representation of the time series of the Gross National Product of the United States (in billions of dollars) from 1889 to 1900.

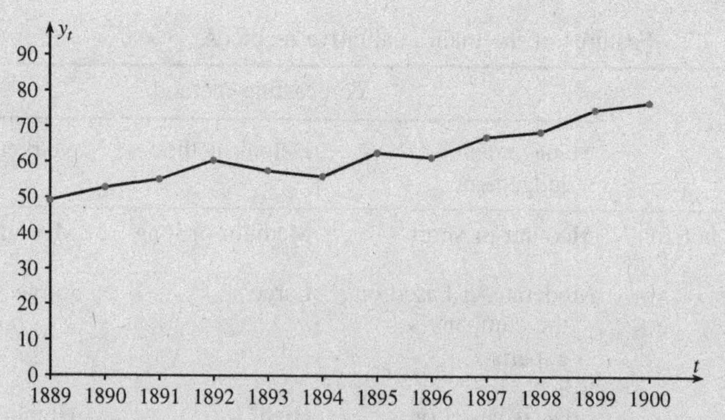

Figure 2.1 Annual trend ($t = 1, \ldots, 12$) of the Gross National Product of the United States (in billions of dollars) from 1889 to 1900.

2.3.2 Classification of time series

Time series can be classified according to the *density index*, defined as the fraction of past observations which are zero. If a time series has a low density index (usually $< 30\%$), then it is called *continuous* (see Figure 2.2).

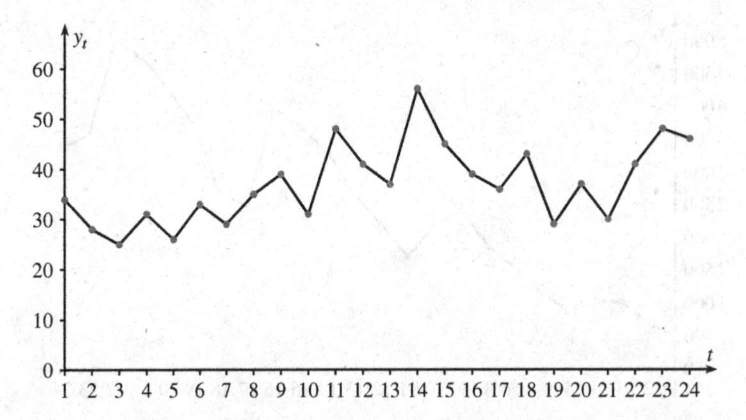

Figure 2.2 A continuous time series.

In contrast, a time series is *sporadic* when it is characterized by a significant proportion (usually more than 30%) of zero values (see Figure 2.3). Typical sporadic time series are those of products whose sales volumes are low. If zero values alternate regularly with strictly positive observations, then the sporadic time series is called *periodic*, otherwise it is called *random*.

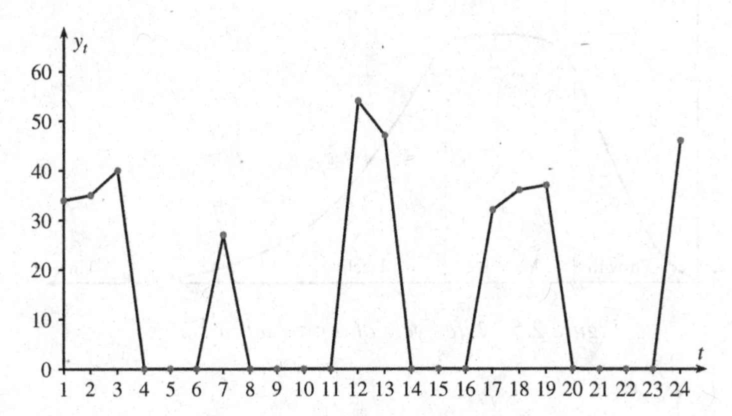

Figure 2.3 A sporadic time series.

Moreover, a time series is said to be *regular* if it can be decomposed into four main components: *trend*, *cyclical variation*, *seasonal variation* and *residual variation* (see Figure 2.4). Otherwise, it is defined as *irregular* (see Figure 2.6).

- *Trend.* The trend is the long-term modification of data patterns over time; it may depend on changes in population and on the product (or service) life cycle (see Figure 2.5).

- *Cyclical variation.* Cyclical variation is caused by the so-called business cycle, which depends on macro-economic issues. It is quite irregular, but its pattern is roughly periodic.

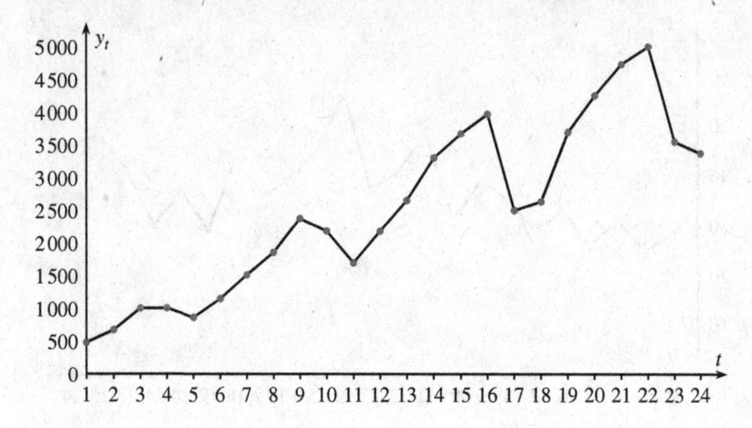

Figure 2.4 A regular time series.

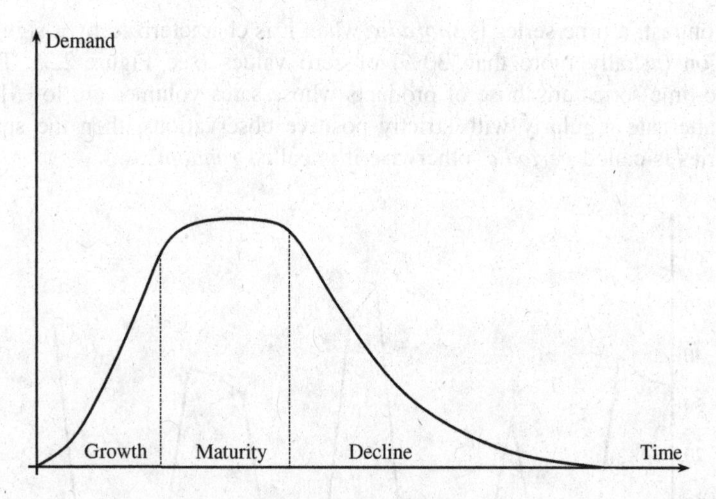

Figure 2.5 Life cycle of a product or service.

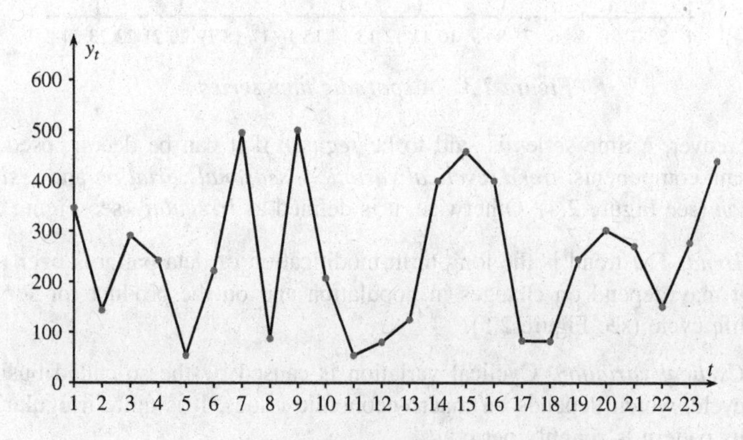

Figure 2.6 An irregular time series.

- *Seasonal variation.* Seasonal variation is caused by the periodicity of several human activities. Typical examples are the ups and downs in the demand of some items over the year. This type of effect can also be observed on a weekly horizon (e.g. some product sales are higher on weekends than on working days). For example, in Figure 2.4 the length of the seasonal cycle is equal to six time periods.
- *Residual variation.* Residual variation is the portion of the data pattern that cannot be interpreted as trend, cyclical or seasonal variation. It is often the result of numerous causes, each of which has a small impact.

2.3.2.1 The forecasting process

The forecasting process is usually divided into three main steps.

1. *Data preprocessing.* Data are seldom ready to be used to make a forecast. Indeed, outliers need to identified, and possibly removed in such a way that they do not affect the predictions.
2. *Choice of the forecasting method.* Here, the most suitable forecasting method is selected among a set of alternative techniques based on the accuracy they would have provided if used in the past. For parametrized methods, this step also includes the determination of the optimal value of each parameter.
3. *Evaluation of the forecasting accuracy.* Once the forecasted variable becomes known, an a posteriori error can be computed. Such errors are then combined in order to assess the accuracy of the method currently used. This measure can be used to finely tune up the parameters of the method or even to select an alternative technique.

These three phases will be described in detail in the following sections.

2.4 Data preprocessing

Past observations are often characterized by

- the absence of some data;
- errors or outliers;
- inconsistency or incoherence.

For these reasons, time series usually need to be preprocessed as follows:

- data cleaning;
- fusion of data from different sources;
- data transformation.

The main preprocessing operations on a time series are illustrated throughout the rest of this section.

2.4.1 Insertion of missing data

In the simplest case, a missing observation can be replaced with the average of the previous and subsequent observations.

Table 2.3 shows the car sales in Argentina during the last 12 months. The missing observation (month 6) can be set as the average of the sales in months 5 and 7: $y_6 = (y_5 + y_7)/2 = (38\,521 + 41\,345)/2 = 39\,933$.

Table 2.3 Monthly sales of cars in Argentina during last year.

Month	Sales	Month	Sales
1	61 945	7	41 345
2	35 379	8	43 866
3	43 535	9	49 379
4	39 317	10	39 533
5	38 521	11	41 611
6	–	12	52 458

2.4.2 Detection of outliers

An outlier is an observation that appears to deviate significantly from the other members of the time series. Outliers can have many causes: an error in data transmission or recording, an instrument error or simply natural deviations in populations (which is often the case, e.g., of *heavy-tailed* distributions). Except in the latter case, one may wish to discard the outliers that otherwise might mislead most forecasting methods. Then, the removed observations can be treated as missing data (see Section 2.4.1).

The identification of the outliers is quite complex in general since it has to consider a number of features of the time series, including an increasing or decreasing trend, a seasonal effect and so on. In case of constant trend and no seasonal effect, the following rule of thumb can be used: the first and third quartiles, Q_1 and Q_3, respectively, of the time series are identified and the data entries outside the interval $[Q_1 - 1.5(Q_3 - Q_1), Q_3 + 1.5(Q_3 - Q_1)]$ are tagged as outliers. The idea behind this rule is that the entries less than $Q_1 - 1.5(Q_3 - Q_1)$, or greater than $Q_3 + 1.5(Q_3 - Q_1)$, deviate significantly from the 50% most central data entries (which are included in $[Q_1, Q_3]$).

Elleshop distributes electrical appliances in Austria. Table 2.4 reports its sales of TV sets in the province of Klagenfurt during the last 12 months. Since the trend is constant and there are no cyclical effects, we can apply the rule of thumb described in this section. The first and third quartiles of the time series are, respectively, $Q_1 = 866.25$, $Q_3 = 977.50$. Consequently, the interval $[Q_1 - 1.5(Q_3 - Q_1), Q_3 + 1.5(Q_3 - Q_1)]$ is [699.375, 1144.375]. This indicates that the observation related to month 8 is an outlier. The value y_8 is therefore eliminated and replaced by 825, the average of the sales volumes in months 7 and 9.

Table 2.4 Number of TV sets delivered during the last 12 months by Elleshop.

Period	Sales	Period	Sales
1	975	7	770
2	1 025	8	200
3	895	9	880
4	1 055	10	870
5	925	11	915
6	985	12	855

2.4.3 Data aggregation

Forecasts made on aggregated data (e.g. sales in a given geographic area) are usually much more accurate than those made on disaggregated data (e.g. sales in each individual retailer in that area). This phenomenon can be explained as follows. Let $X_1, \ldots, X_n$ be n random variables modelling the disaggregated data. For the sake of simplicity, these variables are assumed to be independent and have the same distribution (in particular, the same mean μ_X and the same variance σ_X^2). Moreover, let Y be the random variable modelling the aggregated data: $Y = X_1 + \ldots + X_n$. Then, the mean μ_Y and the variance σ_Y^2 of Y are

$$\mu_Y = n\mu_X;$$

$$\sigma_Y^2 = n\sigma_X^2.$$

Hence, the coefficient of variation of Y (representing the relative dispersion of Y around the mean μ_Y, assuming $\mu_X > 0$)

$$\sigma_Y/\mu_Y = (1/\sqrt{n})(\sigma_X/\mu_X) \tag{2.1}$$

is less than the relative dispersion of each variable $X_1, \dots, X_n$ around the common mean μ_X. Therefore, the forecasts on Y are expected to be more accurate than those of each $X_1, \dots, X_n$ variable.

The demand for champagne during the next year in three regions of France (Burgundy, Alsace and Provence) is known to be modelled by three independent random variables X_1, X_2 and X_3 whose mean (in thousands of bottles), variance and coefficient of variation are shown in Table 2.5. The aggregated demand Y is characterized by the following statistics:

$$\mu_Y = \mu_{X_1} + \mu_{X_2} + \mu_{X_3} = 6\,800;$$

$$\sigma_Y^2 = \sigma_{X_1}^2 + \sigma_{X_2}^2 + \sigma_{X_3}^2 = 600;$$

$$\sigma_Y / \mu_Y = 0.0036.$$

As expected, the relative dispersion of the aggregated demand is less than that in each individual region.

Table 2.5 Demand for champagne in three regions of France.

Geographic area (i)	Mean (μ_{X_i})	Variance ($\sigma_{X_i}^2$)	Coefficient of variation (σ_{X_i}/μ_{X_i})
Burgundy	2 500	300	0.0069
Alsace	1 800	200	0.0079
Provence	2 500	100	0.0040

Data aggregation can be important when dealing with sporadic time series since the presence of zero values in the time series could cause errors most forecasting methods. In order to overcome this problem, the available data can be grouped by product type, by different geographic areas or by time periods.

The daily sales of Sidol75 aluminum-polishing packages in each individual supermarket of the Suomen chain in Lahti province, Finland, are lumpy (see e.g. the past week time series for a sample supermarket shown in Table 2.6). In order to carry out an accurate forecast, the sales manager decided to group the sales data of all supermarkets in the province of Lahti and to consider weekly sales figures instead of the daily data. The resulting time series is shown in Table 2.7. Based on these data, the demand for Sidol75 in the Lahti province for the following week was estimated at 264 packages.

Table 2.6 Daily sales of Sidol75 in a sample Suomen supermarket in the province of Lahti.

Day	Sales
1	4
2	0
3	0
4	3
5	8
6	7
7	0

Table 2.7 Weekly sales of Sidol75 sold in the Suomen supermarkets in the province of Lahti during the last 20 weeks.

Week	Sales	Week	Sales
1	254	11	263
2	262	12	265
3	260	13	271
4	264	14	256
5	255	15	269
6	258	16	262
7	262	17	258
8	267	18	262
9	256	19	264
10	259	20	265

2.4.4 Removing the calendar variations

Some time series that represent a total of some variable for a given time period (e.g. a month) contain calendar effects including variable month length, day-of-the-week effects and holidays. In such cases it is most important to remove the calendar variations. The simplest way to achieve this is to replace each past observation $y_t, t = 1, \ldots, T$, with the adjusted $y'_t = w_t y_t$, where w_t is a suitably determined coefficient. If, for example, y_t is a monthly time series, the coefficient $w_t, t = 1, \ldots, T$, can be calculated as follows:

$$w_t = \bar{n}/n_t, t = 1, \ldots, T,$$

where $\bar{n}$ is the average number of days in a month (which can be assumed to be 365/12), while n_t is the number of (working) days during month t.

Sotam is a Tunisian producer of orangeade. Table 2.8 shows the quantity of oranges (in quintals) used by its plant during the last 10 weeks. The same table also indicates the working days on which the plant was operating. The average number $\bar{n}$ of working days in the 10 weeks was 4.80. Each weight $w_t, t = 1, \ldots, 10$, can therefore be determined as $w_t = \bar{n}/n_t$. Once the weights are known, the modified values of the time series are easily determined, as shown in Table 2.9.

Table 2.8 Quantity of oranges and number of working days for each week in the Sotam problem.

Week	Sales (q)	Number of working days	Week	Sales (q)	Number of working days
1	34 500	4	6	36 090	5
2	36 080	5	7	35 820	4
3	36 380	5	8	36 050	5
4	36 150	5	9	36 240	5
5	36 120	5	10	36 150	5

Table 2.9 Modified time series of the Sotam problem obtained by removing calendar variations.

t	w_t	y'_t	t	w_t	y'_t
1	1.20	41 400.00	6	0.96	34 646.40
2	0.96	34 636.80	7	1.20	42 984.00
3	0.96	34 924.80	8	0.96	34 608.00
4	0.96	34 704.00	9	0.96	34 790.40
5	0.96	34 675.20	10	0.96	34 704.00

2.4.5 Deflating monetary time series

Inflation is a significant component of apparent growth in any monetary time series, that is, time series whose observations are measured in euros, dollars or any other currency (e.g. the sales of a finished product or the prices of a raw material). By adjusting for inflation, the real variations over time can be identified. Let r_k be the average rate of inflation in time period k. *Inflation adjustment*, or *deflation*, is accomplished by dividing each observation y_t of the monetary time series by a price index w_t (such as the Consumer Price Index), which measures the variation in prices from time period 1 to time period t. The deflated time series $y'_t = y_t/w_t$ is said to be measured in *constant dollars* (or euros etc.), whereas the original time series was measured in *nominal* or *current dollars* (or euros etc.). Because of the capitalization effect, the price index is expressed by

$$w_t = \prod_{k=1}^{t-1}(1 + r_k), t = 2, \ldots, T \tag{2.2}$$

and $w_1 = 1$.

Cavis is a wine-making company that sells its products almost exclusively in France. The annual sales during the last 10 years are reported in Table 2.10. The same table also shows the annual rate of inflation recorded in the time period. The deflated data, obtained through (2.2), are reported in Table 2.11.

Table 2.10 Annual Cavis sales and the corresponding annual inflation rates.

Year	Inflation rate (%)	Sales (M€)	Year	Inflation rate (%)	Sales (M€)
1	2.80	1.03	6	2.10	1.31
2	2.50	1.13	7	1.80	1.36
3	2.70	1.20	8	3.30	1.29
4	2.20	1.24	9	0.80	1.33
5	2.00	1.26	10	1.50	1.37

Table 2.11 Annual deflated sales by Cavis.

Year	Price index (%)	Deflated sales (M€)	Year	Price index (%)	Deflated sales (M€)
1	100.00	1.03	6	112.81	1.16
2	102.80	1.10	7	115.18	1.18
3	105.37	1.14	8	117.25	1.10
4	108.21	1.15	9	121.12	1.10
5	110.60	1.14	10	122.09	1.12

2.4.6 Adjusting for population variations

When forecasting future sales (or another economic variable) in a given geographic area, demographic variations need to be taken into account. Let a_t be the reference population in time period $t, t = 1, \ldots, T$. Let $y_t, t = 1, \ldots, T$, be the expenditure time series. Then, the forecasts are carried out on $y'_t = y_t/w_t$, where w_t is

$$w_t = a_t/a_1, t = 1, \ldots, T.$$

Salus is a private company specialized in assistance services to the elderly. It has been operating throughout Lombardy for 10 years. The annual number of customers making use of the services offered by Salus is shown in Table 2.12. Considering the annual population of Lombardy (second column of Table 2.13) during the same 10 years, the modified time series is obtained as shown in Table 2.13.

Table 2.12 Annual number of Salus customers.

Year	Number of customers	Year	Number of customers
1	1 435	6	5 056
2	2 887	7	5 432
3	3 450	8	5 382
4	4 578	9	5 920
5	4 935	10	6 003

Table 2.13 Adjustment of the number of Salus customers for variation of the population.

t	a_t	w_t	y'_t	t	a_t	w_t	y'_t
1	9 121 714	1.000	1 435.000	6	9 475 202	1.039	4 867.378
2	9 033 602	0.990	2 915.159	7	9 545 441	1.046	5 190.871
3	9 108 645	0.999	3 454.950	8	9 642 406	1.057	5 091.371
4	9 246 796	1.014	4 516.073	9	9 742 676	1.068	5 542.681
5	9 393 092	1.030	4 792.422	10	9 826 141	1.077	5 572.650

2.4.7 Normalizing the data

Normalization of a time series y_t, $t = 1, \ldots, T$, is obtained by transforming it into another time series y'_t, $t = 1, \ldots, T$, whose observations belong to a given interval [*min*, *max*] (where, of course, $min < max$). Each value y_t, $t = 1, \ldots, T$, will correspond to:

$$y'_t = \frac{y_t - a}{b - a}(max - min) + min,$$

where a and b represent, respectively, the minimum and maximum values of y_t, $t = 1, \ldots, T$. The most common normalization is the [0, 1] normalization.

The time series of Sotam shown in Table 2.9 (third and sixth columns), when normalized in the interval [2 000, 4 000], will result in Table 2.14.

Table 2.14 Sotam time series normalized in interval [2 000, 4 000].

t	y'_t	t	y'_t
1	3 621.78	6	2 009.17
2	2 006.88	7	4 000.00
3	2 075.64	8	2 000.00
4	2 022.92	9	2 043.55
5	2 016.05	10	2 022.92

2.5 Choice of the forecasting method

2.5.1 Notation

Let

$$p_t(\tau), \tau = 1, \ldots,$$

be the τ periods ahead forecast, made at time period t, that is, the forecast on $y_{t+\tau}$ made at t. For the sake of simplicity, the one-period ahead forecast ($\tau = 1$) is denoted as

$$p_t(1) = p_{t+1}.$$

Once parameter y_t becomes known at time t, it can be compared to the forecast made τ periods before $p_i(\tau)$ (with $i + \tau = t$) in order to compute the error:

$$e_i(\tau) = y_t - p_i(\tau), i + \tau = t.$$

Again, if $\tau = 1 (i = t - 1)$, the notation can be simplified:

$$e_{t-1}(1) = e_t = y_t - p_t.$$

2.5.2 Causal versus extrapolation methods

Quantitative forecasting methods can be classified into two macro-categories.

- *Causal methods*, which are based on the hypothesis that a future pattern depends on the past or current values of some variables.

The demand for small, economical automobiles is related to the business cycle of a country and can, therefore, be correlated to its Gross National Product.

These methods can be applied when it is possible to express the relationship between the present or past value of the causal variables and the future value of the variable to forecast:

$$p_T(\tau) = f(x_{11}, \ldots, x_{1T}, x_{n1}, \ldots, x_{nT}), \tau = 1, \ldots, \tag{2.3}$$

where $x_{i1}, \ldots, x_{iT}$ are the T known values of the independent variable X_i at time period t, $t = 1, \ldots, T$.
Relationship (2.3) can be linear, that is,

$$p_T(\tau) = a_0 + \sum_{i=1}^{n} \sum_{t=1}^{T} a_{it} x_{it}, \tau = 1, \ldots,$$

or nonlinear. Coefficients $a_0, a_{it}, i = 1, \ldots, n, t = 1, \ldots, T$, can be estimated through well-known statistical methods, like *regression analysis* (see Problem 2.5). The greatest advantage of causal methods lies in their capacity to anticipate the variations of the variable to be forecasted, so they are particularly useful in medium- and long-term forecasts. Unfortunately, in many cases, it is difficult to identify causal variables having a high correlation with the dependent variable. For this reason, causal methods are not very popular in logistics and will not be further examined in this volume.

- *Time series extrapolation*, which assumes that some features of the past time data pattern will remain the same. The data pattern is then projected in the future. The reliability of these techniques is therefore greater in the short and medium term. The main extrapolative methods will be illustrated in the following sections of this chapter.

2.5.3 Decomposition method

The decomposition method can be used for continuous and regular time series, which can be decomposed into a trend, a cyclical variation, a seasonal variation and a residual variation. These components can be combined in a *multiplicative* model:

$$y_t = q_t v_t s_t r_t, t = 1, \ldots, T, \tag{2.4}$$

where q_t represents the trend at time period t, v_t is an index representing the business cycle at time t, s_t is the seasonal index at time period t and r_t is the residual index at time period t.

In model (2.4), q_t is expressed in the same unit of measurement as $y_t, t = 1, \ldots, T$. Consequently, v_t, s_t and $r_t, t = 1, \ldots, T$, are dimensionless and greater than 0. Of course, if one of the effects is absent, then the corresponding index should be assumed equal to 1. In addition, let M be the periodicity of the seasonal cycle. Then, the average of the seasonal indices in M consecutive time periods should be 1:

$$\frac{\sum_{t=j+1}^{j+M} s_t}{M} = 1, j = 0, \ldots, T - M.$$

Alternatively, the four effects can be combined as follows:

- additive model:

$$y_t = q_t + v_t + s_t + r_t, t = 1, \ldots, T;$$

- mixed model:

$$y_t = (q_t + v_t)s_t r_t, t = 1, \ldots, T;$$

- logarithmic model:

$$\log y_t = \log q_t + \log v_t + \log s_t + \log r_t, t = 1, \ldots, T.$$

The logarithmic model can be seen as the logarithmic transformation of the multiplicative model, used to transform the multiplicative model into an additive one.

As a rule, if the cyclical, seasonal and residual variations do not depend on the trend, then the most suitable model is the additive one. Otherwise, a multiplicative model is more suitable, which is often the case of most economic time series. In the following, we will refer to the multiplicative model. The decomposition method is divided into the following three steps.

1. the time series $y_t, t = 1, \ldots, T$, is broken down into its four components $q_t, v_t, s_t, r_t, t = 1, \ldots, T$;
2. q, v and s are projected one or more periods ahead;
3. these projections are combined in the following way:

$$p_T(\tau) = q_T(\tau)v_T(\tau)s_T(\tau), \tau = 1, \ldots, \tag{2.5}$$

   to produce the data forecasts for the required future time periods $t + \tau$.

Phase 1 is made up of the following steps.

2.5.3.1 Evaluating the combined trend-cyclical variation

The combined trend-cyclical variation is obtained by removing from $y_t, t = 1, \ldots, T$, the seasonal variation and the residual variation. This can be done by observing that the average value of M consecutive observations does not include any seasonal variation. Moreover, the influence of the residual variation is quite low, especially if M is sufficiently large (e.g. $M = 12$). Thus, the time series

$$\frac{y_1 + \cdots + y_M}{M}, \ldots, \frac{y_{T-M+1} + \cdots + y_T}{M}, \tag{2.6}$$

includes only the trend and the cyclical variation.

If M is an odd number, then (2.6) refers to the median of the interval defined by the first M time periods, that is, $t = \lceil \frac{M}{2} \rceil$. Similarly, for the subsequent time periods the following correspondences hold:

$$(qv)_{\lceil \frac{M}{2} \rceil} = \frac{y_1 + \cdots + y_M}{M},$$

$$\ldots$$

$$(qv)_{T-\lceil \frac{M}{2} \rceil+1} = \frac{y_{T-M+1} + \cdots + y_T}{M}.$$

If M is an even number, the elements of (2.6) do not refer to a well-defined time period. For example, if $M = 12$, the left-hand side of (2.6) would correspond to a time period included between the sixth and the seventh time periods. This drawback can be overcome by weighting the observations $y_{t-M/2}$ and $y_{t+M/2}$ with a coefficient equal to $1/2$, so the following time series can be obtained:

$$(qv)_t = \frac{\frac{1}{2}y_{t-M/2} + y_{t-M/2+1} + \cdots + y_{t+M/2-1} + \frac{1}{2}y_{t+M/2}}{M}, \tag{2.7}$$

$$t = \frac{M}{2} + 1, \ldots, T - \frac{M}{2}.$$

2.5.3.2 Separating the trend and the cyclical variation

In most cases, the trend can be described by a simple functional relationship, such as a linear or quadratic function. It can therefore be obtained by a simple regression analysis applied to time series $(qv)_t, t = j, \ldots, J$ (where $j = \lceil \frac{M}{2} \rceil$ and $J = T - \lceil \frac{M}{2} \rceil + 1$, if M is odd; $j = \frac{M}{2} + 1$, and $J = T - \frac{M}{2}$, if M is even).

For example, if the trend is linear, then

$$q_t = a + bt, t = j, \ldots, J,$$

where the coefficients a and b can be easily obtained using the least squares method.

Known $q_t, t = j, \ldots, J$, the cyclical variation $v_t, t = j, \ldots, J$, can be calculated for every $t = j, \ldots, J$ as

$$v_t = \frac{(qv)_t}{q_t}.$$

2.5.3.3 Evaluating the combined seasonal-residual variation

The combined seasonal-residual variation $(sr)_t, t = j, \ldots, J$, can be obtained as:

$$(sr)_t = \frac{y_t}{(qv)_t}.$$

2.5.3.4 Separating the seasonal variation and the residual variation

The seasonal variation can be expressed by means of M indices $\bar{s}_1, \ldots, \bar{s}_M$, satisfying

$$s_{kM+t} = \bar{s}_t, t = 1, \ldots, M, k = 0, 1, \ldots \tag{2.8}$$

Each index $\bar{s}_t, t = 1, \ldots, M$, can be computed as the average of the items of time series $(sr)_t, t = j, \ldots, J$, corresponding to *homologous* time periods. Time periods are homologous when they are equidistant from each other by M time periods, so that $\bar{s}_t, t = 1, \ldots, M$, is the arithmetic mean of $(sr)_t, (sr)_{M+t}, \ldots$. As observed previously, the arithmetic mean reduces the residual variation noticeably. Hence,

$$\frac{\sum_{t=1}^{M} \bar{s}_t}{M} = 1. \tag{2.9}$$

Whenever the relationship (2.9) is not satisfied, instead of $\bar{s}_t$, the following normalized index can be used:

$$\tilde{s}_t = \frac{M\bar{s}_t}{\sum_{t=1}^{M} \bar{s}_t}, t = 1, \ldots, M.$$

Of course,

$$\frac{\sum_{t=1}^{M} \tilde{s}_t}{M} = 1.$$

Finally, the random variation $r_t, t = j, \ldots, J$, can be obtained by dividing each item of the time series $(sr)_t, t = j, \ldots, J$, by the corresponding seasonal index $s_t, t = j, \ldots, J$:

$$r_t = \frac{(sr)_t}{s_t}.$$

If the decomposition has been done correctly, the mean value of $r_t, t = j, \ldots, J$, is close to 1.

The subsequent phase leads to project q, v and s over one or more future time periods. The cyclical variation is estimated in a qualitative way, on the basis of information about the business cycle, or more simply by letting

$$v_t(\tau) = v_J, \tau = 1, \ldots,$$

that is, setting $v_t(\tau), \tau = 1, \ldots$, equal to the current value of the cyclical variation. This assumption is particularly suitable for short-term forecasts, since variations in the business cycle can be appreciated only after several months or even a few years.

Finally, forecasts can be obtained by combining, according to (2.5), the extrapolation of the trend, cyclical and seasonal variations.

P&A is a consulting firm, headquartered in Hannover, Germany, which has been asked to estimate the demand for electromedical equipment in the Niedersachsen region for the next six months. By examining the historical data, the company was able to assess the monthly sales during the last 163 months. These data (in k€), already preprocessed, are shown in Table 2.15 and in Figure 2.7. The periodicity M of the seasonal cycle is equal to 12.

Time series $(qv)_t$ (Table 2.16, Table 2.17 and Figure 2.8) was obtained using (2.7) for $t = 7, \ldots, 157$. An examination of Figure 2.9 allows us to assume that the trend is linear, that is,

$$q_t = a + bt, t = 7, \ldots, 157,$$

where the coefficients a and b, determined by using the least squares method, are

$$a = 638.51;$$

$$b = 1.43.$$

The indices $\bar{s}_1, \ldots, \bar{s}_{12}$ (Table 2.19 and Figure 2.12) verify condition (2.9), while the mean value of the residual variation is about 1. Therefore, the hypothesis on which the decomposition was carried out can be considered correct.

The forecasts of the demand for the six subsequent time periods (Table 2.21 and Figure 2.15) were obtained by combining the projections of the trend as well as of the cyclical and seasonal variations (see also Figures 2.10, 2.11, 2.13 and Tables 2.18 and 2.19). The cyclical variation was estimated (Figure 2.14) by using a squared regression curve, constructed on the basis of the values of v_t, $t = 150, \ldots, 157$:

$$v_t = a(t - 149)^2 + b(t - 149) + c, t = 150, \ldots$$

Table 2.15 P&A monthly sales (in k€) of electromedical equipment in the Niedersachsen region, Germany.

Month	Sales	Month	Sales	Month	Sales	Month	Sales
1	511.7	42	820.4	83	728.0	124	877.1
2	468.3	43	795.9	84	817.6	125	959.7
3	571.9	44	774.9	85	618.1	126	916.3
4	648.2	45	750.4	86	565.6	127	870.8
5	705.6	46	759.5	87	691.6	128	832.3
6	709.1	47	740.6	88	768.6	129	760.2
7	676.9	48	809.9	89	903.0	130	833.7
8	661.5	49	603.4	90	847.7	131	827.4
9	611.8	50	558.6	91	830.9	132	864.5
10	640.5	51	711.2	92	772.1	133	705.6
11	611.1	52	760.9	93	755.3	134	619.5
12	697.2	53	840.0	94	779.1	135	723.1
13	548.8	54	835.8	95	770.0	136	847.7
14	492.1	55	777.0	96	844.2	137	942.9
15	613.2	56	727.3	97	671.3	138	917.0
16	692.3	57	714.0	98	607.6	139	897.4
17	721.7	58	744.8	99	737.8	140	859.6
18	672.0	59	723.1	100	863.1	141	821.8
19	670.6	60	770.7	101	908.6	142	872.2
20	635.6	61	581.0	102	891.1	143	795.9
21	611.8	62	555.8	103	853.3	144	824.6
22	686.0	63	665.7	104	836.5	145	669.9
23	630.7	64	770.7	105	797.3	146	618.1
24	750.4	65	836.5	106	840.7	147	756.0
25	515.2	66	779.1	107	816.9	148	901.6
26	498.4	67	745.5	108	872.2	149	968.8
27	627.2	68	739.2	109	613.9	150	968.8
28	741.3	69	676.2	110	595.0	151	921.2
29	760.9	70	710.5	111	744.1	152	891.1
30	754.6	71	711.9	112	812.0	153	882.0
31	733.6	72	731.5	113	941.5	154	887.6
32	704.9	73	598.5	114	940.1	155	840.0
33	709.8	74	578.9	115	863.1	156	935.9
34	733.6	75	675.5	116	829.5	157	763.7
35	714.7	76	756.0	117	808.5	158	700.0
36	831.6	77	865.2	118	800.1	159	844.2
37	586.6	78	819.0	119	836.5	160	989.1
38	536.9	79	800.8	120	870.8	161	1 045.8
39	654.5	80	758.1	121	684.6	162	1 012.9
40	767.9	81	737.8	122	644.7	163	970.9
41	848.4	82	774.9	123	721.0		

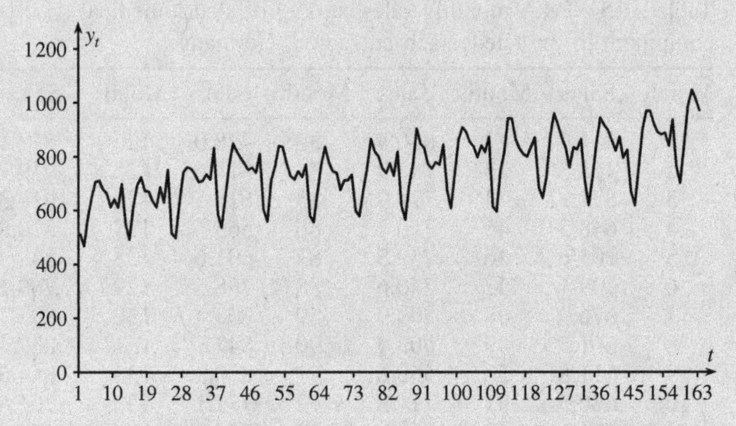

Figure 2.7 P&A monthly sales (in k€) of electromedical equipment.

Table 2.16 Determination of the combined trend and cyclical variation $(qv)_t, t = 7, \ldots, 157$, in the P&A problem.

t	$(qv)_t$	t	$(qv)_t$
7	627.70	...	...
8	630.23	152	864.65
9	632.95	153	871.73
10	636.50	154	879.05
11	639.01	155	885.91
12	638.14	156	890.95
...	...	157	894.86

Table 2.17 Trend $q_t, t = 7, \ldots, 157$, and cyclical variation $v_t, t = 7, \ldots, 157$, in the P&A problem.

t	$(qv)_t$	q_t	v_t	t	$(qv)_t$	q_t	v_t
7	627.70	648.52	0.97	...	...	...	...
8	630.23	650.95	0.97	152	864.65	856.02	1.01
9	632.95	651.39	0.97	153	871.73	857.45	1.02
10	636.50	652.82	0.98	154	879.05	858.88	1.02
11	639.01	654.25	0.98	155	885.91	860.31	1.03
12	638.14	655.68	0.97	156	891.95	861.74	1.03
...	...	...	...	157	894.86	863.17	1.04

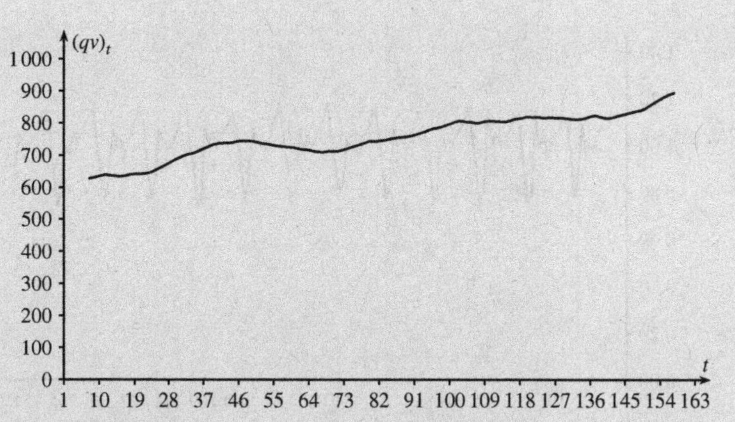

Figure 2.8 *Combined trend and cyclical variation* $(qv)_t$, $t = 7, \ldots, 157$, *in the P&A problem.*

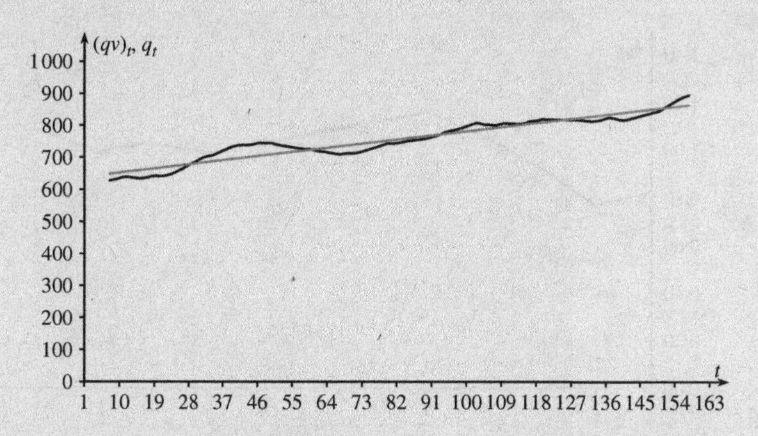

Figure 2.9 *Linear trend (in grey)* q_t, $t = 7, \ldots, 157$, *in the P&A problem.*

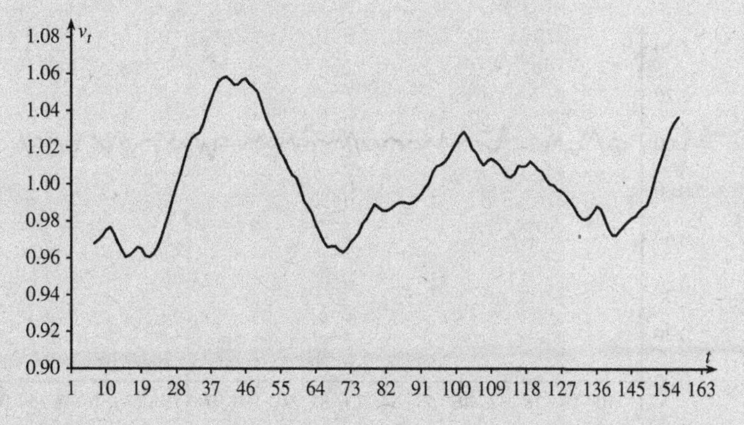

Figure 2.10 *Cyclical variation* v_t, $t = 7, \ldots, 157$, *in the P&A problem.*

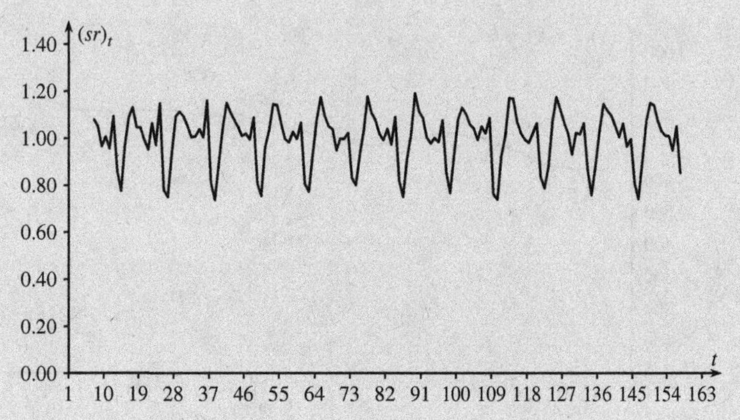

Figure 2.11 *Combined seasonal-residual variation* $(sr)_t$, $t = 7, \ldots, 157$, *in the P&A problem.*

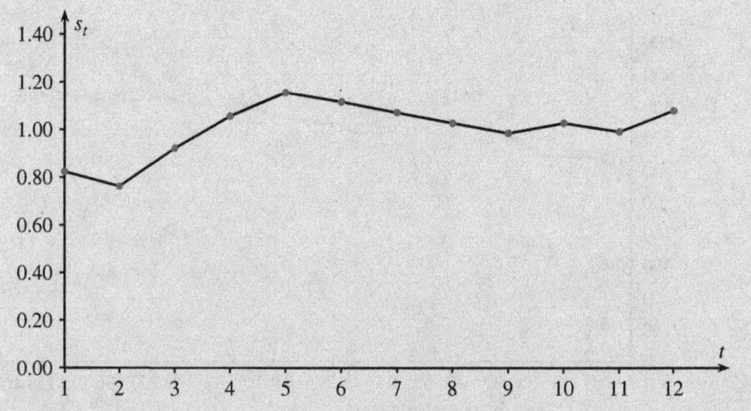

Figure 2.12 *Seasonal variation* $s_t = \bar{s}_t$, $t = 1, \ldots, 12$, *in the P&A problem.*

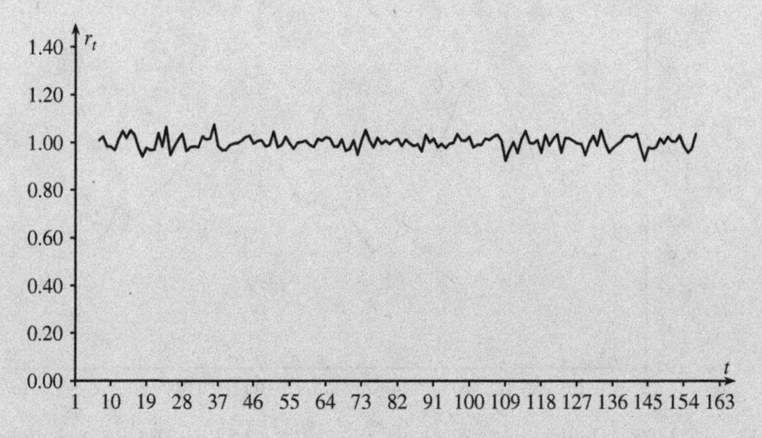

Figure 2.13 *Residual variation* r_t, $t = 7, \ldots, 157$, *in the P&A problem.*

Table 2.18 Combined seasonal-residual $(sr)_t, t = 7, \ldots, 157$, in the P&A problem.

t	y_t	$(qv)_t$	$(sr)_t$	t	y_t	$(qv)_t$	$(sr)_t$
7	676.9	627.70	1.08	...	...	...	...
8	661.5	630.23	1.05	152	891.1	864.6	1.03
9	611.8	632.95	0.97	153	882.0	871.7	1.01
10	640.5	636.50	1.01	154	887.6	879.1	1.01
11	611.1	639.01	0.96	155	840.0	885.9	0.95
12	697.2	638.14	1.09	156	935.9	891.0	1.05
...	...	...	...	157	763.7	894.9	0.85

Table 2.19 Evaluation of the seasonal variation $\bar{s}_t, t = 1, \ldots, 12$, in the P&A problem.

t	$\bar{s}_t$	t	$\bar{s}_t$
1	0.82	7	1.07
2	0.76	8	1.02
3	0.92	9	0.98
4	1.05	10	1.02
5	1.15	11	0.99
6	1.11	12	1.08

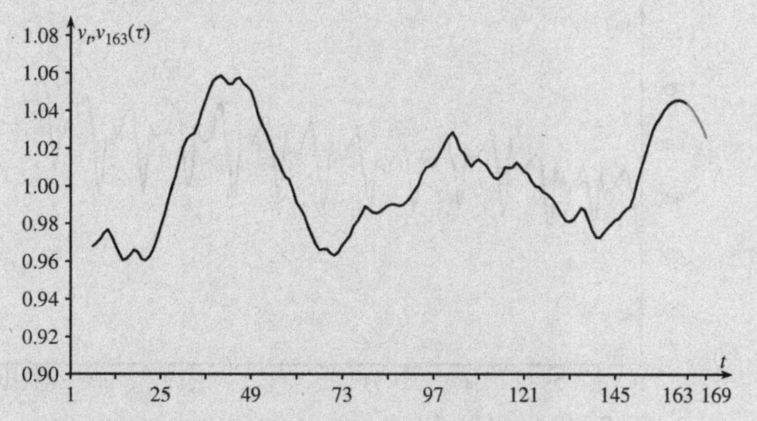

Figure 2.14 Projection of the cyclical variation (in grey) $v_T(\tau)$, $T = 163$, $\tau = 1, \ldots, 6$, in the P&A problem.

Table 2.20 Residual variation $r_t, t = 7, \ldots, 157$, in the P&A problem.

t	$(sr)_t$	s_t	r_t	t	$(sr)_t$	s_t	r_t
7	1.08	1.07	1.01	...	...	...	...
8	1.05	1.02	1.02	152	1.03	1.02	1.01
9	0.97	0.98	0.98	153	1.01	0.98	1.03
10	1.01	1.02	0.98	154	1.01	1.02	0.99
11	0.96	0.99	0.97	155	0.95	0.99	0.96
12	1.09	1.08	1.01	156	1.05	1.08	0.98
...	...	...	...	157	0.85	0.82	1.04

Table 2.21 Demand forecast $p_T(\tau)$, $T = 163$, $\tau = 1, \ldots, 6$, in the P&A problem.

$T + \tau$	$q_T(\tau)$	$v_T(\tau)$	$s_T(\tau)$	$p_T(\tau)$
164	873.19	1.04	1.02	933.84
165	874.62	1.04	0.98	895.28
166	876.05	1.04	1.02	932.52
167	877.48	1.04	0.99	898.46
168	878.92	1.03	1.08	976.03
169	880.35	1.03	0.82	743.76

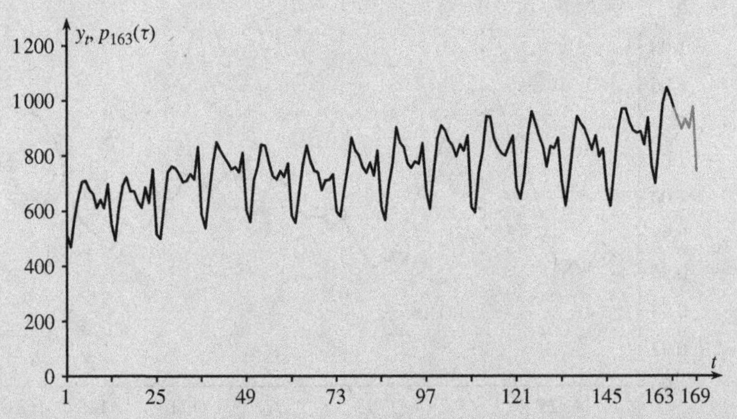

Figure 2.15 P&A monthly demand forecast (in grey) of electromedical equipment in the next six months.

Parameters a, b and c were obtained by minimizing the sum of squared errors in time periods $t = 150, \ldots, 157$:

$$a = -0.0004;$$

$$b = 0.0094;$$

$$c = 0.9856$$

(see also Tables 2.20 and 2.21.)

2.5.4 Further time series extrapolation methods: The constant trend case

We first analyse the case in which the past data pattern does not show any relevant cyclical and seasonal variations, and the trend is constant. We suppose initially that a forecast is generated only for one period ahead.

2.5.4.1 Elementary technique

The forecast for one period ahead is simply given by

$$p_{T+1} = y_T.$$

The method is straightforward. The forecast time series reproduces the data pattern with one-period delay. Consequently, it usually produces rather poor predictions.

Regens Book is an online bookseller with a distribution centre located in Patton, California, United States. The sales of the last 10 weeks are shown in Table 2.22. The data plot in Figure 2.16 shows that there is a constant trend.

By using the elementary technique, we obtain

$$p_{11} = y_{10} = 86.$$

Table 2.22 Weekly sales in the Regens Book problem.

Week	Sales	Week	Sales
1	89	6	80
2	106	7	88
3	92	8	87
4	98	9	92
5	77	10	86

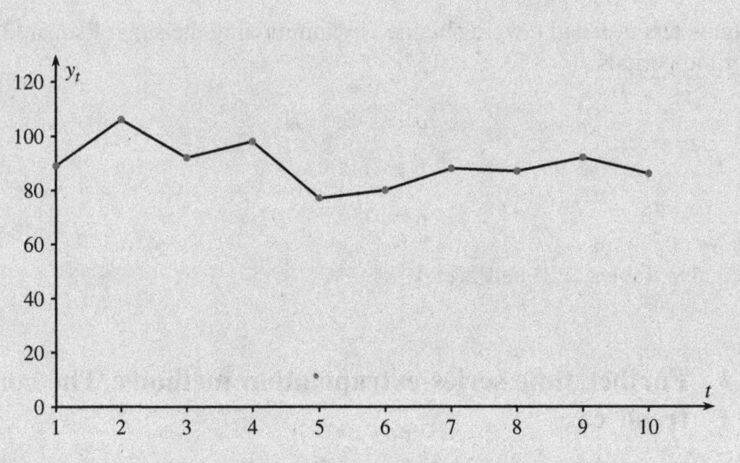

Figure 2.16 Weekly sales in the Regens Book problem.

2.5.4.2 Simple moving average method

The simple moving average method uses the average of the r (≥ 1) most recent data entries as the forecast for one period ahead:

$$p_{T+1} = \sum_{k=0}^{r-1} \frac{y_{T-k}}{r}.$$

If $r = 1$, the simple moving average method reduces to the elementary technique.

A key aspect of the simple moving average method is the choice of parameter r. A small value of r allows a rapid adjustment of the forecast to data pattern fluctuations but, at the same time, increases the influence of residual variations. In contrast, a high value of r effectively filters the residual variations, but produces a slow adaptation to data pattern variations. The optimal choice of r will be discussed in Section 2.7.2.

Using the simple moving average method for solving the Regens Book problem in this section, we obtain the forecasts

$$p_{11} = \frac{y_{10} + y_9}{2} = 89,$$

and

$$p_{11} = \frac{y_{10} + y_9 + y_8}{3} = 88.33,$$

with $r = 2$ and $r = 3$, respectively.

2.5.4.3 Weighted moving average method

This method is a variation of the simple moving average technique in which lower weights are assigned to older data. In particular, the forecast for one period ahead is:

$$p_{T+1} = \frac{\sum_{t=1}^{T} t y_t}{\sum_{t=1}^{T} t}.$$

By using the weighted moving average method to solve the Regens Book problem, we obtain

$$p_{11} = \frac{y_1 + 2y_2 + 3y_3 + 4y_4 + 5y_5 + 6y_6 + 7y_7 + 8y_8 + 9y_9 + 10y_{10}}{1+2+3+4+5+6+7+8+9+10} =$$

$$= \frac{4\,834}{55} = 87.89.$$

2.5.4.4 Exponential smoothing method

The exponential smoothing method (also called the *Brown method*) can be seen as a more sophisticated evolution over the simple moving average technique. The forecast of one period ahead is given by

$$p_{T+1} = \alpha y_T + (1 - \alpha) p_T, \tag{2.10}$$

where $\alpha \in [0, 1]$ is a *smoothing constant*. Here, p_T represents the forecasting value for time period T made at time period $T - 1$.

If we use the exponential smoothing method T with parameter $\alpha = 0.2$ and $y_T = 1\,020$, $p_T = 975$, we get

$$p_{T+1} = 0.2 \times 1\,020 + (1 - 0.2) \times 975 = 984.$$

Rewriting (2.10) as

$$p_{T+1} = p_T + \alpha(y_T - p_T) = p_T + \alpha e_T,$$

we obtain the following interpretation: the forecasting value at time period $T + 1$ corresponds to the sum of the value estimated at time period $T - 1$ and a fraction α of the forecasting error at time period T. This means that if the value of p_T is overestimated with respect to y_T, the forecast p_{T+1} is lower than p_T; likewise, when p_T is an underestimate of y_T, then p_{T+1} is increased.

All the historical data are embedded into p_T, and hence all do not appear explicitly in the previous formula. Applying Equation (2.10) recursively, all the historical data appear explicitly:

$$p_T = \alpha y_{T-1} + (1-\alpha)p_{T-1}.$$

From Equation (2.10), we obtain

$$p_{T+1} = \alpha y_T + (1-\alpha)[\alpha y_{T-1} + (1-\alpha)p_{T-1}]. \tag{2.11}$$

In (2.11) the value of p_{T-1} can be substituted rewriting (2.10) for p_{T-1}, and so on, backward to p_2, whose value can be set equal to y_1, as occurs for the elementary technique and the simple and weighted moving average methods. In this way we will have

$$p_{T+1} = \alpha \sum_{k=0}^{T-2} (1-\alpha)^k y_{T-k} + (1-\alpha)^{T-1} y_1. \tag{2.12}$$

In this equation, past data entries are multiplied by exponentially decreasing weights (this is the reason for the name of the method). Finally, we observe that the sum of all weights in (2.12) is equal to 1 (see Problem 2.15).

If the exponential smoothing method (with $\alpha = 0.1$) is applied to the Regens Book problem, we get

$$p_2 = y_1 = 89,$$

and, hence,

$$p_3 = \alpha y_2 + (1-\alpha)p_2 = 90.70.$$

By iterating the procedure, we get the forecasts reported in (Table 2.23) and, in particular, $p_{11} = 88.79$.

Table 2.23 Forecasts of the weekly sales in the Regens Book problem.

t	p_t	t	p_t
2	89.00	7	89.08
3	90.70	8	88.97
4	90.83	9	88.78
5	91.55	10	89.10
6	90.09	11	88.79

The choice of α plays a fundamental role in the exponential smoothing method. Large values of α involve a greater weight for more recent historical data and therefore a more outstanding capacity to follow the changes of values rapidly as they are gradually made available with the passing of time; however, this corresponds to less filtering of the residual variations of the time series. In contrast, low values of α yield a forecast less subject to random components, but, at the same time, the most recent data variations progressively available are incorporated in the forecast with a longer delay. In practice, the optimal value of α is chosen on the basis of the same considerations that can be used to determine the optimal value of r in the simple moving average method. For this reason, also in this case the reader should see Section 2.7.2.

The forecasts for the subsequent time periods The methods just illustrated can be used for one-period-ahead forecast. In order to forecast for subsequent time periods, it is sufficient to recall that the trend is assumed to be constant. Consequently,

$$p_T(\tau) = p_{T+1}, \tau = 2, \ldots,$$

where the forecasting value p_{T+1} is obtained by using any technique described elsewhere in this section.

In such a context, the forecasting time horizon is said to be *rolling* because, once a new datum becomes available, the time horizon shifts one period ahead.

Three-week-ahead forecasts are needed for the Regens Book problem. By using the simple moving average method with $r = 2$, the forecasts are

$$p_{11}[= p_{10}(1)] = p_{10}(2) = p_{10}(3) = \frac{y_{10} + y_9}{2} = 89.$$

At time period $t = 11$, the rolling forecasting horizon would cover the time periods $t = 12, 13, 14$. For example, if $y_{11} = 90$, then

$$p_{11}(1) = p_{11}(2) = p_{11}(3) = \frac{y_{11} + y_{10}}{2} = 88.$$

Thus, for $t = 12$, a new updated forecast is available ($p_{11}(1) = 88$), which substitutes the previous one ($p_{10}(2) = 89$). Similar considerations are valid for $t = 13$, whereas for time period $t = 14$ a first forecast ($p_{11}(3) = 88$) is determined.

2.5.5 Further time series extrapolation methods: The linear trend case

If the trend is linear and no cyclical or seasonal variation is displayed, the forecasting methods are based on the following computational scheme:

$$p_T(\tau) = a_T + b_T\tau, \tau = 1, \ldots \qquad (2.13)$$

To estimate a_T and b_T, we can use the methods illustrated throughout this section.

2.5.5.1 Elementary technique

This is the simplest forecasting method, on the basis of which we have

$$a_T = y_T;$$

$$b_T = y_T - y_{T-1}.$$

BFT is a Greek telecom company which needs to forecast overall administrative costs over the near future. The historical cost series (in €) during the past 14 months is reported in Table 2.24. As shown in Figure 2.17, the time series shows a linear trend. By utilizing the elementary technique, we get

$$p_{14}(\tau) = a_{14} + b_{14}\tau = y_{14} + (y_{14} - y_{13})\tau = 194\,561 - 217\tau, \tau = 1, \ldots$$

In particular, for the first period ahead ($\tau = 1$), the forecast is

$$p_{15} = € \ 194\,344.$$

Table 2.24 Monthly administrative costs (in €) at BFT.

Month	Costs	Month	Costs
1	189 676	8	193 159
2	190 105	9	192 494
3	190 511	10	193 467
4	190 986	11	193 620
5	192 230	12	193 838
6	192 407	13	194 778
7	192 277	14	194 561

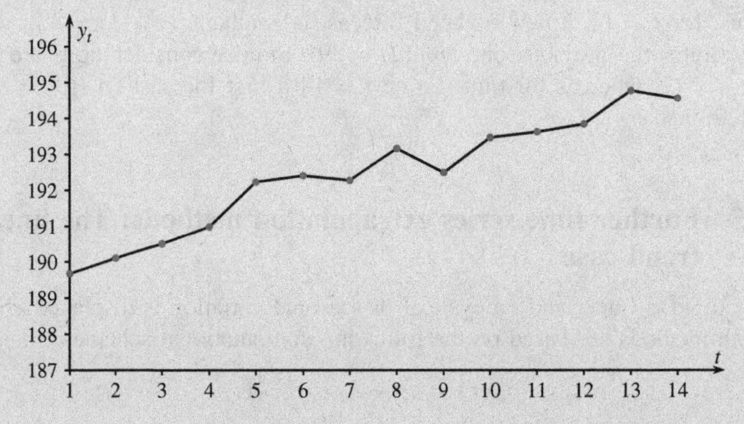

Figure 2.17 Monthly administrative costs (in k€) at BFT.

2.5.5.2 Linear regression method

In order to estimate a_T and b_T, this method determines the regression line which best interpolates the r (≥ 2) most recent data entries (that is, $y_{T-r+1}, \ldots, y_T$):

$$b_T = \frac{\frac{-(r-1)}{2} \sum_{k=0}^{r-1} y_{T-k} + \sum_{k=0}^{r-1} k y_{T-k}}{\frac{r(r-1)^2}{4} - \frac{r(r-1)(2r-1)}{6}};$$

$$a_T = \frac{\sum_{k=0}^{r-1} y_{T-k} + b_T \frac{r(r-1)}{2}}{r}.$$

In order to determine the optimal value of r, we refer the reader to Section 2.7.2.

By applying the linear regression method (with $r = 4$) to the BFT problem, we get

$$\sum_{k=0}^{r-1} y_{T-k} = 776\,797;$$

$$\sum_{k=0}^{r-1} k y_{T-k} = 1\,163\,314.$$

Hence,

$$b_{14} = 376.30;$$

$$a_{14} = 194\,763.70;$$

$$p_{14}(\tau) = a_{14} + b_{14}\tau = 194\,763.70 + 376.30\tau,\ \tau = 1, \ldots$$

In particular, the one-period-ahead forecast ($\tau = 1$) is

$$p_{15} = € \ 195\,140.$$

2.5.5.3 Double moving average method

The method is an extension of the simple moving average method illustrated in Section 2.5.4.2. Let $r (\geq 2)$ be a double moving average parameter. We get

$$a_T = 2\gamma_T - \eta_T;$$

$$b_T = \frac{2}{r-1}(\gamma_T - \eta_T),$$

where γ_T is the average of the r most recent data entries:

$$\gamma_T = \sum_{k=0}^{r-1} \frac{y_{T-k}}{r},$$

and η_T represents the average of the r most recent average data entries, that is,

$$\eta_T = \sum_{k=0}^{r-1} \frac{\gamma_{T-k}}{r}.$$

Whenever r past demand data are not available ($T < r$), the computation of γ_T and η_T can be executed along the guidelines illustrated for the moving average method. Again, we refer the reader to Section 2.7.2 to gain an insight into the determination of optimal value of r.

By using the double moving average method (with $r = 3$) to solve the BFT problem, we get

$$\gamma_{14} = \frac{y_{14} + y_{13} + y_{12}}{3} = 194\,392.33;$$

$$\eta_{14} = \frac{\gamma_{14} + \gamma_{13} + \gamma_{12}}{3},$$

where

$$\gamma_{13} = \frac{y_{13} + y_{12} + y_{11}}{3} = 194\,078.67;$$

$$\gamma_{12} = \frac{y_{12} + y_{11} + y_{10}}{3} = 193\,641.67.$$

Hence,

$$\eta_{14} = 194\,037.56;$$

$$a_{14} = 2\gamma_{14} - \eta_{14} = 194\,747.11;$$

$$b_{14} = \gamma_{14} - \eta_{14} = 354.78;$$

$$p_{14}(\tau) = a_{14} + b_{14}\tau = 194\,747.11 + 354.78\tau, \tau = 1, \ldots.$$

In particular, the one-period-ahead forecast ($\tau = 1$) is:

$$p_{15} = € \; 195\,101.89.$$

2.5.5.4 Holt method

The exponential smoothing method, introduced in Section 2.5.4.4, is unable to deal with a linear trend. The Holt method is a modification of the exponential smoothing method and is based on the following two relations

$$a_T = \alpha y_T + (1-\alpha)(a_{T-1} + b_{T-1}); \tag{2.14}$$

$$b_T = \beta(a_T - a_{T-1}) + (1-\beta)b_{T-1}. \tag{2.15}$$

In equation (2.14), a_T is updated by smoothing b_{T-1} summed to the previously smoothed a_{T-1}. Formula (2.15) allows one to update the slope b_T, by smoothing the trend shown in the last time period (that is, $a_T - a_{T-1}$) summed up to the previous estimation of the slope times $(1-\beta)$.

Applying recursively Equations (2.14) and (2.15), it is possible to express a_T and b_T as a function of the past entries $y_t, t = 1, \ldots, T$. In order to start the procedure, a_1 and b_1 must be specified. They can be chosen in the following way:

$$a_1 = y_1;$$

$$b_1 = 0.$$

In this way, we have $p_2 = p_1(1) = a_1 + b_1 = y_1$, as in the exponential smoothing method. The choice of parameters α and β is carried out according to the same criterion illustrated for the exponential smoothing method.

We apply the Holt method ($\alpha, \beta = 0.3$) to the BFT problem. By recursively applying (2.14) and (2.15), we compute the a_t and b_t values ($t = 1, \ldots, 14$) reported in Table 2.25:

$$p_{14}(\tau) = a_{14} + b_{14}\tau = 194\,825.32 + 360.43\tau, \tau = 1, \ldots.$$

In particular, the one-period-ahead forecast ($\tau = 1$) is

$$p_{15} = € \; 195\,185.75.$$

Table 2.25 Parameters a_t and b_t, $t = 1, \ldots, 14$, in the BFT problem.

t	a_t	b_t
1	189 676.00	0.00
2	189 804.70	38.61
3	190 043.62	98.70
4	190 395.42	174.63
5	191 068.04	324.03
6	191 696.55	415.37
7	192 161.44	430.23
8	192 761.87	481.29
9	193 018.41	413.86
10	193 442.69	416.99
11	193 787.78	395.42
12	194 079.64	364.35
13	194 544.19	394.41
14	194 825.32	360.43

2.5.6 Further time series extrapolation methods: The seasonal variation case

This section describes the main forecasting method when the demand pattern displays a constant or linear trend, and a seasonal variation with periodicity M.

2.5.6.1 Elementary technique

If the trend is constant, then

$$p_T(\tau) = y_{T+\tau-M}, \tau = 1, \ldots, M. \tag{2.16}$$

On the basis of Equation (2.16), the forecast related to the time period $T + \tau$ corresponds to the data entry M time periods back. More generally, for a temporal horizon whose length is superior to one cycle, we have

$$p_T(kM + \tau) = y_{T+\tau-M}, \tau = 1, \ldots, M, k = 1, \ldots$$

Elna distributes and installs air-conditioning systems in the Nayarit region of Mexico. The monthly sales figures in the past two years are reported in Table 2.26. As shown in Figure 2.18, the time series shows a remarkable seasonal effect with periodicity $M = 12$. By using the elementary technique, we get

$$p_{24}(\tau) = y_{24+\tau-12}, \tau = 1, \ldots, 12.$$

For instance, the one- and two-period-ahead forecasts will be

$$p_{25} = y_{25-12} = y_{13} = 815;$$

$$p_{24}(2) = y_{26-12} = y_{14} = 1\,015.$$

Table 2.26 Monthly air-conditioning systems installed by Elna.

Period	Quantity	Period	Quantity
1	915	13	815
2	815	14	1 015
3	1 015	15	915
4	1 115	16	1 315
5	1 415	17	1 215
6	1 615	18	1 615
7	1 515	19	1 315
8	1 415	20	1 115
9	815	21	1 115
10	615	22	915
11	315	23	715
12	815	24	615

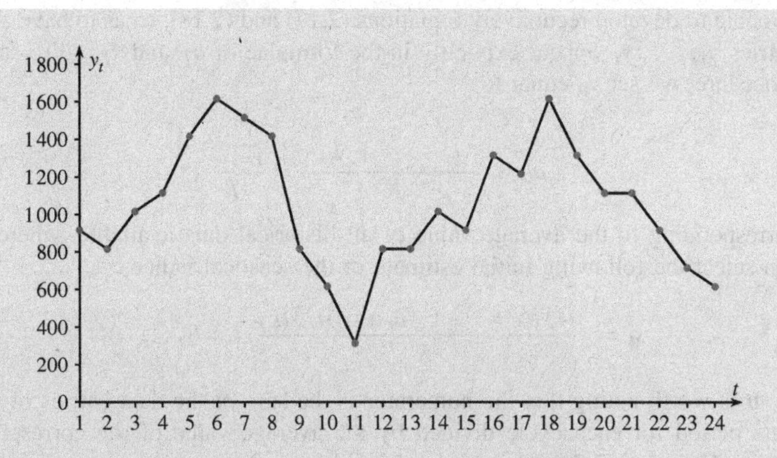

Figure 2.18 Monthly air-conditioning systems installed by Elna.

2.5.6.2 Revised exponential smoothing method

This method can be used whenever the trend is constant. It is based on the following computational scheme:

$$p_T(\tau) = a_T s_{T+\tau}, \tau = 1, \ldots, M,$$

where a_T takes into account the constant trend (and can be interpreted as the forecasted demand without the seasonal variation), whereas $s_{T+\tau}$ (≥ 0) is the seasonal index for time period $T + \tau$.

More generally, for a time horizon whose duration is greater than one cycle, we get

$$p_T(kM + \tau) = a_T s_{T+\tau}, \tau = 1, \ldots, M, k = 1, 2, \ldots$$

Assuming, without loss of generality, that the available historical data are sufficient to cover an integer number $K = T/M$ of cycles, the parameters a_T and $s_{T+\tau}$, $\tau = 1, \ldots, M$, can be computed by the following relations:

$$a_T = \alpha \frac{y_T}{s_T} + (1 - \alpha) a_{T-1}; \tag{2.17}$$

$$s_{T+\tau} = s_{kM+\tau} = \gamma \frac{y_{(K-1)M+\tau}}{a_{(K-1)M+\tau}} + (1 - \gamma) s_{(K-1)M+\tau}, \tau = 1, \ldots, M, \tag{2.18}$$

where α and γ are smoothing constants ($0 \leq \alpha, \gamma \leq 1$). Equation (2.17) expresses a_T as the weighted sum of two components: the first, y_T/s_T, represents the value of the entry at time period T without the seasonal variation, while the second represents the forecast, without the seasonal variation, at time period $T - 1$.

A similar interpretation applies to Equation (2.18). However, in this case, it is necessary to take into account the periodicity of the seasonal variation. It is possible to develop recursively Equations (2.17) and (2.18), so as to have all data entries $y_1, \ldots, y_T$ appear explicitly in the formulae of a_T and $s_{T+\tau}$. To start the procedure, we set a_0 equal to

$$a_0 = \frac{\bar{y}_{(1)} + \cdots + \bar{y}_{(K)}}{K} = \frac{\sum_{t=1}^{T} y_t}{T}, \tag{2.19}$$

corresponding to the average value of all historical data available, whereas we can select the following initial estimate of the seasonal indices:

$$s_t = \frac{y_t/\bar{y}_{(1)} + \cdots + y_{t+(k-1)M}/\bar{y}_{(K)}}{K}, t = 1, \ldots, M. \tag{2.20}$$

It is worth noting that the numerator is the sum of the data entries of the tth time period for each cycle divided by the average value of the corresponding cycles. Equation (2.20) implies

$$\sum_{t=1}^{M} s_t = \frac{T}{K} = M,$$

that is, the average seasonal index for the first cycle is equal to 1. However, this condition might not be satisfied for the subsequent cycles, and for this reason it is necessary to normalize the indices s_t, $t = (k-1)M + 1, \ldots, kM$, $k = 2, \ldots$.

We refer the reader to Section 2.7.2 to get an insight into the determination of optimal values of α and γ.

We apply the revised exponential smoothing method (with parameters $\alpha = \gamma = 0.05$) to the Elna problem. Firstly, we compute the average sales in the first and second cycles:

$$\bar{y}_{(1)} = \frac{y_1 + \cdots + y_{12}}{12} = 1\,031.67;$$

$$\bar{y}_{(2)} = \frac{y_{13} + \cdots + y_{24}}{12} = 1\,056.67,$$

which allows us to calculate

$$a_0 = \frac{\bar{y}_{(1)} + \bar{y}_{(2)}}{2} = 1\,044.17.$$

Then, we compute the seasonal indices s_t, $t = 1, \ldots, 12$, by using formula (2.20) (see Table 2.27). It is worth noting that

$$\bar{s} = \frac{\sum_{t=1}^{12} s_t}{12} = 1.$$

Table 2.27 Time series $s_t, t = 1, \ldots, 12$, for the Elna air-conditioning systems forecasting problem.

t	s_t	T	s_t
1	0.83	7	1.36
2	0.88	8	1.21
3	0.92	9	0.92
4	1.16	10	0.73
5	1.26	11	0.49
6	1.55	12	0.69

By using relationships (2.17) and (2.18), we get Tables 2.28 and 2.29, where the seasonal indices $s_t, t = 13, \ldots, 36$, have already been normalized. These parameters allow us to compute, among others, the one- and two-period-ahead forecasts:

$$p_{25} = p_{24}(1) = 874.84;$$

$$p_{24}(2) = 924.17.$$

Table 2.28 Time series $a_t, t = 1, \ldots, 12$, for the Elna air-conditioning systems forecasting problem.

t	a_t	t	a_t
1	1 047.14	13	1 021.65
2	1 041.34	14	1 028.83
3	1 044.14	15	1 026.70
4	1 039.89	16	1 032.11
5	1 044.01	17	1 028.49
6	1 044.01	18	1 029.24
7	1 047.65	19	1 026.06
8	1 053.58	20	1 020.43
9	1 045.07	21	1 030.28
10	1 034.88	22	1 041.91
11	1 015.21	23	1 063.95
12	1 023.86	24	1 055.19

Figure 2.19 shows both the time series and its extrapolation over the next 12 months.

Table 2.29 Times series $s_t, t = 13, \ldots, 36$, for the Elna air-conditioning systems problem.

t	s_t	t	s_t
13	0.83	25	0.83
14	0.87	26	0.88
15	0.93	27	0.92
16	1.16	28	1.16
17	1.27	29	1.26
18	1.55	30	1.55
19	1.36	31	1.36
20	1.22	32	1.21
21	0.92	33	0.92
22	0.72	34	0.73
23	0.48	35	0.49
24	0.69	36	0.69

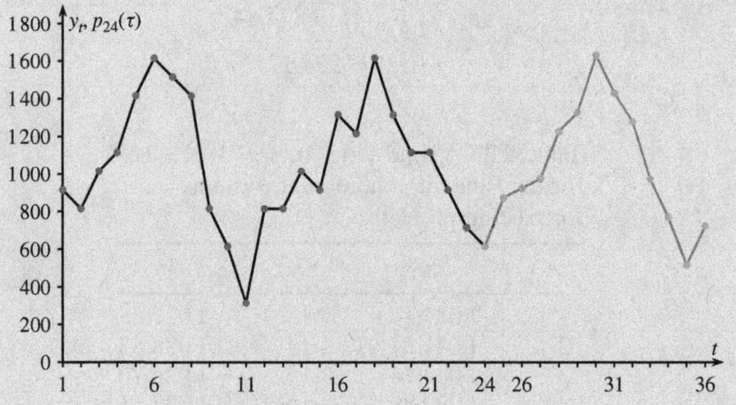

Figure 2.19 Monthly sales series during the past two years and its projection over the next 12 months (in grey) for the Elna air-conditioning systems forecasting problem.

2.5.6.3 Winters method

The Winters method can be used whenever there is a linear trend and a seasonal variation

$$p_T(\tau) = (a_T + b_T(\tau))s_{T+\tau}, \tau = 1, \ldots, M$$

and

$$p_T(kM + \tau) = (a_T + b_T(kM + \tau))s_{T+\tau}, \tau = 1, \ldots, M, k = 1, \ldots$$

As in the revised exponential smoothing method, we assume that the historical data available are enough to have an integer number $K = T/M$ of cycles. The Winters method is based on the following relationships for the computation of a_T, b_T and $s_{T+\tau}$, $\tau = 1, \ldots, M$:

$$a_T = \alpha\left(\frac{y_T}{s_T}\right) + (1-\alpha)(a_{T-1} + b_{T-1}); \quad (2.21)$$

$$b_T = \beta(a_T - a_{T-1}) + (1-\beta)b_{T-1}; \quad (2.22)$$

$$s_{T+\tau} = s_{KM+\tau} = \gamma\left(\frac{y_{(K-1)M+\tau}}{a_{(K-1)M+\tau}}\right) + (1-\gamma)s_{(K-1)M+\tau}, \tau = 1, \ldots, M, \quad (2.23)$$

where α, β and γ are smoothing constants chosen in the interval $[0, 1]$. We refer the reader to Section 2.7.2 to get an insight into the determination of their optimal values.

Equations (2.21) and (2.22) are derived from the equations (2.14) and (2.15) of the Holt method, except that y_T is deseasonalized (similarly to the revised exponential smoothing method). Moreover, relationship (2.23) is the same as (2.18) of the revised exponential smoothing method.

The Winters method can be initialized as follows:

$$a_0 = \frac{\bar{y}_{(1)} + \cdots + \bar{y}_{(K)}}{K} = \frac{\sum_{t=1}^{T} y_t}{T}; \quad (2.24)$$

$$b_0 = \frac{1}{K-1}\left(\frac{\bar{y}_{(2)} - \bar{y}_{(1)}}{M} + \cdots + \frac{\bar{y}_{(K)} - \bar{y}_{(K-1)}}{M}\right) = \frac{\bar{y}_{(K)} - \bar{y}_{(1)}}{(K-1)M} = \frac{\bar{y}_{(K)} - \bar{y}_{(1)}}{T-M}; \quad (2.25)$$

$$s_t = \frac{y_t/\bar{y}_{(1)} + \cdots + y_{t+(k-1)M}/\bar{y}_K}{K}, t = 1, \ldots, M. \quad (2.26)$$

Equations (2.24) and (2.26) are the same as (2.19) and (2.20) used in the revised exponential smoothing method. In (2.25), the terms in brackets are $K - 1$. The quantity $\frac{\bar{y}_{(K)} - \bar{y}_{(1)}}{T-M}$ expresses the variation, in $T - M$ time periods, of the average of historical data, from the first to the last cycle. This is because the average of the historical data of the first cycle is centred with respect to the time period $t = (M + 1)/2$, whereas the average of the historical data of the last cycle is centred with respect to the time period $t = (K - 1)M + (M + 1)/2$. The values determinated according to (2.26) are already normalized. For this reason, an eventual normalization procedure should be executed only for the seasonal indices determined by (2.23) and corresponding to cycles greater that one.

Elna also distributes microwave ovens in the Nayarit region. The monthly sales volumes over the past two years are reported in Table 2.30. As shown in

Table 2.30 Elna monthly sales of microwave ovens during the past 24 months.

Period	Quantity	Period	Quantity
1	682	13	416
2	416	14	1 746
3	1 613	15	2 411
4	1 613	16	2 544
5	1 746	17	4 140
6	2 677	18	4 539
7	4 672	19	7 997
8	5 603	20	8 263
9	3 741	21	7 465
10	1 480	22	3 209
11	682	23	1 081
12	150	24	283

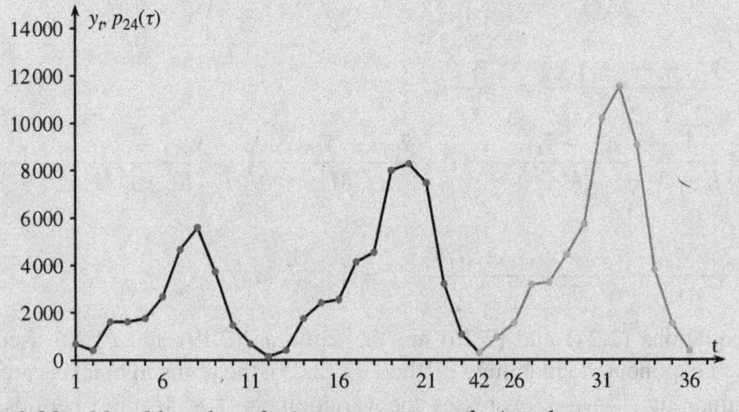

Figure 2.20 Monthly sales of microwave ovens during the past two years at Elna and their projection over the next 12 *months (in grey).*

Figure 2.20, the time series is characterized by a seasonal effect with periodicity $M = 12$ as well as by a linear trend.

To forecast the sales over the next 12 months with the Winters method, we first compute the average sales in the first $K = 2$ cycles:

$$\bar{y}_{(1)} = \frac{y_1 + \cdots + y_{12}}{12} = 2\,089.58;$$

$$\bar{y}_{(2)} = \frac{y_{13} + \cdots + y_{24}}{12} = 3\,674.50.$$

Then, we calculate

$$a_0 = \frac{\bar{y}_{(1)} + \bar{y}_{(2)}}{2} = 2\,882.04;$$

$$b_0 = \frac{\bar{y}_{(2)} - \bar{y}_{(1)}}{24 - 12} = 132.08,$$

and, subsequently, the seasonal indices $s_t, t = 1, \ldots, M$, reported in Table 2.31. Then, by using formulae (2.21)–(2.23) with $\alpha = 0.2$, $\gamma = 0.1$ and $\beta = 0.3$, we get the parameters in Tables 2.32 and 2.33 (where $s_t, t = 13, \ldots, 36$, are normalized). Then, the 12-month-ahead forecasts are computed (see Table 2.34 and Figure 2.20).

Table 2.31 Time series $s_t, t = 1, \ldots, 12$, for the Elna microwave ovens forecasting problem.

t	s_t	t	s_t
1	0.22	7	2.21
2	0.34	8	2.47
3	0.71	9	1.91
4	0.73	10	0.79
5	0.98	11	0.31
6	1.26	12	0.07

Table 2.32 Parameters a_t and b_t, $t = 1, \ldots, 24$, for the Elna microwave ovens forecasting problem.

t	a_t	b_t	t	a_t	b_t
1	3 031.87	137.40	13	1 814.88	−53.99
2	2 782.21	21.28	14	2 500.19	167.80
3	2 694.59	−11.39	15	2 817.44	212.63
4	2 587.20	−40.19	16	3 126.02	241.42
5	2 393.52	−86.24	17	3 555.85	297.94
6	2 271.36	−97.01	18	3 805.57	283.48
7	2 163.03	−100.41	19	3 994.18	255.02
8	2 104.69	−87.79	20	4 061.19	198.61
9	2 005.06	−91.34	21	4 187.14	176.81
10	1 905.28	−93.87	22	4 300.20	157.69
11	1 888.72	−70.68	23	4 248.50	94.87
12	1 857.65	−58.79	24	4 225.30	59.45

Table 2.33 Time series of normalized $s_t, t = 13, \ldots, 36$, for the Elna microwave ovens forecasting problem.

t	s_t	t	s_t
13	0.22	25	0.22
14	0.32	26	0.36
15	0.71	27	0.72
16	0.72	28	0.73
17	0.96	29	0.98
18	1.26	30	1.25
19	2.21	31	2.19
20	2.50	32	2.45
21	1.92	33	1.90
22	0.79	34	0.79
23	0.32	35	0.31
24	0.08	36	0.07

Table 2.34 Forecasts over the next 12 months for the Elna microwave ovens forecasting problem.

τ	$p_{24}(\tau)$	τ	$p_{24}(\tau)$
1	953.27	7	10 184.04
2	1 557.05	8	11 535.31
3	3 179.26	9	9 068.49
4	3 279.20	10	3 805.78
5	4 443.11	11	1 517.74
6	5 735.14	12	368.72

2.5.7 Further time series extrapolation methods: The irregular time series case

The simplest forecasting method for the irregular time series is the *shift method* which is based on the assumption that the time series will repeat in the future, possibly scaled up or down. The model is

$$p_T(\tau) = \alpha_{T+\tau} y_{T+\tau-L}, \tau = 1, \ldots,$$

where $\alpha_{T+\tau}$ is the *scaling factor* at time $T + \tau$. This means that at time $T + \tau$, the forecast will be equal to $\alpha_{T+\tau}$ times the data entry at time $T + \tau - L$. It is worth noting that L is not necessarily related to a possible seasonality of the time series. Its optimal value can be determined by following the guidelines illustrated in Section 2.7.2.

The scaling factor belongs to the interval [0, 2] and can be computed on the basis of the data entries reported in the last two periods (i.e. $T+\tau-L$ and $T+\tau-2L$):

$$\alpha_{T+\tau} = 1 + \frac{y_{T+\tau-L} - y_{T+\tau-2L}}{\max\{y_{T+\tau-L}, y_{T+\tau-2L}\}}, \tag{2.27}$$

where we assume that $\alpha_{T+\tau} = 1$, whenever (2.27) turns out to be 0/0. The choice of $\alpha_{T+\tau}$ according to (2.27) ensures that the forecasting method can also be applied to sporadic time series (not only irregular).

MCE is an international freight forwarder headquartered in New Zealand with more than 500 terminals and representatives around the world. The daily sales (in NZ$) at the Wellington centre during the past 24 days are shown in Table 2.35. As illustrated by Figure 2.21, the time series is irregular. In order to forecast the sales during the next six working days, the shift method can be used. By setting $L = 6$, from formula (2.27) we get the values of $\alpha_{T+\tau}$ and $p_T(\tau)$, $\tau, = 1, \dots, 6$, reported in Table 2.36. Figure 2.22 illustrates the required sales forecasts.

Table 2.35 Daily sales (in NZ$) at the Wellington MCE centre.

Period	Revenue	Period	Revenue
1	215	13	2 366
2	1 768	14	1 119
3	10 331	15	9 991
4	287	16	10 252
5	10 689	17	8 974
6	4 003	18	9 499
7	2 801	19	500
8	4 056	20	340
9	10 989	21	6 991
10	6 520	22	11 732
11	5 790	23	8 414
12	9 685	24	9 514

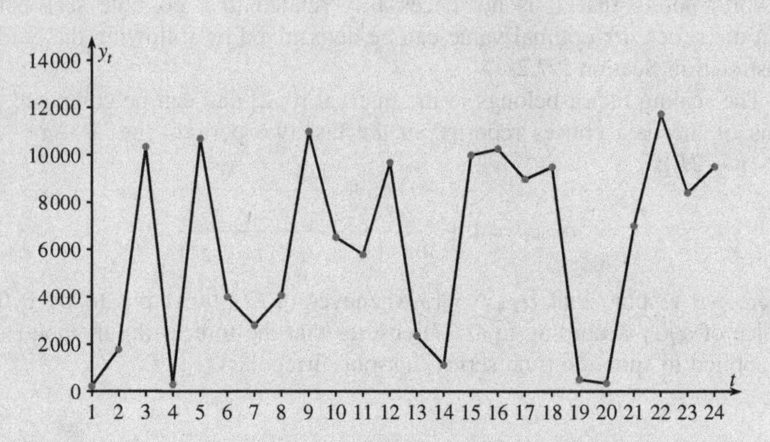

Figure 2.21 Daily sales (in NZ$) at the Wellington MCE centre.

Table 2.36 Time series $\alpha_{T+\tau}$ and $p_T(\tau)$, $\tau = 1, \ldots, 6$, for the MCE problem.

τ	$T+\tau$	$\alpha_{T+\tau}$	$p_T(\tau)$
1	25	0.21	105.66
2	26	0.30	103.31
3	27	0.70	4 891.81
4	28	1.13	13 212.00
5	29	0.94	7 888.95
6	30	1.00	9 529.00

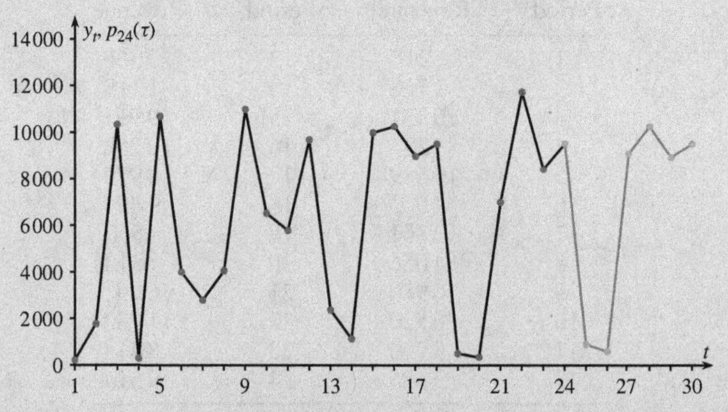

Figure 2.22 Daily sales and forecasts (in grey) at the Wellington MCE centre.

2.5.8 Sporadic time series

If the time series is sporadic, two cases may occur.

- The time series is random; the *Croston method*, described in the following, can be used or the shift method, if the time series is also irregular.
- The time series is periodic; this is a simpler case, and one of the methods described in Sections 2.5.3–2.5.7 can be adapted.

2.5.8.1 Croston method

The method is made up of three steps.

1. We remove all the zero data entries from the time series $y_t, t = 1, \ldots, T$. Let $y'_t, t = 1, \ldots, T'$, be the corresponding compact time series. Make the one-period-ahead forecast $p_{T'+1}$ through, for example, the exponential smoothing method.
2. We assign the previous forecast to a time period $T + \tau$, where τ is determined as follows. Let $z_k, k = 1, \ldots, K$, be the time series of the inter-arrival times between nonzero data entries in $y_t, t = 1, \ldots, T$. Time lag τ can be determined by applying a forecasting method (e.g. the exponential smoothing method) to time series $z_k, k = 1, \ldots, K$.
3. Once computed $p_T(\tau)$, the forecasts for the subsequent time periods are zero, except for time periods $T + 2\tau, T + 3\tau, \ldots$ when the forecast is set equal to $p_T(\tau)$.

It is worth noting that the forecasts defined here are periodic even if the time series $y_t, t = 1, \ldots, T$, is not periodic.

The logistician of Ipergent, a Belgian supermarket chain, wants to determine the sales of a 0.5 kg dog food package of a renowned brand at the sales point in Louvain. The sales recorded during the past 21 days are reported in Table 2.37 and in Figure 2.23. Since the time series is both sporadic and random, the Croston method is used. Firstly, the manager computes the time series $y'_t, t = 1, \ldots, T'$ (see Table 2.38), where $T' = 9$. The forecast of the first nonzero entry $p_{T'+1}$ is calculated by applying the exponential smoothing method with $\alpha = 0.2$ to $y'_t, t = 1, \ldots, T'$:

$$p_{10} = 11.11.$$

Then, the manager determines the auxiliary time series $z_k, k = 1, \ldots, K$ (as shown in Table 2.39) which is then used to forecast τ with the exponential smoothing method with $\alpha = 0.2$:

$$z_6 = 2.54.$$

Table 2.37 Daily sales of dog food (in 0.5 kg packages) at the Ipergent supermarket in Louvain.

Day	Quantity	Day	Quantity	Day	Quantity
1	10	8	12	15	11
2	0	9	11	16	0
3	0	10	9	17	0
4	14	11	0	18	0
5	0	12	0	19	12
6	0	13	0	20	0
7	0	14	14	21	10

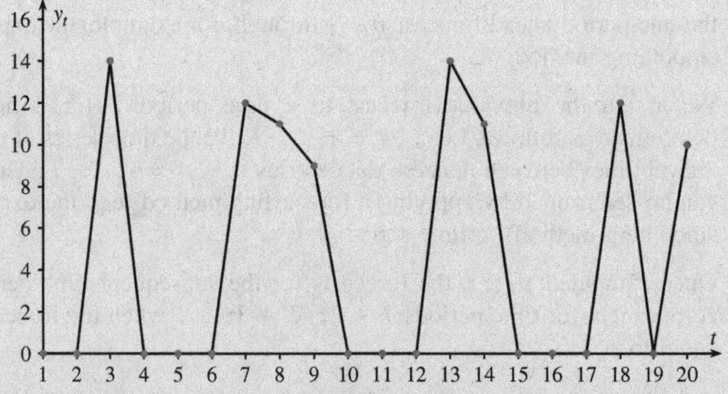

Figure 2.23 Daily sales of dog food (in 0.5 kg packages) at the Ipergent supermarket in Louvain.

Table 2.38 Auxiliary time series y'_t, $t = 1, \ldots, T'$, in the Ipergent problem.

t	y'_t	t	y'_t
1	10	6	14
2	14	7	11
3	12	8	12
4	11	9	10
5	9		

Table 2.39 Auxiliary time series z_k, $k = 1, \ldots, 6$, in the Ipergent problem.

k	z_k	k	z_k
1	1	4	4
2	3	5	4
3	4	6	2

The time lag τ is then approximated to three days. Hence,

$$p_{21}(1) = p_{21}(2) = 0,$$

and

$$p_{21}(3) = 11.11.$$

2.5.8.2 Sporadic and periodic time series

Let M be the periodicity of the time series $y_t, t = 1, \ldots, T$. On the basis of Section 2.5.8.1, the following forecasting framework can be used:

1. the zero data entries are removed from $y_t, t = 1, \ldots, T$, and the corresponding compact time series $y'_t, t = 1, \ldots, T'$, is obtained;
2. the forecasts for one or more periods ahead are computed by applying to $y'_t, t = 1, \ldots, T'$, one of the methods devised for the regular time series or the shift method in case y'_t is irregular;
3. Then the one-, two-, ..., periods-ahead forecasts of y'_t are the forecasts of y_t in time periods $T + M, T + 2M, \ldots$, while the forecasts associated to the other time periods are set equal to 0.

Belem is a pastry producer located in Lisbon. The sales of the 1 kg package of its cake called Torta de Viana during the past three years are reported in Table 2.40. As shown in Figure 2.24, the time series is sporadic and periodic. The compact time series y'_t (Table 2.41) shows a seasonal variation with periodicity $M = 5$ as illustrated in Figure 2.25. To estimate the sales during the next five months, we use the Winters method, with parameters $K = 3$, $\alpha = \gamma = 0.1$ and $\beta = 0.3$. By taking

Table 2.40 Monthly sales of the Torta de Viana cake (1 kg packages) in the Belem problem.

Month	Quantity	Month	Quantity	Month	Quantity
1	200	13	189	25	187
2	11	14	7	26	9
3	0	15	0	27	0
4	0	16	0	28	0
5	0	17	0	29	0
6	0	18	0	30	0
7	0	19	0	31	0
8	0	20	0	32	0
9	0	21	0	33	0
10	14	22	18	34	17
11	121	23	118	35	114
12	450	24	489	36	498

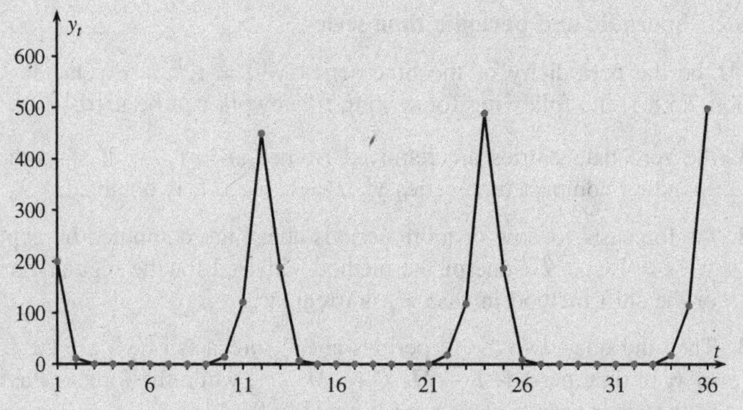

Figure 2.24 Monthly sales of the Torta de Viana cake (1 kg packages) in the Belem problem.

Table 2.41 Compact time series y'_t, $t = 1, \ldots, 15$, for the Belem problem.

t	y'_t	t	y'_t	t	y'_t
1	200	6	189	11	187
2	11	7	7	12	9
3	14	8	18	13	17
4	121	9	118	14	114
5	450	10	489	15	498

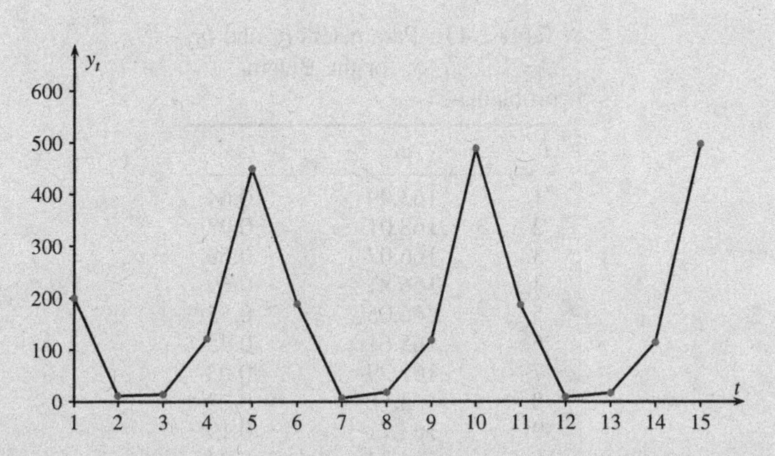

Figure 2.25 Compact time series $y'_t, t = 1, \ldots, 15$, for the Belem problem.

into account that

$$\bar{y}_{(1)} = \frac{y_1 + \cdots + y_5}{5} = 159.20;$$

$$\bar{y}_{(2)} = \frac{y_6 + \cdots + y_{10}}{5} = 164.20;$$

$$\bar{y}_{(3)} = \frac{y_{11} + \cdots + y_{15}}{5} = 165,$$

a_0 and b_0 are computed as follows:

$$a_0 = \frac{\bar{y}_{(1)} + \bar{y}_{(2)} + \bar{y}_{(3)}}{3} = 162.80;$$

$$b_0 = \frac{\bar{y}_{(3)} - \bar{y}_{(1)}}{10} = 0.58,$$

while Table 2.42 reports the seasonal indices $s_t, t = 1, \ldots, 5$. Table 2.43 reports the parameters a_t and $b_t, t = 1, \ldots, 15$, while Table 2.44 shows

Table 2.42 Parameters s_t, $t = 1, \ldots, 5$, for the Belem problem.

t	s_t
1	1.18
2	0.06
3	0.10
4	0.72
5	2.94

Table 2.43 Parameters a_t and b_t, $t = 1, \ldots, 15$, for the Belem problem.

t	a_t	b_t
1	163.99	0.64
2	168.01	0.98
3	166.07	0.69
4	166.81	0.69
5	166.05	0.55
6	165.61	0.45
7	161.29	−0.03
8	163.78	0.22
9	163.71	0.19
10	164.34	0.24
11	163.87	0.17
12	164.18	0.18
13	164.81	0.23
14	164.17	0.14
15	164.88	0.20

Table 2.44 Normalized indices s_t, $t = 6, \ldots, 20$, for the Belem problem.

t	s_t	t	s_t	t	s_t
6	1.21	11	1.19	16	1.17
7	0.06	12	0.05	17	0.05
8	0.10	13	0.10	18	0.10
9	0.73	14	0.73	19	0.72
10	2.91	15	2.93	20	2.95

Table 2.45 Forecasts of the Torta de Viana cake for five months ahead in the Belem problem.

Month	Forecast
1	193.52
2	9.00
3	16.77
4	118.98
5	489.92

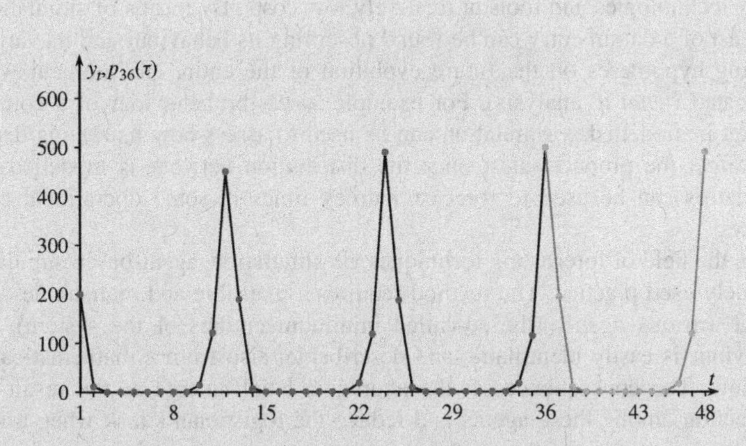

Figure 2.26 Monthly sales and forecasts (in grey) in the Belem problem.

the normalized indices s_t, $t = 6, \ldots, 20$. Then, Table 2.45 shows the forecasts for the five periods ahead for y'_t while Figure 2.26 illustrates the corresponding forecasts for y_t.

2.6 Advanced forecasting method

For the sake of completeness, some advanced forecasting techniques are outlined in this section. As stated in Section 2.2, the first approach can be classified as a time series extrapolation technique, while the subsequent ones are causal methods.

Box-Jenkins method The Box-Jenkins method is made up of three phases: *identification, parameter evaluation* and *diagnostic check*. In the first phase, the most appropriate forecasting method is selected from a set of techniques. To this end, the historical data entries (at least 50) are used to generate a set of autocorrelation functions. In the second phase, the coefficients of the forecasting method are selected so as to minimize the mean squared error. Finally, in the third phase, an autocorrelation function of the error is determined to verify the adequacy of the method chosen and its corresponding parameters. In case of a negative result, the entire procedure is executed again by discarding the forecasting method previously chosen. Of course, when new data entries become available, the whole procedure is run again.

Computer simulation Simulation is a widely used tool for carrying out forecasts thanks to the level of progress achieved in the ICT field that makes available many technologies and tools at relatively low costs. By means of simulation, the forecast of a datum entry can be found observing its behaviour and its variations, making hypotheses on the future evolution of the entire environment where it is located (*what if* analysis). For example, once the behaviour of a company's market is modelled, a simulation can be used to assess how marketing decisions can affect the product sales; once the distribution network is modelled, traffic simulators can be used to forecast journey times in some operational circumstances.

In the field of forecasting techniques via simulation, agent-based simulation is a widely used practice. The method requires a plausible and realistic description of the various *agents* (the so-called minimum entities of the system), whose behaviour is easily identifiable and describable, also from a mathematical point of view. The consequences at the aggregate level emerge as the result of the interaction among these agents and reduce the logistician's task when assessing the phenomena observed at a global level.

Artifical neural network An artificial neural network (ANN) is a computational model inspired by the structure of biological neural networks. ANNs implement a sophisticated regression technique and are usually used to model complex relationships between inputs and outputs or to find patterns in data. ANNs are particularly advantageous for the following reasons:

- they can carry out nonlinear mappings between the input and the output variables;
- the forecast, after the learning phase, is carried out in a very short time;
- the approach is nonparametric and, therefore, it works without any hypothesis about the data structure underlying the model.

In particular, ANNs appear to be the most suitable tool for forecasting the journey times also in the short term, thanks to their rapidity of response and their capacity to connect among themselves the different causes that affect the same output. A critical point in the forecasting process based on ANNs concerns the structure and the dimension of the network itself: an erroneous choice can lead to poor forecasts. If an ANN is chosen is too small, the model will not be able to represent the desired mapping between inputs and outputs (*underfitting*). If, instead, an ANN is chosen too large, it will be unable to generalize past observations, thus providing incorrect forecasts (*overfitting*).

2.7 Accuracy measure and forecasting monitoring

Although it may sound odd, forecasts are almost always wrong. This is because of the inherent random nature of the variables to be predicted. Moreover, errors can be caused by the fact that the hypotheses on which the forecasts are based no longer hold. This being said, a relevant question is to determine by *how much* forecasts can be wrong. In order to deal with this issue, forecasting methods have been evaluated through suitable accuracy measures calculated on the basis of errors made in the past. Such measures can be employed to both evaluate the impact of the forecasting error on the business and select the most precise approach. Moreover, in the case of periodic predictions (like those required by inventory management), forecasting errors should be monitored in order to adjust parameters if needed. For the sake of brevity, we examine these issues for the case of a regular time series where an one-period-ahead forecast has to be generated.

2.7.1 Accuracy measures

To evaluate the accuracy of a forecasting method, the errors that would have been made in the past have to be computed. Then a number of measures (the *mean absolute deviation* (MAD), the *mean absolute percentage deviation* ($MAPD$) and the *mean squared error* (MSE)) at time period T can be defined:

$$MAD_T = \frac{\sum_{t=2}^{T} |e_t|}{T-1}; \quad (2.28)$$

$$MAPD_T = \frac{\sum_{t=2}^{T} \frac{|e_t|}{y_t}}{T-1}; \quad (2.29)$$

$$MSE_T = \frac{\sum_{t=2}^{T} e_t^2}{T-2}. \quad (2.30)$$

where $T > 1$ for Equations (2.28) and (2.29), and $T > 2$ for Equation (2.30).

These three accuracy measures can be used at time period $t = T$ to establish a comparison between different forecasting methods. In particular, $MAPD$ can be used to evaluate the quality of a forecasting method (see Table 2.46).

Table 2.46 Evaluation of the quality of a forecasting method by means of the mean absolute percentage deviation.

Mean absolute percentage deviation	Quality of forecasting
$\leq 10\%$	Very good
$> 10\%, \leq 20\%$	Good
$> 20\%, \leq 30\%$	Moderate
$> 30\%$	Poor

We want to assess the accuracy of the elementary technique, of the moving average (with $r = 2$), of the weighted moving average and of exponential smoothing ($\alpha = 0.1$) methods in the Regens Book example, introduced in Section 2.5.4.1. The results of the four forecasting methods are reported in Table 2.47. On the basis of the accuracy measures shown in Table 2.48, we can state that all the methods provide good-quality forecasts. In particular, the most accurate technique turned out to be the exponential smoothing method (with $\alpha = 0.1$).

Table 2.47 Forecast errors in the Regens Book example.

		Elementary technique		Simple moving average ($r = 2$)		Weighted moving average		Exponential smoothing ($\alpha = 0.1$)	
t	y_t	p_t	e_t	p_t	e_t	p_t	e_t	p_t	e_t
1	89	–	–	–	–	–	–	–	–
2	106	89	17.00	89.00	17.00	89.00	17.00	89.00	17.00
3	92	106	−14.00	97.50	−5.50	100.33	−8.33	90.70	1.30
4	98	92	6.00	99.00	−1.00	96.17	1.83	90.83	7.17
5	77	98	−21.00	95.00	−18.00	96.90	−19.90	91.55	−14.55
6	80	77	3.00	87.50	−7.50	90.27	−10.27	90.09	−10.09
7	88	80	8.00	78.50	9.50	87.33	0.67	89.08	−1.08
8	87	88	−1.00	84.00	3.00	87.50	−0.50	88.97	−1.97
9	92	87	5.00	87.50	4.50	87.39	4.61	88.78	3.22
10	86	92	−6.00	89.50	−3.50	88.31	−2.31	89.10	−3.10
11	–	86	–	89.00	–	87.89	–	88.79	–

Table 2.48 Accuracy values of the forecasting methods used by Regens Book.

	Elementary technique	Simple moving average ($r = 2$)	Weighted moving average	Exponential smoothing ($\alpha = 0.1$)
MAD_{10}	9.00	7.72	7.27	6.61
$MAPD_{10}$	10.12%	8.78%	8.30%	7.43%
MSE_{10}	137.13	104.03	111.31	85.08

As a rule of thumb, a forecasting method made on a sporadic time series can be assumed to be accurate if its *MAPD* is less than 30%.

2.7.2 Tuning of the forecasting methods

The accuracy measures can be used to tune the forecasting methods depending on one or more parameters, like exponential smoothing and Winters techniques.

The basic idea is to assign the parameters the values that would have maximized the accuracy of the forecasts in the past. For instance, by using the mean squared error, the most suitable parameter α^* of the exponential smoothing method at time period T can be determined as the solution of the optimization problem:

$$\text{Minimize} \quad MSE_T(\alpha)$$
$$\text{subject to} \quad 0 \leq \alpha \leq 1.$$

Since the number of parameters is usually small and the parameters themselves are usually bounded in nature, a good approximated solution can be found through a discretization. For the sake of simplicity, we illustrate this procedure for a forecasting method with a single parameter δ and an accuracy measure $A(\delta)$:

1. Let δ_{min} and δ_{max} be the minimum and the maximum feasible values of parameter δ. Moreover, let Δ be a discretization step. Set $h = 1$, $\delta_h = \delta_{min}$ and $min = \infty$.
2. If $\delta_h \leq \delta_{max}$, determine the accuracy measure $A(\delta_h)$ corresponding to parameter value δ_h. If $A(\delta_h) < min$, set $min = A(\delta_h)$ and $\bar{\delta} = \delta_h$.
3. If $\delta_h > \delta_{max}$, *STOP*, set $\bar{\delta}$ as the most suitable value of parameter δ, otherwise set $\delta_{h+1} = \delta + \Delta$, $h = h + 1$ and go back to Step 2.

Of course, in the case of multiple parameters there may be a different discretization step for each parameter.

We want to tune the exponential smoothing method for the Regens Book problem, by utilizing the mean absolute deviation as an accuracy measure.

By using $\alpha_{min} = 0$, $\alpha_{max} = 1$ and $\Delta = 0.1$, we get the deviations reported in Table 2.49. On the basis of these results, the most suitable value of the forecasting parameter is $\bar{\alpha} = 0.2$.

Table 2.49 Mean absolute deviations $MAD_{10}(\alpha)$ corresponding to various α values in the Regens Book problem.

α	$MAD_{10}(\alpha)$	α	$MAD_{10}(\alpha)$
0.0	6.56	0.6	7.64
0.1	6.61	0.7	7.80
0.2	6.42	0.8	7.91
0.3	6.65	0.9	8.26
0.4	7.07	1.0	9.00
0.5	7.40		

2.7.3 Forecast control

A forecasting method works correctly if the errors are random and not systematic. Typical systematic errors occur when the variable to be predicted is constantly underestimated or overestimated, or a seasonal variation is not taken into account. Forecast control aims at identifying systematic errors in periodic predictions (caused e.g. by a shift in trend) in order to adjust parameters if needed. Forecasting control can be done through a *tracking signal* or a *control chart.*

2.7.3.1 Tracking signal

The tracking signal K_T at time period T ($T > 1$) is defined as the ratio between the cumulative error at time period T and MAD_T, that is,

$$K_T = \frac{\sum_{t=2}^{T} e_t}{MAD_T}.$$

The tracking signal is greater than zero if the forecast systematically underestimates the datum entry; in contrast, a negative value of K_T indicates a systematic overestimate of the datum entry. For this reason, a forecast is assumed to be unbiased if the tracking signal falls in the range of $\pm\bar{K}$. The value of $\bar{K}$ is established heuristically, and usually varies between 2 and 5. If the tracking signal is outside this interval, the parameters of the forecasting method should be modified or a different forecasting method should be selected.

In this sense, the tracking signal can be also used as a different measure to evaluate the accuracy of a forecasting method.

Three months ago, the logistician of the supermarket chain Plaza (in Bolivia) was in charge of devising and monitoring monthly forecasts for the company's sportswear items. The sales of the previous 12 months (in k$) are reported in Table 2.50. After a preliminary study of the time series, he decided to make use of the exponential smoothing method. The optimal smoothing parameter α was obtained by minimizing $MAD_{12}(\alpha)$. The corresponding solution was $\alpha^* = 0.26$ which was associated with $MAD_{12}(\alpha^*) = 15.95$. The forecasts $p_t, t = 2, \ldots, 12$, are reported in the third column of Table 2.50.

Hence, the manager verified that the tracking signal K_{12} was in the range $[-4, 4]$ ($K_{12} = -1.65$). Then, he devised the forecast for the next month $p_{13} = \$\ 970.08\,\text{k}$.

For the subsequent month ($T = 13$), the sales turned out to be equal to $y_{13} = \$\ 1\,024\,\text{k}$. The associated tracking signal value was $K_{13} = 1.44$, while the new forecast was $p_{14} = \$\ 984.22\,\text{k}$.

Two months later ($T = 14$), the sales were known to be $y_{14} = \$\ 1\,035\,\text{k}$; the tracking signal was $K_{14} = 3.63$ and the forecast was $p_{15} = \$\ 997.55\,\text{k}$. Three months later ($T = 15$), the sales were $y_{15} = \$\ 1\,047\,\text{k}$ and the relative

Table 2.50 Monthly sales of sportsware (in k$) in the Plaza problem and the corresponding exponential smoothing forecasts with $\alpha = 0.26$.

Month	Sales	Forecast	Month	Sales	Forecast
1	977	–	7	948	968.00
2	958	977.00	8	996	962.76
3	989	972.01	9	955	971.48
4	966	976.47	10	959	967.15
5	956	973.72	11	960	965.02
6	965	969.07	12	988	963.70

Table 2.51 Monthly sales of sportsware (in k$) during the first 14 months in the Plaza problem and the corresponding exponential smoothing forecasts (with $\alpha = 0.89$).

Month	Sales	Forecast	Month	Sales	Forecast
1	977	–	9	955	990.84
2	958	977.00	10	959	959.00
3	989	960.12	11	960	959.00
4	966	985.78	12	988	959.89
5	956	968.21	13	1 024	984.86
6	965	957.36	14	1 035	1 019.63
7	948	964.15	15	1 047	1 033.29
8	996	949.80			

tracking signal was $K_{15} = 5.43$, out of the feasible range. Thus, the exponential smoothing parameter was replaced by $\alpha^* = 0.89$, associated with MAD_{15} instead of $MAD_{15}(\alpha^*) = 20.22$. The new forecasts are shown in Table 2.51. The new value of the tracking signal was $K_{15} = 3.81$ and the new forecast was $p_{16} = \$\ 1\,045.47\,\text{k}$.

2.7.3.2 Control charts

Unlike tracking signals, control charts are based on the plot of single errors $e_t, t = 2, \ldots, T$. Let us assume that such errors are realizations of a normal random variable E with expected value μ_E and standard deviation σ_E. A forecast is deemed to be effective if each error e_t is in the confidence interval $\pm k\sigma_E$, where k is quite often set equal to 3 ('3-σ' control chart).

The standard deviation σ_E is usually substituted by its sample mean, computed on the basis of the data entry available at time period T:

$$S_E = \sqrt{\frac{\sum_{t=2}^{T} (e_t - \bar{E})^2}{T-2}}. \tag{2.31}$$

It is worth noting that the expected error $\bar{E}$ is equal to 0 in case there are no systematic errors, in which case (2.31) becomes:

$$S_E \approx \sqrt{\frac{\sum_{t=2}^{T} (e_t - 0)^2}{T-2}} = \sqrt{MSE_T}. \tag{2.32}$$

In case $\bar{E}$ is significantly different from 0, the forecast is *biased* (see Figure 2.27), which would suggest updating the parameters of the forecasting method (or, eventually, changing it with a more appropriate one).

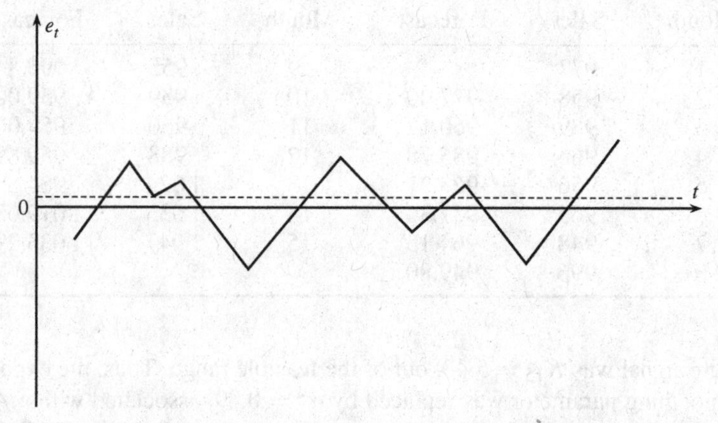

Figure 2.27 Visual examination of the control chart: error with a sample mean not equal to 0.

In addition to the previous analytical check, it is useful to verify visually whether the error pattern reveals the possibility of improving the forecast by introducing suitable modifications:

1. the error pattern shows a positive or negative trend; in this case, the accuracy of the forecasting method is progressively diminishing;
2. the error pattern is periodic; this can happen if an existing seasonal effect has not been identified.

Softline, a company manufacturing leather sofas, uses the moving average method to forecast ($r = 2$) the monthly unit production cost (assuming sales volume is constant) of its top product called Trinity. To assess the forecasting process, the logistician uses a '3-σ' control chart. Table 2.52 reports the unit production costs as well as the forecasts and the errors made during the past two years. The corresponding sample mean error was

$$\bar{E} = € \ 1.40.$$

Table 2.52 Unit production costs, forecasts and errors (in €) in the Softline problem.

t	y_t	p_t	e_t	t	y_t	p_t	e_t
1	452.00	–	–	13	468.50	458.90	9.60
2	463.50	452.00	11.50	14	462.00	465.65	−3.65
3	478.00	457.75	20.25	15	470.00	465.25	4.75
4	466.50	470.75	−4.25	16	473.80	466.00	7.80
5	457.00	472.25	−15.25	17	472.00	471.90	0.10
6	465.20	461.75	3.45	18	478.00	472.90	5.10
7	460.00	461.10	−1.10	19	468.40	475.00	−6.60
8	457.50	462.60	−5.10	20	473.40	473.20	−0.20
9	463.50	458.75	4.75	21	468.00	470.90	−2.90
10	456.00	460.50	−4.50	22	463.00	470.70	−7.70
11	455.00	459.75	−4.75	23	470.50	465.50	5.00
12	462.80	455.50	7.30	24	475.00	466.75	8.25
				25	–	472.75	–

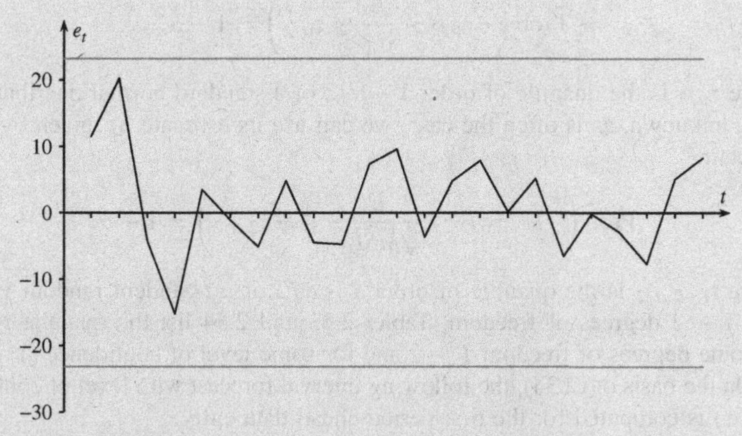

Figure 2.28 Control chart corresponding to the sales forecasts in the Softline problem.

Since this value was a small fraction of a typical monthly unit production cost, the forecast was deemed to be unbiased. The sample standard deviation of the error was computed through (2.32):

$$S_E = \sqrt{MSE_{24}} = € \; 7.89.$$

All the errors were in the range of $\pm 3 \times$ € 7.89 = € 23.67. Hence the forecasting process was deemed to be under control. Moreover, the visual examination of the chart in Figure 2.28 did not show any systematic forecasting error.

2.8 Interval forecasts

The previous sections of this chapter are devoted to *point forecasts*, that is, forecasts expressed by a single possible outcome. Here, we want to devise a range of outcomes that the variable to be forecasted is likely to assume with a given level of confidence (*interval forecast*). We will focus on the one-period-ahead forecast case and assume that the forecasting errors e_t, $t = 2, \ldots, T$, are independent realizations of a normal random variable E with mean μ_E equal to zero and standard deviation σ_E (usually unknown). It is worth noting that if μ_E were different from zero, there would be a systematic error and the forecasts could be improved by subtracting this constant offset μ_E. As a result, we can assume $\mu_E = 0$ without any loss of generality.

Given a level of confidence $(1 - \alpha)$, if σ_E were known, the normalized error $\frac{E-\mu_E}{\sigma_E}$ will satisfy the relationship

$$\text{Prob}\left(-z_{\alpha/2} \leq \frac{E}{\sigma_E} \leq z_{\alpha/2}\right) = 1 - \alpha, \tag{2.33}$$

where $z_{\alpha/2}$ is the quantile of order $1 - \alpha/2$ of a standard normal distribution. If σ_E is unknown, as is often the case, we can use its estimate S_E given by (2.32) to obtain

$$\text{Prob}\left(-t_{T-2,\alpha/2} \leq \frac{E}{\sqrt{MSE_T}} \leq t_{T-2,\alpha/2}\right) = 1 - \alpha, \tag{2.34}$$

where $t_{T-2,\alpha/2}$ is the quantile of order $1 - \alpha/2$ of a t-Student random variable with $T - 2$ degrees of freedom. Tables 2.53 and 2.54 list the quantile $t_{T-2;\alpha/2}$ for some degrees of freedom $T - 2$ and for some level of confidence $(1 - \alpha/2)$.

On the basis of (2.34), the following interval forecast with level of confidence $(1 - \alpha)$ is computed for the first-period-ahead data entry:

$$[p_{T+1} - t_{T-2,\alpha/2}\sqrt{MSE_T};\, p_{T+1} + t_{T-2,\alpha/2}\sqrt{MSE_T}], \tag{2.35}$$

with a level of confidence equal to $(1 - \alpha)$.

Table 2.53 Part I – quantiles of Student's t-distribution.

$t_{T-2,\alpha/2}$	Level of confidence $(1-\alpha/2)$				
	0.900	0.950	0.975	0.990	0.995
1	3.0777	6.3137	12.7062	31.8210	63.6559
2	1.8856	2.9200	4.3027	6.9645	9.9250
3	1.6377	2.3534	3.1824	4.5407	5.8408
4	1.5332	2.1318	2.7765	3.7469	4.6041
5	1.4759	2.0150	2.5706	3.3649	4.0321
6	1.4398	1.9432	2.4469	3.1427	3.7074
7	1.4149	1.8946	2.3646	2.9979	3.4995
8	1.3968	1.8595	2.3060	2.8965	3.3554
9	1.3830	1.8331	2.2622	2.8214	3.2498
10	1.3722	1.8125	2.2281	2.7638	3.1693
11	1.3634	1.7959	2.2010	2.7181	3.1058
12	1.3562	1.7823	2.1788	2.6810	3.0545
13	1.3502	1.7709	2.1604	2.6503	3.0123
14	1.3450	1.7613	2.1448	2.6245	2.9768
15	1.3406	1.7531	2.1315	2.6025	2.9467
16	1.3368	1.7459	2.1199	2.5835	2.9208
17	1.3334	1.7396	2.1098	2.5669	2.8982
18	1.3304	1.7341	2.1009	2.5524	2.8784
19	1.3277	1.7291	2.0930	2.5395	2.8609
20	1.3253	1.7247	2.0860	2.5280	2.8453
21	1.3232	1.7207	2.0796	2.5176	2.8314
22	1.3212	1.7171	2.0739	2.5083	2.8188
23	1.3195	1.7139	2.0687	2.4999	2.8073
24	1.3178	1.7109	2.0639	2.4922	2.7970
25	1.3163	1.7081	2.0595	2.4851	2.7874
26	1.3150	1.7056	2.0555	2.4786	2.7787
27	1.3137	1.7033	2.0518	2.4727	2.7707
28	1.3125	1.7011	2.0484	2.4671	2.7633
29	1.3114	1.6991	2.0452	2.4620	2.7564
30	1.3104	1.6973	2.0423	2.4573	2.7500
31	1.3095	1.6955	2.0395	2.4528	2.7440
32	1.3086	1.6939	2.0369	2.4487	2.7385
33	1.3077	1.6924	2.0345	2.4448	2.7333
34	1.3070	1.6909	2.0322	2.4411	2.7284
35	1.3062	1.6896	2.0301	2.4377	2.7238
36	1.3055	1.6883	2.0281	2.4345	2.7195
37	1.3049	1.6871	2.0262	2.4314	2.7154
38	1.3042	1.6860	2.0244	2.4286	2.7116
39	1.3036	1.6849	2.0227	2.4258	2.7079
40	1.3031	1.6839	2.0211	2.4233	2.7045
41	1.3025	1.6829	2.0195	2.4208	2.7012
42	1.3020	1.6820	2.0181	2.4185	2.6981
43	1.3016	1.6811	2.0167	2.4163	2.6951
44	1.3011	1.6802	2.0154	2.4141	2.6923

Table 2.54 Part II – quantiles of Student's t-distribution.

$t_{T-2,\alpha/2}$	Level of confidence $(1-\alpha/2)$				
	0.900	0.950	0.975	0.990	0.995
45	1.3007	1.6794	2.0141	2.4121	2.6896
46	1.3002	1.6787	2.0129	2.4102	2.6870
47	1.2998	1.6779	2.0117	2.4083	2.6846
48	1.2994	1.6772	2.0106	2.4066	2.6822
49	1.2991	1.6766	2.0096	2.4049	2.6800
50	1.2987	1.6759	2.0086	2.4033	2.6778
51	1.2984	1.6753	2.0076	2.4017	2.6757
52	1.2980	1.6747	2.0066	2.4002	2.6737
53	1.2977	1.6741	2.0057	2.3988	2.6718
54	1.2974	1.6736	2.0049	2.3974	2.6700
55	1.2971	1.6730	2.0040	2.3961	2.6682
56	1.2969	1.6725	2.0032	2.3948	2.6665
57	1.2966	1.6720	2.0025	2.3936	2.6649
58	1.2963	1.6716	2.0017	2.3924	2.6633
59	1.2961	1.6711	2.0010	2.3912	2.6618
60	1.2958	1.6706	2.0003	2.3901	2.6603
70	1.2938	1.6669	1.9944	2.3808	2.6479
80	1.2922	1.6641	1.9901	2.3739	2.6387
90	1.2910	1.6620	1.9867	2.3685	2.6316
100	1.2901	1.6602	1.9840	2.3642	2.6259

We want to devise, at time period $T = 24$, an interval forecast for the one-month-ahead unit production cost of the Trinity sofa in the Softline problem, with a level of confidence $(1-\alpha) = 0.90$. The quantiles involved into the computation are: $t_{T-2,\alpha/2} = t_{22,\alpha/2} = 1.71$ (see Table 2.53). By utilizing (2.35) and taking into account that $p_{25} =$ € 472.75 and $\sqrt{MSE_{24}} =$ € 7.89, we get:

$$[\text{€ } (472.75 - 1.7171 \times 7.89); \text{€ } (472.75 + 1.71 \times 7.89)]$$
$$= [\text{€ } 459.20; \text{€ } 486.30].$$

Hence, the sales of the Trinity sofa at time period $t = 25$ will be included between € 459.20 and € 486.30 with a level of confidence equal to 90%.

Again, we emphasize that this interval forecast relies on the hypothesis that the underlying unit production cost process will not change during the next month.

2.9 Case study: Forecasting methods at Adriatica Accumulatori

Adriatica Accumulatori is an electromechanical firm headquartered in Termoli, Italy, manufacturing car spare parts for the Italian market. In 2008, the results of a survey showed that, although Adriatica Accumulatori car battery sales constantly increased during the previous decade, the company progressively lost market share (see Table 2.55). Until 2008, the company had traditionally based its production and marketing plans on sales forecasts, provided by a time series extrapolation technique which, applied to the data shown in Table 2.55, results in the following expression:

$$p_{10}(\tau) = 285\,875.06 + 15\,951.08\tau, \tau = 1, 2, \ldots,$$

which would provide the following demand forecasts: 301 826 units in 2009 ($\tau = 1$) (with a 9.4% increase with respect to 2008) and 317 777 units in 2010 ($\tau = 12$) (with a 15.2% increase with respect to 2008). However, the results of the survey convinced the company management that during the previous decade Adriatica Accumulatori had lost an opportunity to sell more, mainly because its forecasts were not related to market demand. Based on this reasoning, it was decided to predict sales by first estimating the Italian market demand and then evaluating different scenarios corresponding to the current market share and increased shares achievable through appropriate marketing initiatives.

Table 2.55 Number of spare batteries sold.

Year	Italian market sales	Adriatica Accumulatori sales	Adriatica Accumulatori market share (%)
1999	693 326	138 665	20
2000	803 666	152 696	19
2001	947 243	170 503	18
2002	1 136 433	193 192	17
2003	1 406 432	210 964	15
2004	1 666 011	233 241	14
2005	1 869 683	243 058	13
2006	2 136 463	256 375	12
2007	2 316 402	266 386	11
2008	2 507 929	275 872	11

In order to forecast the Italian market sales, a causal method was used (see Section 2.5.2). The time series of national battery sales was correlated to the number of cars sold two years before (see Table 2.56). Then, the Italian demand

Table 2.56 Car sales in Italy over 10 years.

Year	Number	Year	Number
1997	253 321	2003	886 297
1998	381 385	2004	1 014 975
1999	491 755	2005	1 162 246
2000	634 706	2006	1 167 614
2001	951 704	2007	1 217 929
2002	830 175	2008	1 363 594

of spare batteries was forecast for the years 2009 and 2010, as 2 396 003 and 2 676 295 units, respectively. Then, the company management generated several scenarios based on different market shares. In the case where the firm maintained a market share equal to 11%, the demand would be equal to 263 560 units in 2009 (with a 4.5% increase with respect to 2008), and 294 392 units in 2010 (with a 6.7% increase with respect to 2008).

The two forecasting methods used by the company provided different results. Therefore, the company decided to better analyse the logic underlying the two approaches. Because the Italian economy was undergoing a period of quick and dramatic change, the latter method was deemed to provide more accurate predictions than the former, which is more suitable when the past demand pattern is likely to be replicated in the future.

2.10 Case study: Sales forecasting at Orlea

Orlea is an Australian firm which produces and distributes bicycles. The company makes more than 150 models, divided into four classes: Mountain Bike, Racing, City and Junior. Each class is further articulated into families (e.g. class City consists of four families). The models differ also in colour and size, with more than 500 versions overall. The range of bicycles is renewed every sales period, which starts from September to August of the subsequent year. The company is active on the Australian market (about 60% of the company sales), with a a network of more than 450 dealers across the country, and on some foreign markets, in collaboration with affiliate companies and distributors. In Australia, the logistics system is the MTS type (see Section 1.2.2), since dealers, due to the limited availability of funds and the difficulties in sales forecasting, work with low inventory levels and issue orders of low size which the company has to fulfil within a short time. Conversely, the affiliate companies and the foreign distributors require large lots of bicycles with a quite long delivery time (up to four months). In this case, the company works with an MTO-type logistics system. In order to keep inventory management costs low, the company decided in 2008 to adopt a sales-forecasting system for the short term (a few months).

For products of the Dream family (which includes nine models and belongs to the City class), historical data of monthly sales show (see Figure 2.29) some

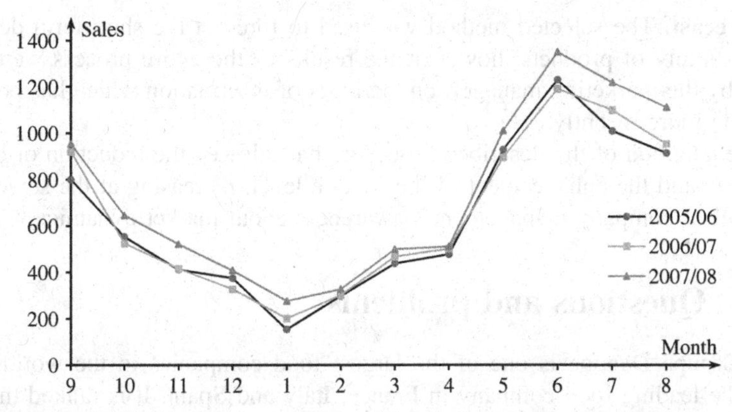

Figure 2.29 Monthly trend of sales for bicycles of the Dream family over three sales periods.

regularity (and, in particular, a strong seasonal variation) in the trend for the available three sales periods (2005/06, 2006/07 and 2007/08).

The sales-forecasting method has been articulated into three phases. In the first, a sales forecast for the product family level has been defined, by using extrapolation methods based on time series of monthly sales for all the models in the Dream family. In the next phase, the forecasts of the demand obtained in the previous step are disaggregated among the nine models of the family, by proportionally dividing the overall value and using specific coefficients which are monthly updated. Finally, the demand for each version has been estimated qualitatively by the marketing managers, based on their experience, starting from the data related to the corresponding model. For the first phase, the following process was adopted.

Data preprocessing. The available historical values (from the company information system) were cleaned of outliers due, for example, to the different number of working days in the month or to exceptional events, on the basis of recently available information.

Initialization. Data from the 2005/06 period were used to this end.

Parameter calibration. The values of the parameters which characterize the adopted forecasting methods (like the Holt method and the Winters method) were selected by simulating the forecast of the demand trend for all the 2006/07 period; from the comparison between forecast data and real demand values, forecasting errors were evaluated and, thus, by means of an accuracy measure (the *MSE*), the best parameter values were chosen.

Choice of the forecasting method. Once the optimal parameter values were determined for each of the adopted forecasting methods, these techniques were compared by generating the forecasts for the 2007/08 period; by using the *MSE* for comparing the accuracy of such techniques, the most suitable forecasting method was chosen.

Forecast. The selected method was used to forecast the short-term demand for the family of products; however, the results of the entire process were corrected by the marketing manager, on the basis of information which has become available more recently.

The adoption of the described procedure has allowed the reduction of operative costs and the enhancement of the service level, increasing at the same time the level of company management's awareness about market dynamics.

2.11 Questions and problems

2.1 Groupe Danone is one of the largest food companies in the world. It is the leading food company in France, Italy and Spain. It is ranked third in Europe and seventh worldwide. In Italy, Danone has invested more in probiotic products, such as Activia and Actimel. During the first three months of last year, the company carried out an opinion poll by launching a campaign called 'Choose the Flavour' on the Facebook fanpage Activia Italia. The goal was to let consumers directly select the new yoghurt flavours. In this way, Facebook was used as a tool to carry out a market survey. Discuss the limits and opportunities offered by such a qualitative forecasting method.

2.2 In freight transportation, a typical logistics problem is the allocation of empty containers to depots, in order to satisfy future demands of transportation services. How can a carrier predict demands of containers for the transport of a given category of products?

2.3 To what extent are the forecasting practices different in an MTS and in an MTA system?

2.4 How would you predict the future demand of a new product?

2.5 Hot Spot is a firm based in the United States whose core business is the maintenance of home heaters. The company usually forecasts service requests on the basis of the number of installed heaters. The number of installed home heaters and the number of service requests received over past years in New Jersey are reported in Table 2.57.
Forecast the service request p_9 by using, to this purpose:

- a single regression analysis (service requests versus the total number of heaters installed);
- a multiple regression analysis (service requests versus the number of heaters installed less than, or equal to, two years ago and at least two years ago).

Which technique is the most accurate? Why?

Table 2.57 Number of Hot Spot heater installations and service requests.

Year	Heater installations ≤ two years ago	Heater installations > two years ago	Total	Service requests
1	260 000	69 500	329 500	18 672
2	265 000	74 200	339 200	19 076
3	287 800	82 850	370 650	20 994
4	313 750	90 550	404 300	23 249
5	345 350	97 150	442 500	25 025
6	379 050	105 950	485 000	28 111
7	416 950	111 550	528 500	30 985
8	459 100	117 000	576 100	33 397
9	502 550	123 200	625 750	

2.6 Sunshine is one of the world's leading suppliers of fast-moving goods in household care and personal product categories. According to the company management, sales of the Facial Cleansing Blemish product mostly depend on promotion expenditure made by the company and its competitor. Table 2.58 reports facial soap sales for the previous 10 quarters, and quarterly levels of sales in Canada for that detergent with respect to (a) Sunshine promotion expenditure, and (b) the competitor's promotion expenditure divided by the Sunshine promotion expenditure. Make a sales forecast for the next two time periods under the hypothesis that Sunshine will increase its promotion expenditure to CAN$ 14.5 million and the competitor's promotion expenditure will remain the same as in the actual quarter.

Table 2.58 Data for the forecasting problem of Sunshine.

Quarter	Sunshine promotion expenditure (Millions CAN$)	Competitor's promotion expenditure / Sunshine promotion expenditure	Sales of Blemish in Canada (Millions CAN$)
1	6.0	1.2	46.8
2	6.8	1.2	52.7
3	7.5	1.4	60.5
4	7.5	1.5	56.6
5	9.0	1.5	64.4
6	10.5	1.7	74.1
7	12.0	1.8	72.2
8	12.0	1.5	78.0
9	13.5	1.4	87.8
10	13.5	1.5	95.6

2.7 Carlinek is a Polish company, based in Poznan, which retails Aqua-Floor, a well-known water-resistant laminated floor covering. The company needs to forecast the demand of this product, by using historical data on the sales of the latest nine months, reported in Table 2.59.

Table 2.59 Aqua-Floor product quantities (in m^2) sold by Carlinek in Poznan during the last nine months.

Month	Quantity	Month	Quantity
1	450	6	437
2	670	7	456
3	332	8	–
4	123	9	231
5	343		

After checking the presence of possible outliers in the historical database, use the weighted moving average method to determine the one-month-ahead sales forecast.

2.8 Table 2.60 reports an estimate of the annual mean demand and the corresponding variance of fruit and vegetable products of Flanders, Belgium. The products are divided into five categories: fall–winter, spring–summer, exotic fruits, citrus fruits and dried fruits.
Verify whether the aggregate forecast is more or less accurate than the estimated demand for each category.

Table 2.60 Annual mean demand and variance of fruit and vegetable products of Flanders.

Category	Mean (t)	Variance (t^2)
Fall–winter	123 000	24 570
Spring–summer	245 000	35 650
Exotic fruits	9 860	320
Citrus fruits	98 000	2 980
Dried fruits	2 450	456

2.9 Vaal Engineering is a South African company which produces and distributes industrial slotting machines in Africa. The number of 800xp machines sold during the last 60 months is reported in Table 2.61.

Table 2.61 Number of 800xp slotting machines sold by Vaal Engineering during the last 60 months.

Period	Quantity	Period	Quantity	Period	Quantity	Period	Quantity
1	60	16	104	31	32	46	144
2	83	17	205	32	81	47	185
3	130	18	235	33	149	48	280
4	120	19	32	34	141	49	89
5	187	20	81	35	177	50	73
6	248	21	149	36	281	51	137
7	51	22	141	37	75	52	146
8	76	23	177	38	78	53	191
9	125	24	281	39	130	54	268
10	134	25	54	40	112	55	31
11	196	26	94	41	202	56	63
12	262	27	132	42	262	57	126
13	54	28	104	43	44	58	133
14	94	29	205	44	80	59	213
15	132	30	235	45	128	60	281

After plotting the data, clean the time series from possible seasonal and residual variations and then determine the trend component.

2.10 A time series $w_t, t = 1, \ldots, 24$, is reported in Table 2.62. Can the time series correspond to a seasonal variation over time, with periodicity $M = 12$, obtained through the application of a decomposition method (a multiplicative way to combine the components)? If so, explain the meaning of w_1, otherwise what can we do?

Table 2.62 Time series $w_t, t = 1, \ldots, 24$, used in Problem 2.10.

t	w_t	t	w_t	t	w_t
1	1.012	9	0.876	17	1.174
2	1.123	10	0.904	18	1.055
3	1.088	11	0.812	19	1.101
4	1.122	12	0.714	20	1.086
5	1.097	13	1.057	21	0.909
6	1.023	14	1.134	22	0.987
7	1.001	15	1.099	23	0.893
8	0.987	16	1.180	24	0.811

2.11 The time series decomposition method (additive) into its four components of trend, cyclical, seasonal and residual variations is used to estimate monthly sales (in kg) of a product. Characterize the seasonal component $s_{\bar{t}}$ for some $\bar{t}$, in terms of units, magnitude (can it be positive, negative or zero? Explain its meaning) and the average value of M consecutive time periods (M is the periodicity).

2.12 Mitsumishi is a Korean company whose number of M5 light trucks sold during the last 42 months is reported in Table 2.63. The company invested significant financial resources in promotion, and the M5 sales increased in some months, as shown in Table 2.64.

(a) Forecast sales for the next six months using an appropriate forecasting method.

(b) Plot a control chart. Are you able to detect any anomalies?

Table 2.63 Number of M5 trucks sold by Mitsumishi.

Month	Quantity	Month	Quantity	Month	Quantity
1	22 882	15	20 967	29	28 414
2	19 981	16	19 759	30	22 537
3	18 811	17	22 200	31	22 845
4	19 352	18	24 162	32	9 451
5	27 226	19	20 275	33	15 842
6	18 932	20	7 949	34	16 409
7	18 931	21	14 328	35	13 881
8	8 523	22	16 691	36	11 230
9	13 064	23	13 784	37	24 765
10	13 733	24	10 986	38	21 739
11	12 597	25	24 768	39	25 153
12	7 645	26	19 351	40	20 515
13	23 478	27	23 953	41	24 038
14	17 019	28	18 855	42	25 151

Table 2.64 M5 sales improvement (in %) achieved during some months by Mitsumishi.

Month	Sales improvement	Month	Sales improvement
5	30	19	8
6	13	29	30
7	3	30	12
17	20	31	5
18	15		

2.13 Sit is an American company which produces and sells high-quality printers. The number of XC2100 printers monthly sold in Berkeley, California, during the last 12 months is reported in Table 2.65. The company has always forecasted the monthly one-period-ahead demand by using the moving average method ($r = 2$), but the logistician is now evaluating the option of adopting a new forecasting technique, called Q. Knowing the forecasting errors e_t, $t = 2, \ldots, 12$, obtained by using the Q method for the one-period-ahead forecasts (and reported in Table 2.66), determine the value that e_8 should take to induce the logistician to replace the moving average method with Q.

Table 2.65 Number of XC2100 printers sold during the last 12 months by Sit.

Period	Quantity	Period	Quantity
1	835	7	810
2	798	8	814
3	831	9	800
4	772	10	793
5	750	11	805
6	783	12	829

Table 2.66 Forecasting errors e_t, $t = 2, \ldots, 12$, generated by using the Q method.

Period	Error	Period	Error
1	–	7	16
2	24	8	x
3	−13	9	−8
4	19	10	41
5	−35	11	31
6	27	12	−28

2.14 Aldes is a food and beverage company located in the Republic of Mauritius that is specialized in the export of canned food, such as tuna fish, mainly sold in Africa, the Middle East and Western Europe, in cartons of 48 cans of 185 g each.

The monthly exportation value of tuna fish to Jordan during the last eight months is summarized in Table 2.67.

Estimate the two-period-ahead exports of tuna fish by using the exponential smoothing method with $\alpha = 0.20$, $\alpha = 0.25$ and $\alpha = 0.35$. By comparing the obtained results, what is the best value of α?

If the company decides to promote the product through a discount policy to increase the sales, how convenient it is to modify the choice of α? Is it possible to use the selected method for six-month-ahead forecasting?

Table 2.67 Aldes exports (in $) of tuna fish to Jordan during the last eight months.

Month	Sales	Month	Sales
1	100 000	5	92 000
2	98 000	6	87 500
3	70 500	7	95 000
4	90 000	8	99 500

2.15 Prove that the sum of weights in Equation (2.12) is equal to 1.

2.16 Given the time series $y_t, t = 1, \ldots, 12$, reported in Table 2.68, use the exponential smoothing method to determine the forecasting value p_{13}, assuming a time-variable parameter $\alpha_t = \max_t \left\{0.24; \sqrt[3]{\frac{3}{5t^2}}\right\} t = 2, \ldots, 12$. Modify Equation (2.12) when the time-variable parameter is used and verify that the sum of weights still remains equal to 1. Determine whether the time-variable parameter implies a more accurate forecasting than the value of $\alpha = 0.3$.

Table 2.68 Time series $y_t, t = 1, \ldots, 12$, used in Problem 2.16.

t	y_t	t	y_t
1	1 100	7	920
2	935	8	985
3	1 120	9	1 070
4	1 040	10	940
5	1 060	11	1 100
6	1 100	12	970

2.17 Let $y_t, t = 1, \ldots, T$, a time series, characterized by a linear trend, no cyclical and seasonal variations. Prove that the linear regression method corresponds to the elementary technique when $r = 2$.

2.18 EuroPack is a leading European firm of packaging products. The plant located in Denmark realizes, in a continuous cycle, more than 150 products in 60 g/m^2 polyethylene and nylon film.

(a) Estimate the amount of polyethylene and nylon film needed to ensure daily production in the next week by using historical data related to the last 30 days (shown in Table 2.69).

(b) Determine the tracking signal (band ± 4) at the current time period.

(c) Determine the interval forecast of polyethylene and nylon films for one day ahead with respect to $T = 30$, with a level of confidence $(1 - \alpha) = 0.95$.

Table 2.69 Quantity (in thousands of kg) of kg/m^2 polyethylene and nylon films used by EuroPack in the last 30 days.

Period	Quantity	Period	Quantity
1	133	16	196
2	155	17	220
3	179	18	207
4	191	19	213
5	176	20	219
6	150	21	219
7	204	22	170
8	145	23	213
9	204	24	227
10	209	25	205
11	149	26	203
12	210	27	187
13	185	28	225
14	204	29	235
15	203	30	194

2.19 Oasis is a pineapple soft drink sold in Germany. The monthly sales data over the past 44 months are reported in Table 2.70.

(a) Plot the data on a graph. What important observations can you make about the demand pattern? Which data are relevant and should be used for forecasting purposes?

(b) Use two different forecasting methods to predict sales over the next four months.

Table 2.70 Sales (in kl) of Oasis soft drink in Germany over the last 44 months.

Month	Quantity	Month	Quantity	Month	Quantity	Month	Quantity
1	9 050	13	10 000	25	9 150	37	8 900
2	8 050	14	9 750	26	8 750	38	9 450
3	7 000	15	9 300	27	8 800	39	8 750
4	6 120	16	10 100	28	9 400	40	9 500
5	8 250	17	10 400	29	12 000	41	13 400
6	11 450	18	15 650	30	13 450	42	14 000
7	10 900	19	16 350	31	14 900	43	16 850
8	12 850	20	17 000	32	18 760	44	21 000
9	10 650	21	13 600	33	12 250		
10	11 000	22	11 250	34	11 000		
11	9 200	23	9 500	35	9 600		
12	8 900	24	9 200	36	9 100		

(c) Estimate *MAPD* of both methods using the data of the last six months. Which approach seems to work best?

(d) Select the most accurate method (see Problem 2.19(c)) and then determine, with a level of confidence $(1 - \alpha) = 0.90$, the interval forecast for the one-month-ahead sales at $T = 44$.

2.20 Hollaflowers is a Dutch cut-flowers company with its own distribution network that covers five European countries. In order to maintain a product quality control in the last phase of the distribution system, the company uses its resources for timely delivery to customers. To this end, it is required to determine the necessary resources to satisfy future demands of customers. Based on daily data over the last five weeks, which are reported in Table 2.71, forecast the number of vans for the one-week-ahead deliveries.

Table 2.71 Number of vans daily used over the last five weeks by Hollaflowers.

	Week				
Day	1	2	3	4	5
Monday	3	24	21	25	24
Tuesday	8	16	25	23	16
Wednesday	10	12	12	19	8
Thursday	2	18	15	12	22
Friday	6	6	26	12	14

2.21 The National Health Service of Belgium is responsible for the distribution of Neurozam to all national hospitals. Neurozam is a drug used to fight a rare neurological disease. Its high cost and perishability forced the National Health Service to have small amount of stocks, used to satisfy demand of not more than three days. Plan the supply of the drug of the National Health Service taking into account the number of packages of the drug distributed in the last 20 days (see Table 2.72).

2.22 A company is planning to add extra capacity to a plant currently manufacturing 110 000 items per year. After an accurate sales forecast over the next few years, one is quite sure that the most likely value of the annual demand is 140 000 items and that the MSE is equal to 10^8. It is known that the company loses \$ 3 for each unit of unused capacity and \$ 7 for each unit of unsatisfied demand. How much capacity should the company buy? (Hint: the forecasting error can be assumed to be normally distributed.)

2.23 Given the time series $y_t, t = 1, \ldots, 42$, reported in Table 2.63, it is assumed that $y_t = 0$, for $t = 4, 5, 6, 16, 17, 18, 28, 29, 30, 40, 41, 42$.

Table 2.72 Number of packages (three vials each) of Neurozam distributed over the last 20 days from the Belgian National Health Service.

Day	Quantity	Day	Quantity
1	23	11	0
2	25	12	0
3	19	13	0
4	0	14	16
5	14	15	14
6	0	16	6
7	0	17	0
8	13	18	0
9	16	19	12
10	0	20	0

Determine $p_{42}(\tau), \tau = 1, \ldots, 6$, using, if it is necessary, the same forecasting method selected in Problem 2.12.

2.24 Monitor the forecasts for the monthly sales of sportswear of the Plaza supermarkets (see Section 2.7.3.1) by using, as an accuracy measure, the tracking signal instead of *MAD*.

2.25 To control a one-period-ahead forecasting process determined with the exponential smoothing method ($\alpha = 0.3$) on the time series reported in Table 2.62 (see Problem 2.14), use a '3-σ' control chart. What you can say about the accuracy of the forecasting method? Construct a 95% level of confidence for the one-period-ahead forecast.

3

Locating facilities in logistics systems

3.1 Introduction

Facility location problems deal with the number, position, equipment and size of new facilities, as well as the divestment, displacement or downsizing of existing facilities. In business logistics, the location planning process consists of designing the set of facilities through which commodities flow from suppliers to demand points, whereas in the public sector it consists of determining the set of facilities from which users are serviced. In this chapter some of the most important facility location problems are examined. To put this analysis in the right perspective, a number of relevant issues are first introduced and discussed.

When location decisions are needed Facility location decisions must obviously be made when a logistics system is started from scratch. They are also required as a consequence of variations in the demand pattern or spatial distribution, or following modifications of materials, energy or labour cost. In particular, location decisions are often made when new products or services are launched, or outdated products are withdrawn from the market.

Location decisions may be strategic or tactical Whenever facilities are purchased or built, location decisions involve sizeable investments. In this case, changing sites or equipment is unlikely in the short or medium run. This may be true even if facilities are leased. On the other hand, if space and equipment are rented (e.g. from a public warehouse) or operations are subcontracted, facility location decisions can be reversible in the medium term.

Introduction to Logistics Systems Management, Second Edition. Gianpaolo Ghiani, Gilbert Laporte and Roberto Musmanno.

Location and allocation decisions are intertwined Location decisions are strictly related to those of defining facility area boundaries (e.g. allocating demand to facilities). For example, in a two-echelon distribution logistics system (see Figure 3.1), opening a new RDC must be accompanied by a redefinition of the sales districts along with a different allocation of the RDCs to the CDCs and of the CDCs to the production plants. For this reason location problems are sometimes referred to as *location-allocation* problems.

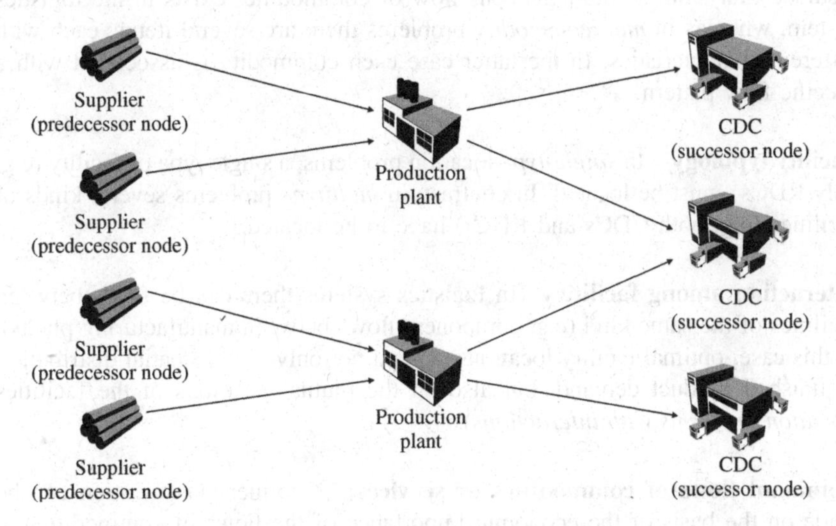

Figure 3.1 A two-echelon single-type location problem.

Location decisions may affect demand Facility location may affect the demand volume. For example, opening a new RDC may lead to the acquisition of customers who previously could not be served at a satisfactory service level because they were located too far.

Facility location problems can be classified with respect to a number of criteria. The classification proposed in this chapter is logistics-oriented.

Key factors in decisions In business logistics the key factors that affect location choices (of manufacturing plants, distribution centres etc.) are of an economic nature (costs and or profits). In service logistics, the key factor when locating public utility facilities (fire service, places of worship, dumps, police stations, bank offices etc.) is *accessibility*, in order to achieve equity among servicing users.

Number of facilities Location problems may concern a single facility or multiple facilities. The location of a single facility is a simpler problem since it is not necessary to consider further decision aspects connected to location (i.e. demand allocation).

Decision space Location problems can be *continuous* if the candidate locations are points of a continuous space, or *discrete*, which is the most common case, whenever the facilities are chosen from a list of potential sites. Moreover, some location problems can be modelled on a graph (directed, undirected or mixed), where facilities can be located on the vertices or on the arcs or edges.

Homogeneity of the material flows In *single-commodity* problems it can be assumed that a single homogeneous flow of commodities exists in the logistics system, whereas in *multicommodity* problems there are several items, each with different characteristics. In the latter case each commodity is associated with a specific flow pattern.

Facility typology In *single-type* location problems, a single type of facility (e.g. only RDCs) must be located. In contrast, in *multitype* problems several kinds of facilities (e.g. both CDCs and RDCs) have to be located.

Interaction among facilities In logistics systems there can be flows between facilities of the same kind (e.g. component flows between manufacturing plants). In this case, optimal facility locations depend not only on the spatial distribution of finished product demand, but also on the mutual positions of the facilities (*location problems with interactions*).

Dominant flows of commodities or services A further classification can be made on the basis of the economic importance of the flows of commodities or services between the facilities to be located and those directly connected to them. *Single-echelon* location problems are single-type problems such that either the flow coming out from the facilities or the flow entering the facilities to be located is negligible. A typical case of a single-echelon problem arises in manufacturing plant location (e.g. in the steel industry), in which the weight of the finished products is significantly lower than that of the raw materials used (iron and coal); under these assumptions, transport costs associated with incoming flows to manufacturing plants are dominant. Another example of a single-echelon problem is the location of the DCs of a commercial business company that purchases goods at pre-established prices (inclusive of the transport cost up to the DCs) and also supplies sales points spread over a wide geographic area. In this case, the goal is to minimize the sum of the facility cost of DCs and the transport cost from these to the sales points.

In *two-echelon* single-type problems, both inbound and outbound flows of commodities or services are relevant. This is the case, for example, when manufacturing plants have to be located taking into account both the transport cost from the supply points (predecessor nodes) to manufacturing plants and the transport cost from the manufacturing plants to the DCs (successor nodes) (Figure 3.1).

In two-echelon problems, constraints aiming at balancing inbound and outbound flows have to be considered.

Divisibility of flows In some logistics systems, flows of products or services from predecessor nodes or flows directed to successor nodes are not divisible (for administrative or book-keeping reasons). For example, for the logistics system depicted in Figure 3.1, one could impose that every DC be served by a single manufacturing plant; the corresponding location problem should take into account this additional constraint.

Influence of transport on location decisions Most facility location models assume direct connections between facilities of a logistics system (door-to-door connections); this assumption is correct if vehicles travel fully loaded. However, when it is necessary to delivery or pick up small loads, the same vehicle can typically serve multiple points (e.g. retailers in a district) and transport costs are affected by the order in which the successor or predecessor nodes are served; the optimal location of logistics facilities is therefore dependent on vehicles routes (*location-routing* models). In such cases, traditional facility location models cannot be used properly. To illustrate this concept, consider Figure 3.2 where a DC serves three sales points located at the vertices of triangle *ABC*.

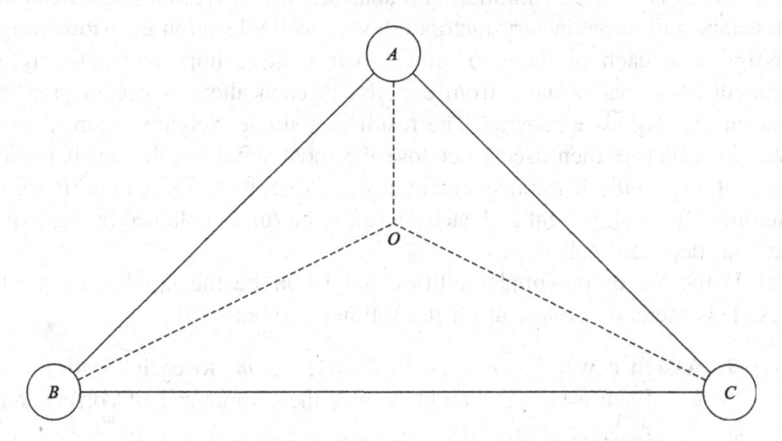

Figure 3.2 Steiner point O in the triangle ABC.

Under the hypothesis that the facility costs are independent of where the DC is located, there can be two extreme cases:

- each sales point requires a full load supply and, therefore, the optimal location of the DC corresponds to the Steiner point O (i.e. the internal point of the triangle so that the three angles formed by the intersections of the three segments starting from the vertices are all equal to 120°);
- the sum of the demands of the three sales points is less than or equal to the vehicle capacity and, therefore, the DC can be located at any point of the triangle ABC perimeter.

To solve location problems, two different methods can be used. The first, which is qualitative, simple, intuitive and applicable to discrete location problems, is illustrated in Section 3.2. The second method is quantitative and based on the formulation of the location problem as an optimization model. Some quantitative methods for the most relevant facility location problems are illustrated in Section 3.3. It is also possible to use a hybrid method, obtained by combining qualitative and quantitative factors in the decision-making process; this method will be briefly described in Section 3.4.

3.2 Qualitative methods

Qualitative methods are suitable whenever the number of candidate solutions is discrete (and usually fairly small) and the location decision is affected by some criteria that are difficult to assess in monetary terms (e.g. proximity to transport infrastructures such as parking areas, motorways, railways, ports and airports; proximity to shopping areas, competitors etc.; or labour relations, a fall or a rise in population etc.). An intuitive and easy-to-apply method is the *weighted scoring method*, which is a form of multicriteria analysis. It involves the identification of all monetary and nonmonetary factors relevant to the location decision; weights are assigned to each of them to reflect their relative importance; finally, the assignment of scores (usually from experts) to each alternative expresses their impact on the various attributes. The result is a single weighted score for each alternative, which is then used to choose the most suitable solution. It is worth noting that, especially if the investment cost is large, it is of key importance that the rationale for each weight and each score can be fully explained by the experts to the final decision maker.

Let V the set of potential facilities and let m be the number of location factors. This method is made up of the following three steps.

Step 1. Assign a weight $w_k \in (0, 1), k = 1, \dots, m$, to each location factor k [1]. Each assigned weight defines the importance of corresponding location factor (note that generally it is required that $\sum_{k=1}^{m} w_k = 1$).

Step 2. Assign a score $s_{ik}, i \in V, k = 1, \dots, m$ (typically, from 0 to 10) to location factor k of site i.

Step 3. For every site $i \in V$, the sum of the weighted score is calculated as $r_i = \sum_{k=1}^{m} w_k s_{ik}$. The preferred site i^* is the one that achieves the highest overall score, that is, $i^* = \arg\max_{i \in V} \{r_i\}$.

[1] Different techniques can be used to determine the weights to attribute to location factors. Such a procedure is illustrated in Chapter 4 (see Section 4.3).

Jet Market has to decide where to locate a retail outlet in Berne. External consultants have selected seven factors which are considered the most important for the location decision. Each factor has a weight from 0 to 1 (Table 3.1).

Table 3.1 Weights associated to each location factor in the Jet Market problem.

ID	Location factor	Weight
1	Renting cost	0.40
2	Availability of skilled workforce	0.14
3	Proximity to transport infrastructure	0.06
4	Proximity to parking areas	0.20
5	Number of shop windows	0.05
6	Proximity of retail competitors	0.10
7	Proximity of complementary shops	0.05
		1.00

By applying the weighted scoring method, three possible commercial areas are evaluated (1, 2 and 3). Scores vary between 0 and 10 and are reported in Table 3.2. The weighted scores for each site are 4.48, 4.14 and 4.65, respectively. On the basis of this analysis, the third site has the highest score and thus it is chosen.

Table 3.2 Evaluation of the location factors for the three potential sites in the Jet Market problem.

	Score		
Location factor	Site 1	Site 2	Site 3
1	5	4	4
2	3	5	5
3	6	4	5
4	4	3	5
5	7	8	7
6	4	5	5
7	3	2	4

3.3 Quantitative methods

This section illustrates some quantitative methods to tackle the most relevant location problems arising both in business and in service logistics systems. Far from being exhaustive, our treatment of the topic covers several models and techniques that can be suitably adapted to solve a much larger set of location problems in the most diverse settings.

3.3.1 Single-commodity single-echelon continuous location problems

The problem considered in this section amounts to finding the optimal location of a single facility in a two-dimensional Cartesian plane. For the sake of simplicity, we assume that this facility has to supply a single commodity to a set of successor nodes (e.g. retailers). We also suppose that the transport cost between the facility to be located and every successor node is proportional to the Euclidean distance between these two entities.

To formulate the problem, let V be the set of successor nodes. For each successor node $i \in V$, both the Cartesian coordinates (x_i, y_i) and an estimate of the demand d_i during the planning horizon are known. Moreover, let (x, y) be the (unknown) Cartesian coordinates of the facility to be located. If its facility cost is independent from its location, the problem consists in finding the coordinates (x^*, y^*) to

$$\text{minimize } f(x, y) = \sum_{i \in V} cd_i \left(\sqrt{(x_i - x)^2 + (y_i - y)^2} \right), \tag{3.1}$$

where c represents the transport cost per distance unit travelled and per unit of commodity transported. This cost is assumed identical for every successor node $i \in V$. Consequently, the objective function (3.1) expresses the overall transport cost to be minimized during the planning horizon. Models like this, seemingly of limited practical interest, are, however, often used when data are not available to perform a study with more sophisticated models, or when one must to determine not the optimal position of the facility (which could be geographically infeasible), but, rather, a location area of interest. In the latter case, the optimal solution (x^*, y^*) of (3.1) would be the centre of gravity of such area. Within this area, the facility would be later selected after taking into account additional nonmonetary factors, by using, for example, the weighted scoring method illustrated in Section 3.2.

Since function (3.1) is convex, its minimizer (x^*, y^*) corresponds to the stationary point, whose coordinates can be determined by imposing

$$\left.\frac{\partial f(x, y)}{\partial x}\right|_{x=x^*} = 0;$$
$$\left.\frac{\partial f(x, y)}{\partial y}\right|_{y=y^*} = 0.$$

This means that, for determining x^*, the following relation must hold:

$$-\frac{1}{2}c\sum_{i\in V}\frac{d_i[2(x_i-x^*)]}{\sqrt{(x_i-x^*)^2+(y_i-y^*)^2}}=0.$$

Simplifying, we obtain

$$\sum_{i\in V}\frac{d_i x_i}{(x_i-x^*)^2+(y_i-y^*)^2}-\sum_{i\in V}\frac{d_i x^*}{(x_i-x^*)^2+(y_i-y^*)^2}=0,$$

from which

$$x^*=\frac{\sum_{i\in V}\left[\frac{d_i x_i}{\sqrt{(x_i-x^*)^2+(y_i-y^*)^2}}\right]}{\sum_{i\in V}\left[\frac{d_i}{\sqrt{(x_i-x^*)^2+(y_i-y^*)^2}}\right]}. \quad (3.2)$$

Similarly,

$$y^*=\frac{\sum_{i\in V}\left[\frac{d_i y_i}{\sqrt{(x_i-x^*)^2+(y_i-y^*)^2}}\right]}{\sum_{i\in V}\left[\frac{d_i}{\sqrt{(x_i-x^*)^2+(y_i-y^*)^2}}\right]}. \quad (3.3)$$

Equations (3.2) and (3.3) are nonlinear, and their solution is not known in closed form. As a results, the following heuristic, proposed by Weiszfeld, is often used:

Step 0. Let ε be a user-defined tolerance parameter. Set $h=0$. Let $x^{(h)}$ and $y^{(h)}$ be the coordinates of the centre of gravity, computed in the following way:

$$x^{(h)}=\frac{\sum_{i\in V}d_i x_i}{\sum_{i\in V}d_i};$$

$$y^{(h)}=\frac{\sum_{i\in V}d_i y_i}{\sum_{i\in V}d_i}.$$

Let

$$f(x^{(h)},y^{(h)})=\sum_{i\in V}cd_i\left(\sqrt{(x_i-x^{(h)})^2+(y_i-y^{(h)})^2}\right).$$

Step 1. Set $h = h + 1$.
Determine

$$x^{(h)} = \frac{\sum_{i \in V} \left[\frac{d_i x_i}{\sqrt{(x_i - x^{(h-1)})^2 + (y_i - y^{(h-1)})^2}} \right]}{\sum_{i \in V} \left[\frac{d_i}{\sqrt{(x_i - x^{(h-1)})^2 + (y_i - y^{(h-1)})^2}} \right]};$$

$$y^{(h)} = \frac{\sum_{i \in V} \left[\frac{d_i y_i}{\sqrt{(x_i - x^{(h-1)})^2 + (y_i - y^{(h-1)})^2}} \right]}{\sum_{i \in V} \left[\frac{d_i}{\sqrt{(x_i - x^{(h-1)})^2 + (y_i - y^{(h-1)})^2}} \right]},$$

and

$$f(x^{(h)}, y^{(h)}) = \sum_{i \in V} c d_i \left(\sqrt{(x_i - x^{(h)})^2 + (y_i - y^{(h)})^2} \right).$$

Step 2. If $f(x^{(h-1)}, y^{(h-1)}) - f(x^{(h)}, y^{(h)}) \leq \varepsilon$, *STOP*, $x^{(h)}$ and $y^{(h)}$ represent a good approximation of the optimal coordinates x^* and y^*; otherwise, go back to Step 1.

Setting $\varepsilon = 0$, the described algorithm here becomes exact. Indeed, it can be proved that the sequence $\left\{ f(x^{(h)}, y^{(h)}) \right\}_{h=0}^{\infty}$ is monotonically decreasing and that the method converges to the optimal solution (x^*, y^*).

It is worth noting that this algorithm can also be adapted to solve a single-commodity, single-echelon continuous location problem in which the nodes in V represent the predecessor nodes of the facility to be located (e.g. the manufacturing plants if the facility to locate is a DC). In this case, it is sufficient to replace the demand d_i of each successor node $i \in V$ with the supply o_i of each predecessor node $i \in V$.

The Karakum Desert lies in Turkmenistan, in central Asia. Nine water wells have been drilled in this desert. Their average flow rates o_i, $i = 1, \ldots, 9$, expressed in litres per minute, are summarized in Table 3.3. The same table also shows the coordinates (in kilometres) of these wells with respect to a two-dimensional Cartesian system.

Transport costs are estimated to be € 0.00002 per kilometre of pipeline and for every litre of water flown.

An aqueduct was built to connect the nine wells. The position of the aqueduct (Figure 3.3) was determined by solving the continuous location problem (3.1). In this, d_i should be replaced by o_i for each $i = 1, \ldots, 9$. The optimal solution is the following:

$$(x^*, y^*) = (15.644, 10.336);$$

$$f(x^*, y^*) = 72.016 \text{ €/min}.$$

Table 3.3 Coordinates and average flow rate of the water wells in the Karakum Desert.

Water well	Abscissa (km)	Ordinate (km)	Flow rate (l/min)
1	0.000	0.000	28 700
2	13.543	11.273	27 500
3	8.578	24.432	28 600
4	42.438	14.583	42 500
5	20.652	5.232	38 700
6	3.782	21.567	25 500
7	14.720	9.565	18 300
8	18.768	3.678	20 500
9	35.650	25.678	21 000

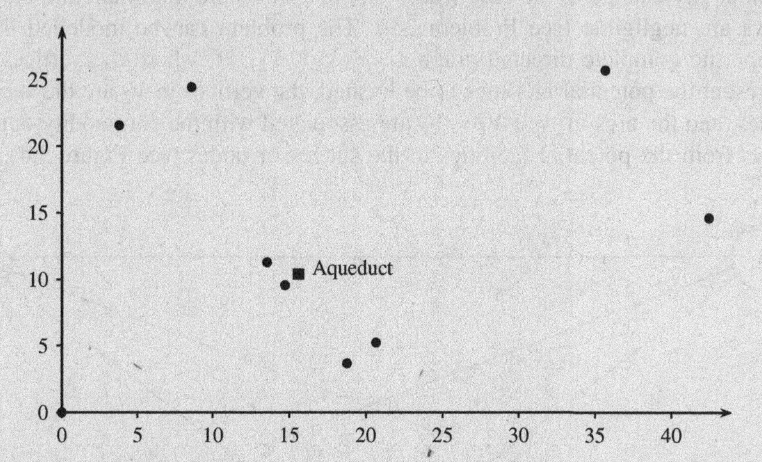

Figure 3.3 Location of the water wells and the aqueduct in the Karakum Desert. The coordinates are expressed in km.

Table 3.4 Sequence of aqueduct coordinates generated by the Weiszfeld heuristic in the Karakum Desert problem.

Iteration	Abscissa (km)	Ordinate (km)	Object function (€/min)	Difference (€/min)
0	18.782	12.617	74.358	–
1	17.189	10.875	72.424	1.934
2	16.522	10.387	72.125	0.299
3	16.185	10.297	72.058	0.066

Using the Weiszfeld heuristic with $\varepsilon = €\ 0.1$/min, we obtain the sequence of Cartesian coordinates of the facility reported in Table 3.4.

The procedure stops at the point of coordinates (16.185, 10.297) which is a good approximation of the optimal solution.

3.3.2 Single-commodity single-echelon discrete location problems

In *single-commodity single-echelon* (SCSE) discrete location problems, we assume that the facilities to be located are homogeneous (e.g. they are all regional warehouses). For the sake of simplicity, we examine the case where inbound flows are negligible, although the same methodology can be applied without any change to the case where inbound flows are important and outbound flows are negligible (see Problem 3.9). The problem can be modelled through a bipartite complete directed graph $G = (V_1 \cup V_2, A)$, where the vertices in V_1 represent the potential facilities to be located, the vertices in V_2 are the successor nodes, and the arcs in $A = V_1 \times V_2$ are associated with the commodity outbound flows from the potential facilities to the successor nodes (see Figure 3.4).

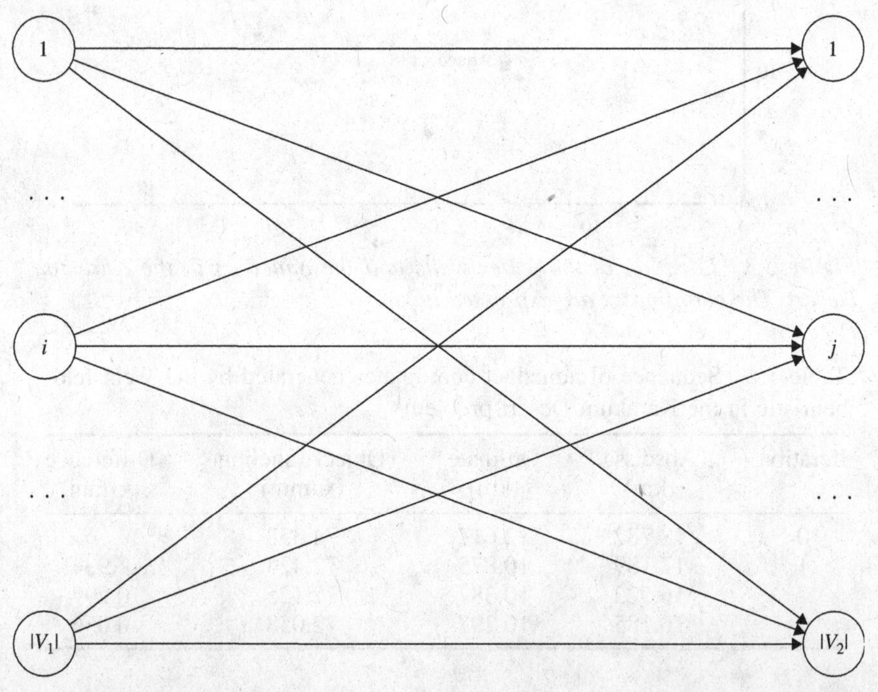

Figure 3.4 Representation of the SCSE problem on a bipartite complete directed graph.

Let $t = 1, \ldots, T$, be the time periods of the planning horizon. Let d_{jt}, $j \in V_2$, $t = 1, \ldots, T$, be the demand estimate of successor node j at time period t and q_{it}, $i \in V_1$, $t = 1, \ldots, T$, the maximum throughput of potential facility i at time period t. In the section, we consider the case where d_{jt}, $j \in V_2$, $t = 1, \ldots, T$, and q_{it}, $i \in V_1$, $t = 1, \ldots, T$, are constant during the planning horizon, that is,

$$d_{jt} = d_j, j \in V_2, t = 1, \ldots, T; \tag{3.4}$$

$$q_{it} = q_i, i \in V_1, t = 1, \ldots, T. \tag{3.5}$$

The more general formulation is left to the reader as an exercise (see Problem 3.12). Under assumptions (3.4)–(3.5), we can choose the following decision variables: u_i, $i \in V_1$, representing the level of activity at facility i in a time period; s_{ij}, $i \in V_1$, $j \in V_2$, representing the quantity of commodity delivered in a time period from facility i to the successor node j. Note that the values of the decision variables remain unchanged when the time periods vary because of assumptions (3.4)–(3.5). Furthermore, let $C_{ij}(s_{ij})$, $i \in V_1$, $j \in V_2$, be the cost of transporting s_{ij} units of commodity from facility i to successor node j in a time period, and let $F_i(u_i)$, $i \in V_1$, be the cost of operating potential facility i at level u_i in a time period.

Assuming that the commodity flow for every successor node is divisible (see Section 3.1), the SCSE problem can be modelled in the following way:

$$\text{Minimize} \sum_{i \in V_1} \sum_{j \in V_2} C_{ij}(s_{ij}) + \sum_{i \in V_1} F_i(u_i) \tag{3.6}$$

subject to

$$\sum_{j \in V_2} s_{ij} = u_i, i \in V_1 \tag{3.7}$$

$$\sum_{i \in V_1} s_{ij} = d_j, j \in V_2 \tag{3.8}$$

$$\sum_{j \in V_2} s_{ij} \leq q_i, i \in V_1 \tag{3.9}$$

$$s_{ij} \geq 0, i \in V_1, j \in V_2 \tag{3.10}$$

$$u_i \geq 0, i \in V_1. \tag{3.11}$$

The values of decision variables u_i, $i \in V_1$, implicitly define a location decision since a facility $i \in V_1$ is opened if only if u_i is strictly positive. The values of decision variables s_{ij}, $i \in V_1$, $j \in V_2$, determine the commodity flows allocation to the successor nodes. The objective function (3.6) is the sum of the facility operating costs, plus the transport cost between facilities and successor nodes in a time period. Constraints (3.7) state that the sum of the flows outgoing a facility equals its activity level in a time period. Constraints (3.8) ensure the demand satisfaction of every successor node, while constraints (3.9) force the activity level of a facility not to exceed the corresponding maximum throughput in a time period.

Model (3.6)–(3.11) is quite general and can be easily adapted to the case where, in order to have an acceptable service level, some arcs $(i, j) \in A$ having a travel time larger than a given threshold cannot be used (here we remove the corresponding decision variables s_{ij} for those arcs).

In this subsection, a particular case of the SCSE location problem is examined in detail. We assume that the transport cost per unit of commodity is constant in a time period, and that facility operating costs are piecewise linear and concave, that is,

$$C_{ij}(s_{ij}) = a_{ij}s_{ij}, i \in V_1, j \in V_2,$$

$$F_i(u_i) = \begin{cases} f_i + g_i u_i, & if\ u_i > 0 \\ 0, & if\ u_i = 0 \end{cases}, i \in V_1. \tag{3.12}$$

In Equation (3.12), f_i, $i \in V_1$, is the average fixed cost of facility i in a time period (obtained by considering the average of the facility set-up costs over the whole planning horizon) and g_i, $i \in V_1$, is the marginal unit cost associated with the activity of facility i in a time period.

Three years ago, Baja bought from a competing company a warehouse in Debrecen, used for the distribution of food products in Hungary. The purchasing cost was 850 000 current dollars. The annual fixed costs due to the warehouse operation were 173 000, 168 000 and 125 000 current dollars in the three subsequent years, respectively. Then, Baja decided to close the warehouse and sells it to another company. The selling price was 680 000 current dollars. The warehouse operated for 48 weeks for each year. Hence, the average weekly fixed cost of the warehouse was

$$(850\,000 + 173\,000 + 168\,000 + 125\,000 - 680\,000)/(48 \times 3) =$$

$$= 4416.67 \text{ current dollars.}$$

The facility cost function defined by (3.12) can be modelled by introducing a binary decision variable y_i, for each $i \in V_1$, set equal to 1 if a potential facility i is opened, 0 otherwise. In addition, constraints (3.7) enable us to avoid the explicit use of the continuous decision variable u_i for each $i \in V_1$. In particular, the objective function (3.6) can be rewritten as

$$\sum_{i \in V_1} \sum_{j \in V_2} a_{ij}s_{ij} + \sum_{i \in V_1} \left(f_i y_i + g_i \sum_{j \in V_2} s_{ij} \right) = \sum_{i \in V_1} \sum_{j \in V_2} (a_{ij} + g_i) s_{ij} + \sum_{i \in V_1} f_i y_i,$$

where $a_{ij} + g_i = b_{ij}$, for each arc $(i, j) \in A$, represents the sum of the transport cost per unit of flow from facility i to customer j and of the marginal unit cost associated with the activity of facility i.

The SCSE model therefore becomes

$$\text{Minimize} \sum_{i \in V_1} \sum_{j \in V_2} b_{ij} s_{ij} + \sum_{i \in V_1} f_i y_i \tag{3.13}$$

subject to

$$\sum_{i \in V_1} s_{ij} = d_j, j \in V_2 \tag{3.14}$$

$$\sum_{j \in V_2} s_{ij} \le q_i y_i, i \in V_1 \tag{3.15}$$

$$s_{ij} \ge 0, i \in V_1, j \in V_2 \tag{3.16}$$

$$y_i \in \{0, 1\}, i \in V_1. \tag{3.17}$$

Constraints (3.15) are facility maximum throughput constraints and are used to express the relationship between the values of the decision variables $s_{ij}, i \in V_1, j \in V_2$, and the binary decision variables $y_i, i \in V_1$.

An equivalent model can be formulated alternatively by replacing the decision variables s_{ij} with $x_{ij}, i \in V_1, j \in V_2$, according to the following relations:

$$s_{ij} = d_j x_{ij}, i \in V_1, j \in V_2.$$

Hence, $x_{ij}, i \in V_1, j \in V_2$, represents the fraction of the demand of successor node j satisfied by facility i in a time period. The equivalent formulation to model (3.13)–(3.17) is reported below.

$$\text{Minimize} \sum_{i \in V_1} \sum_{j \in V_2} c_{ij} x_{ij} + \sum_{i \in V_1} f_i y_i \tag{3.18}$$

subject to

$$\sum_{i \in V_1} x_{ij} = 1, j \in V_2 \tag{3.19}$$

$$\sum_{j \in V_2} d_j x_{ij} \le q_i y_i, i \in V_1 \tag{3.20}$$

$$x_{ij} \ge 0, i \in V_1, j \in V_2 \tag{3.21}$$

$$y_i \in \{0, 1\}, i \in V_1, \tag{3.22}$$

where

$$c_{ij} = b_{ij} d_j = a_{ij} d_j + g_i d_j, i \in V_1, j \in V_2.$$

The SCSE problem (3.18)–(3.22) corresponds to the well-known *capacitated plant location* (CPL) problem.

Milatog is a Russian company producing cattle forage. In the province of Domedovsky, there are seven farms which have an average daily forage demand (in quintals) equal to 36, 42, 34, 50, 27, 30 and 43, respectively.

Milatog intends to purchase some silos, to supply the seven farms. Six different potential sites in the area have been identified, with a maximum daily forage throughput (expressed in quintals) equal to, respectively, 80, 90, 110, 120, 100 and 120. For the next four years, Milatog has estimated the following fixed costs (in €): 321 420, 350 640, 379 860, 401 775, 350 640 and 336 030, respectively. The daily average marginal facility cost (in €) per quintal of forage, for each potential site, is equal to 0.15, 0.18, 0.20, 0.18, 0.15 and 0.17, respectively.

The transport cost per quintal of forage and per kilometre travelled is equal to € 0.06. The kilometric distances for each origin-destination pair are shown in Table 3.5. The daily transport costs are computed by considering that every journey is made up of both an outward and a return journey.

Table 3.5 Kilometric distances between each potential site and each farm for the Milatog problem.

	Farm						
Potential silo	1	2	3	4	5	6	7
1	18	23	19	21	24	17	9
2	21	18	17	23	11	18	20
3	27	18	17	20	23	9	18
4	16	23	9	31	21	23	10
5	31	20	18	19	10	17	18
6	18	17	29	21	22	18	8

Milatog is planning to keep the warehouses in operation for four years (corresponding to $365 \times 3 + 366 = 1\,461$ days). On a daily basis, the CPL model can be formulated as follows:

$$V_1 = \{1, 2, 3, 4, 5, 6\} \text{ is the set of potential sites;}$$

$$V_2 = \{1, 2, 3, 4, 5, 6, 7\} \text{ is the set of farms;}$$

$$f_1 = 321{,}420/1\,461 = € \ 220.$$

Similarly, the other costs f_i, $i = 2, \ldots, 6$, can be computed. Also

$$c_{11} = 0.06 \times 2 \times 18 \times 36 + 0.15 \times 36 = € \ 83.16.$$

A similar procedure is used to calculate the other costs c_{ij}, $i = 1, \ldots, 6$, $j = 1, \ldots, 7$. Furthermore, let $y_i, i = 1, \ldots, 6$, be the binary decision variable associated to potential site i (with value 1 if silo i is purchased by Milatog, 0 otherwise), and $x_{ij}, i = 1, \ldots, 6, j = 1, \ldots, 7$, be the decision variable that expresses the fraction of the average daily demand of farm j and satisfied by silo i.

$$\begin{aligned}\text{Minimize } & 83.16x_{11} + 122.22x_{12} + 82.62x_{13} + 133.50x_{14} + 81.81x_{15} + \\ & + 65.70x_{16} + 52.89x_{17} + 97.20x_{21} + 98.28x_{22} + 75.48x_{23} + \\ & + 147.00x_{24} + 40.50x_{25} + 70.20x_{26} + 110.94x_{27} + 123.84x_{31} + \\ & + 99.12x_{32} + 76.16x_{33} + 130.00x_{34} + 79.92x_{35} + 38.40x_{36} + \\ & + 101.48x_{37} + 75.60x_{41} + 123.48x_{42} + 42.84x_{43} + 195.00x_{44} + \\ & + 72.90x_{45} + 88.20x_{46} + 59.34x_{47} + 139.32x_{51} + 107.10x_{52} + \\ & + 78.54x_{53} + 121.50x_{54} + 36.45x_{55} + 65.70x_{56} + 99.33x_{57} + \\ & + 83.88x_{61} + 92.82x_{62} + 124.10x_{63} + 134.50x_{64} + 75.87x_{65} + \\ & + 69.90x_{66} + 48.59x_{67} + 220y_1 + 240y_2 + 260y_3 + 275y_4 \\ & + 240y_5 + 230y_6\end{aligned}$$

subject to

$$\begin{aligned}& x_{11} + x_{21} + x_{31} + x_{41} + x_{51} + x_{61} = 1 \\ & x_{12} + x_{22} + x_{32} + x_{42} + x_{52} + x_{62} = 1 \\ & x_{13} + x_{23} + x_{33} + x_{43} + x_{53} + x_{63} = 1 \\ & x_{14} + x_{24} + x_{34} + x_{44} + x_{54} + x_{64} = 1 \\ & x_{15} + x_{25} + x_{35} + x_{45} + x_{55} + x_{65} = 1 \\ & x_{16} + x_{26} + x_{36} + x_{46} + x_{56} + x_{66} = 1 \\ & x_{17} + x_{27} + x_{37} + x_{47} + x_{57} + x_{67} = 1 \\ & 36x_{11} + 42x_{12} + 34x_{13} + 50x_{14} + 27x_{15} + 30x_{16} + 43x_{17} \leq 80y_1 \\ & 36x_{21} + 42x_{22} + 34x_{23} + 50x_{24} + 27x_{25} + 30x_{26} + 43x_{27} \leq 90y_2 \\ & 36x_{31} + 42x_{32} + 34x_{33} + 50x_{34} + 27x_{35} + 30x_{36} + 43x_{37} \leq 110y_3 \\ & 36x_{41} + 42x_{42} + 34x_{43} + 50x_{44} + 27x_{45} + 30x_{46} + 43x_{47} \leq 120y_4 \\ & 36x_{51} + 42x_{52} + 34x_{53} + 50x_{54} + 27x_{55} + 30x_{56} + 43x_{57} \leq 100y_5 \\ & 36x_{61} + 42x_{62} + 34x_{63} + 50x_{64} + 27x_{65} + 30x_{66} + 43x_{67} \leq 120y_6\end{aligned}$$

$$x_{ij} \geq 0, i = 1, \ldots, 6, j = 1, \ldots, 7$$

$$y_i \in \{0, 1\}, i = 1, \ldots, 6.$$

The optimal solution of the problem provides for the purchasing of silos 1, 5 and 6 and an overall daily cost equal to € 1 218.08. The daily demand of the seven farms is satisfied as can be deduced from Table 3.6.

Table 3.6 Fraction of the daily forage demand of every farm satisfied by the purchased silos for the Milatog problem.

	Farm						
Purchased silo	1	2	3	4	5	6	7
1	1	0	11/34	0	0	1	0
5	0	0	23/34	1	1	0	0
6	0	1	0	0	0	0	1

The CPL model can be easily adapted to take into account further conditions of practical interest. For example, a potential facility cannot be run economically if its average level of activity is lower than a value q_i^- or higher than a threshold q_i^+. For intermediate values, the operating cost grows linearly. To take this condition into account, the CPL model can be modified by substituting the corresponding constraints (3.20) with the following pair of relations:

$$\sum_{j \in V_2} d_j x_{ij} \leq q_i^+ y_i,$$

$$\sum_{j \in V_2} d_j x_{ij} \geq q_i^- y_i,$$

for all facilities $i \in V_1$ where these conditions apply.

With reference to the Milatog location problem, we assume now that the sixth silo has a minimum daily throughput equal to 90 quintals, whereas the maximum throughput remains unchanged at 120. Consequently, the CPL problem illustrated in the previous box requires the following additional constraint

$$36x_{61} + 42x_{62} + 34x_{63} + 50x_{64} + 27x_{65} + 30x_{66} + 43x_{67} \geq 90y_6.$$

Imposing this constraint implies that the optimal solution of the Milatog problem changes: the purchased silos will always be the first, the fifth and the sixth, but the daily cost results equal to € 1 218.18 and the average daily demands of the seven farms are satisfied in a different way, as reported in Table 3.7.

Table 3.7 Fraction of the daily forage demand of each farm satisfied by the purchased silos for the modified Milatog problem.

	Farm						
Purchased silo	1	2	3	4	5	6	7
1	31/36	0	11/34	0	0	1	0
5	0	0	23/34	1	1	0	0
6	5/36	1	0	0	0	0	1

Another interesting case occurs when the operating cost $F_i(u_i)$ of a potential facility $i \in V_1$ can be represented through a general concave piecewise linear function of its activity level because of economies of scale. In the simplest case, there are only two piecewise lines (Figure 3.5). Then

$$F_i(u_i) = \begin{cases} 0, & \text{if } u_i = 0, \\ f_i' + g_i' u_i, & \text{if } 0 < u_i \leq \bar{u}_i \\ f_i'' + g_i'' u_i, & \text{if } \bar{u}_i < u_i \leq \hat{u}_i \end{cases}, i \in V_1 \tag{3.23}$$

where it results that $f_i' \leq f_i''$, $g_i' \geq g_i''$ and $f_i' + g_i'\bar{u}_i = f_i'' + g_i''\bar{u}_i$.

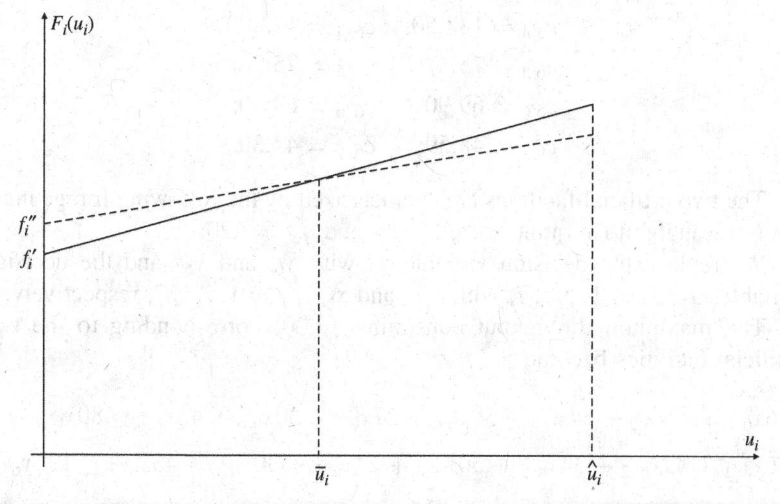

Figure 3.5 Concave two-piecewise linear cost $F_i(u_i)$ of potential facility $i \in V_1$ versus activity level u_i.

In order to model this problem, each potential facility is replaced by as many artificial facilities as the number of piecewise lines of its cost function. For instance, if Equation (3.23) holds, facility $i \in V_1$ is replaced by two artificial facilities i' and i'' whose operating costs are characterized, respectively, by fixed costs equal to $f_{i'} = f_i'$ and $f_{i''} = f_i''$ and by marginal unit costs equal to $g_{i'} = g_i'$ and $g_{i''} = g_i''$. Moreover, it is necessary to add the constraint

$$y_{i'} + y_{i''} \leq 1,$$

to ensure the nonsimultaneous activation of the two artificial facilities i' and i'' (it is easy to verify that this constraint is redundant in case of $\hat{u}_i$ being sufficiently large).

With reference to the Milatog location problem, we assume that the daily average facility unit marginal cost of the sixth silo decreases by € 0.03 per quintal of forage and that the facility fixed costs are increased up to € 339 536 when the level of activity is greater than 80 quintals. The above CPL formulation is modified by replacing facility 6 with two artificial facilities 6′ and 6″, characterized by fixed and marginal unit daily costs equal to, respectively, $f_{i'} =$ € 230, $g_{i'} =$ € 0.17, and $f_{i''} =$ € 232.4, $g_{i''} =$ € 0.14.

Both the artificial facilities are at the same distances to the seven farms as the original facility sixth potential facility, so that the transport and facility unit marginal cost (in €) $c_{6'j}$ and $c_{6''j}$, $j = 1, \ldots, 7$, are

$$c_{6'1} = 83.88; \quad c_{6''1} = 82.80;$$
$$c_{6'2} = 92.82; \quad c_{6''2} = 91.56;$$
$$c_{6'3} = 124.10; \quad c_{6''3} = 123.08;$$
$$c_{6'4} = 134.50; \quad c_{6''4} = 133.00;$$
$$c_{6'5} = 75.87; \quad c_{6''5} = 75.06;$$
$$c_{6'6} = 69.90; \quad c_{6''6} = 69.00;$$
$$c_{6'7} = 48.59; \quad c_{6''7} = 47.30.$$

The two artificial facilities are characterized by the following forage maximum throughput (in quintals): $q_{6'} = 80$ and $q_{6''} = 120$.

We replace the decision variable y_6 with $y_{6'}$ and $y_{6''}$ and the decision variables x_{6j}, $j = 1, \ldots, 7$, with $x_{6'j}$ and $x_{6''j}$, $j = 1, \ldots, 7$, respectively.

The maximum throughput constraints (3.20) corresponding to the two artificial facilities become

$$36x_{6'1} + 42x_{6'2} + 34x_{6'3} + 50x_{6'4} + 27x_{6'5} + 30x_{6'6} + 43x_{6'7} \leq 80y_{6'},$$
$$36x_{6''1} + 42x_{6''2} + 34x_{6''3} + 50x_{6''4} + 27x_{6''5} + 30x_{6''6} + 43x_{6''7} \leq 120y_{6''}.$$

It is also necessary to impose $y_{6'} + y_{6''} \leq 1$ to avoid the simultaneous activation of the two artificial facilities.

The optimal solution yields $y_1^* = y_5^* = y_{6''}^* = 1$ and $y_2^* = y_3^* = y_4^* = y_{6'}^* = 0$. Consequently, silos 1, 5 and 6 will be purchased by Milatog, in particular silo 6 with a maximum daily throughput equal to 120 quintals of forage. The optimal daily cost is equal to € 1 217.58. Table 3.8 shows how the daily forage demand of each farm is satisfied.

Table 3.8 Fraction of the daily forage demand of every farm satisfied by the purchased silos for the Milatog problem in case of two artificial facilities.

	Farm						
Purchased silo	1	2	3	4	5	6	7
1	1/36	0	11/34	0	0	1	0
5	0	0	23/34	1	1	0	0
6	35/36	1	0	0	0	0	1

If facilities to be activated have no throughput constraints, then constraints (3.20) should be replaced by constraints which can be used only to express the existing relationship between decision variables $x_{ij}, i \in V_1, j \in V_2$, and $y_i, i \in V_1$, that is,

$$\sum_{j \in V_2} x_{ij} \leq |V_2| \, y_i, i \in V_1.$$

The resulting model corresponds to the *simple plant location* (SPL) model.

With reference to the Milatog location problem, we assume that the silos have no throughput constraints (i.e. the maximum average daily throughput for each silo is sufficient to satisfy the whole daily average forage demand for all the farms). The original throughput constraints are replaced with

$$x_{11} + x_{12} + x_{13} + x_{14} + x_{15} + x_{16} + x_{17} \leq 7y_1;$$

$$x_{21} + x_{22} + x_{23} + x_{24} + x_{25} + x_{26} + x_{27} \leq 7y_2;$$

$$x_{31} + x_{32} + x_{33} + x_{34} + x_{35} + x_{36} + x_{37} \leq 7y_3;$$

$$x_{41} + x_{42} + x_{43} + x_{44} + x_{45} + x_{46} + x_{47} \leq 7y_4;$$

$$x_{51} + x_{52} + x_{53} + x_{54} + x_{55} + x_{56} + x_{57} \leq 7y_5;$$

$$x_{61} + x_{62} + x_{63} + x_{64} + x_{65} + x_{66} + x_{67} \leq 7y_6.$$

The optimal solution of the problem entails the activation of just one silo (the first), which will supply all the farms, with an overall daily cost equal to € 841.90.

Other SCSE problems can be formulated by modifying the CPL model. In particular, if p facilities should be activated, then the following constraint has to be added to the CPL model:

$$\sum_{i \in V_1} y_i = p. \tag{3.24}$$

If it is requested that a specific subset of facilities $V_1' \subseteq V_1$ be necessarily activated, then the following constraints can be imposed:

$$y_i = 1, i \in V_1'.$$

Considering the SCSE formulation (3.18)–(3.22), (3.24) and assuming that

- the fixed costs are the same for each facility (i.e. $f_i = f$, for each $i \in V_1$; this means that in the objective function (3.18) operating costs of facilities are constant and equal to fp and, hence, it can be omitted);
- $d_j = 1, j \in V_2$;
- $q_i = |V_2|, i \in V_1$,

we obtain the so-called *p-median* model.

United Bank, a Bulgarian credit institute, has used the p-median model to determine the optimal position of two bank offices at Gabrovo. The city area was divided into eight different districts (which form the set V_2), whereas there are identified six different potential sites (forming the set V_1) for the location of the bank offices. The six potential sites have identical facility fixed costs. The connection costs between the potential sites and the centroids of the eight districts are proportional to the corresponding kilometric distances, which are reported in Table 3.9. Let $y_i, i \in V_1$, be the binary decision variable assuming value 1 if a bank office is located at site i, 0 otherwise; $x_{ij}, i \in V_1, j \in V_2$, is the binary decision variable assuming value 1 if the bank office located at site i serves district j, 0 otherwise. The p-median problem (with $p = 2$) is the following:

$$\text{Minimize} \sum_{i \in V_1} \sum_{j \in V_2} c_{ij} x_{ij}$$

subject to

$$\sum_{i \in V_1} x_{ij} = 1, \ j \in V_2$$

$$\sum_{j \in V_2} x_{ij} \leq |V_2| \, y_i, \ i \in V_1$$

$$\sum_{i \in V_1} y_i = 2$$

$$x_{ij} \in \{0, 1\}, \ i \in V_1, j \in V_2 \qquad (3.25)$$

$$y_j \in \{0, 1\}, \ j \in V_2,$$

where $c_{ij}, i \in V_1, j \in V_2$, is the kilometric distance between site i and district j, reported in Table 3.9.

Table 3.9 Kilometric distances between the potential sites and the centroids of the eight districts of Gabrovo for the United Bank problem.

	District							
Potential site	1	2	3	4	5	6	7	8
1	2.1	1.7	2.8	0.3	0.8	2.2	1.8	0.7
2	1.5	2.2	3.1	2.2	0.2	1.9	2.3	1.3
3	0.9	1.6	2.3	0.3	1.7	1.6	0.9	2.7
4	1.8	3.1	2.7	2.6	3.1	0.6	0.2	0.7
5	0.1	2.5	1.8	3.1	0.4	1.2	0.7	1.1
6	0.5	1.4	3.1	0.5	0.2	1.5	2.2	0.8

Note that, due to the particular structure of the problem constraints, relations (3.25) can be relaxed and expressed as

$$x_{ij} \geq 0, i \in V_1, j \in V_2.$$

The relaxed problem is equivalent to the original one, in the sense that the optimal solution of the relaxed problem will automatically satisfy constraints (3.25).

The optimal United Bank problem solution leads to the location of the two bank offices at sites 5 and 6. Districts 1, 3, 6 and 7 are assigned to the bank office located at site 5, while districts 2, 4, 5 and 8 are assigned to the bank office located at site 6.

The CPL model is a location-allocation problem. If the set $\bar{V}_1 \subseteq V_1$ of open facilities is known, it is clear that an optimal allocation of the average quantities demanded by the successor nodes can be determined by solving the following linear programming (LP) problem:

$$\text{Minimize} \sum_{i \in \bar{V}_1} \sum_{j \in V_2} c_{ij} x_{ij} + \sum_{i \in \bar{V}_1} f_i \tag{3.26}$$

subject to

$$\sum_{i \in \bar{V}_1} x_{ij} = 1, \; j \in V_2 \tag{3.27}$$

$$\sum_{j \in V_2} d_j x_{ij} \leq q_i, \; i \in \bar{V}_1 \tag{3.28}$$

$$0 \leq x_{ij} \leq 1, i \in \bar{V}_1, \; j \in V_2. \tag{3.29}$$

With reference to the Milatog location problem, it is assumed that the set $\bar{V}_1 = \{1, 5, 6\}$ of the activated silos (at a daily activation cost equal to € 690) is known; the optimal fraction of the daily forage demand satisfied by the activated silos (shown in Table 3.6) can be obtained by solving the following LP problem:

$$\begin{aligned}
\text{Minimize} \quad & 83.16x_{11} + 122.22x_{12} + 82.62x_{13} + 133.50x_{14} + 81.81x_{15} + \\
& + 65.70x_{16} + 52.89x_{17} + 139.32x_{51} + 107.10x_{52} + 78.54x_{53} + \\
& + 121.50x_{54} + 36.45x_{55} + 65.70x_{56} + 99.33x_{57} + 83.88x_{61} + \\
& + 92.82x_{62} + 124.10x_{63} + 134.50x_{64} + 75.87x_{65} + 69.90x_{66} + \\
& + 48.59x_{67} + 690
\end{aligned}$$

subject to

$$x_{11} + x_{51} + x_{61} = 1$$

$$x_{12} + x_{52} + x_{62} = 1$$

$$x_{13} + x_{53} + x_{63} = 1$$

$$x_{14} + x_{54} + x_{64} = 1$$

$$x_{15} + x_{55} + x_{65} = 1$$

$$x_{16} + x_{56} + x_{66} = 1$$

$$x_{17} + x_{57} + x_{67} = 1$$

$$36x_{11} + 42x_{12} + 34x_{13} + 50x_{14} + 27x_{15} + 30x_{16} + 43x_{17} \leq 80$$

$$36x_{51} + 42x_{52} + 34x_{53} + 50x_{54} + 27x_{55} + 30x_{56} + 43x_{57} \leq 100$$

$$36x_{61} + 42x_{62} + 34x_{63} + 50x_{64} + 27x_{65} + 30x_{66} + 43x_{67} \leq 120$$

$$x_{ij} \geq 0, i = 1, 5, 6, j = 1, \ldots, 7.$$

It can happen that the optimal solution of the allocation problem (3.26)–(3.29) is such that the demand of a successor node $j \in V_2$ can be satisfied by more than one open facility $i \in V_1$ (i.e. some x^*_{ij} values may be fractional), because of maximum throughput constraints (3.28). However, in the SCSE models without constraints (3.20) (as in the SPL and p-median models), there exists at least one optimal solution such that the demand of each successor node $j \in V_2$ is satisfied by a single facility $i \in V_1$ (*single assignment* property). This solution can be obtained as follows. Let $i_j \in \bar{V}_1$ be a facility such that

$$i_j = \arg\min_{i \in \bar{V}_1} \{c_{ij}\}.$$

Then, the values of the allocation decision variables can be found as follows:

$$x^*_{ij} = \begin{cases} 1, & \text{if } i = i_j \\ 0, & \text{otherwise.} \end{cases}$$

To determine the optimal solution (already reported in a previous box) of the Milatog SPL problem, it is sufficient to execute the following steps.

1. Solve the problem assuming the activation of a single silo. By exploiting the single assignment property, we look for the silo $i^* \in V_1$ yielding

$$\min_{i \in V_1} \left\{ f_i + \sum_{j \in V_2} c_{ij} \right\}.$$

The minimum value is reached for the first silo, so that $f_1 + \sum_{j \in V_2} c_{1j} =$ € 841.90. Let $z^{(1)}$ be this cost.

2. Verify that a lower bound $\mathrm{LB}^{(p)}$ on the optimal cost of the Milatog SPL problem is greater than or equal to $z^{(1)}$, supposing that p silos are activated, with $p \geq 2$. The value of $\mathrm{LB}^{(p)}$ can be easily determined in the following way. Since the single assignment property is valid, a lower bound on the daily cost of the demand allocation of each farm $j = 1, \ldots, 7$, can be determined. In particular, for farm

1 at least € 75.60 per day has to be paid; for the other farms at least € 92.82, € 42.84, € 121.50, € 36.45, € 38.40 and € 48.59 per day has to be paid, respectively. A lower bound on the overall daily demand allocation costs is therefore equal to € 456.20. In order to obtain $\mathrm{LB}^{(p)}$, it is necessary to add to these costs the fixed facility costs for the p activated sites, whose lower bound is easily determined by sorting fixed costs $f_i, i = 1, \ldots, 6$, in nondecreasing order and taking the first p sorted values. In this way, $\mathrm{LB}^{(2)} = 456.20 + (220 + 230) =$ € 906.20, and similarly, $\mathrm{LB}^{(3)} = 456.20 + (220 + 230 + 240) =$ € 1 146.20. Consequently, $\mathrm{LB}^{(p)} \geq \mathrm{LB}^{(2)}$, for $p = 3, \ldots, 6$. Since $z^{(1)} < \mathrm{LB}^{(2)}$, the optimal cost of the Milatog SPL problem corresponds to $z^{(1)}$.

3.3.2.1 A Lagrangian heuristic for the capacitated plant location problem

SCSE problems are nondeterministic polynomial-time hard (NP-hard) mixed-integer programming (MIP) problems. An optimal solution can in principle be determined by means of a general-purpose or tailored branch-and-bound or branch-and-cut algorithm. As a rule, capacitated problems are harder than uncapacitated ones. Nowadays, thanks to the development of such algorithms, it is possible to solve fairly large instances by using a COTS solver. Here, we illustrate a heuristic capable of determining a good feasible solution within a reasonable amount of time. Such a heuristic is valuable for more difficult location problems (i.e. multicommodity and multiechelon problems) whose large instances are still out of reach for branch-and-bound or branch-and-cut algorithms, or require unacceptable computing times.

To evaluate whether a heuristic solution is a tight upper bound (UB) on the optimal solution value, it is useful to determine a lower bound (LB) on the optimal solution value. This yields a ratio $(\mathrm{UB} - \mathrm{LB})/\mathrm{LB}$ which represents an overestimate of the relative deviation of the heuristic solution value from the optimum. Lagrangian relaxation techniques usually provide high-quality upper and lower bounds within a few iterations. In the sequel, a Lagrangian heuristic is illustrated for the CPL problem, although this algorithm may be used to solve other SCSE problems.

The fundamental step of the heuristic is the determination of a lower bound, obtained by relaxing demand satisfaction constraints (3.19) in a Lagrangian fashion. Let $\lambda_j \in \Re$ be the Lagrangian multiplier associated with the jth constraint (3.19). Then the relaxed problem is

$$\text{Minimize} \sum_{i \in V_1} \sum_{j \in V_2} c_{ij} x_{ij} + \sum_{i \in V_1} f_i y_i + \sum_{j \in V_2} \lambda_j \left(\sum_{i \in V_1} x_{ij} - 1 \right)$$

$$= \sum_{i \in V_1} \sum_{j \in V_2} (c_{ij} + \lambda_j) x_{ij} + \sum_{i \in V_1} f_i y_i - \sum_{j \in V_2} \lambda_j \qquad (3.30)$$

subject to

$$\sum_{j \in V_2} d_j x_{ij} \leq q_i y_i, i, \in V_1 \qquad (3.31)$$

$$0 \leq x_{ij} \leq 1, i \in V_1, j \in V_2 \qquad (3.32)$$

$$y_i \in \{0, 1\}, i \in V_1, \qquad (3.33)$$

whose optimal objective function value is denoted as $LB_{CPL}(\lambda)$. It should be observed that the elimination of constraints (3.19) imposes the introduction of constraints (3.32) which fix the upper bound of 1 to the value of each of the decision variables $x_{ij}, i \in V_1, j \in V_2$.

With reference to the Milatog CPL problem, the following vector $\lambda \in \Re^7$ of the Lagrangian multipliers is assumed:

$$\lambda = [-186; -170; -140; 60; -115; -166; -112]^{\mathrm{T}}.$$

The relaxed Lagrangian problem is the following:

$$\begin{aligned}
\text{Minimize } & -102.84x_{11} - 47.78x_{12} - 57.38x_{13} + 193.50x_{14} - 33.19x_{15} + \\
& -100.30x_{16} - 59.11x_{17} - 88.80x_{21} - 71.72x_{22} - 64.52x_{23} + \\
& +207.00x_{24} - 74.50x_{25} - 95.80x_{26} - 1.06x_{27} - 62.16x_{31} + \\
& -70.88x_{32} - 63.84x_{33} + 190.00x_{34} - 35.08x_{35} - 127.60x_{36} + \\
& -10.52x_{37} - 110.40x_{41} - 46.52x_{42} - 97.16x_{43} + 255.00x_{44} + \\
& -42.10x_{45} - 77.80x_{46} - 52.66x_{47} - 46.68x_{51} - 62.90x_{52} + \\
& -61.46x_{53} + 181.50x_{54} - 78.55x_{55} - 100.30x_{56} - 12.67x_{57} + \\
& -102.12x_{61} - 77.18x_{62} - 15.90x_{63} + 194.50x_{64} - 39.13x_{65} + \\
& -96.10x_{66} - 63.41x_{67} + 220y_1 + 240y_2 + 260y_3 + 275y_4 + \\
& +240y_5 + 230y_6 + 829
\end{aligned}$$

subject to

$$36x_{11} + 42x_{12} + 34x_{13} + 50x_{14} + 27x_{15} + 30x_{16} + 43x_{17} \leq 80y_1$$

$$36x_{21} + 42x_{22} + 34x_{23} + 50x_{24} + 27x_{25} + 30x_{26} + 43x_{27} \leq 90y_2$$

$$36x_{31} + 42x_{32} + 34x_{33} + 50x_{34} + 27x_{35} + 30x_{36} + 43x_{37} \leq 110y_3$$

$$36x_{41} + 42x_{42} + 34x_{43} + 50x_{44} + 27x_{45} + 30x_{46} + 43x_{47} \leq 120y_4$$
$$36x_{51} + 42x_{52} + 34x_{53} + 50x_{54} + 27x_{55} + 30x_{56} + 43x_{57} \leq 100y_5$$
$$36x_{61} + 42x_{62} + 34x_{63} + 50x_{64} + 27x_{65} + 30x_{66} + 43x_{67} \leq 120y_6$$
$$0 \leq x_{ij} \leq 1, i = 1, \ldots, 6, j = 1, \ldots, 7$$
$$y_i \in \{0, 1\}, i = 1, \ldots, 6.$$

It is easy to check that problem (3.30)–(3.33) can be decomposed into $|V_1|$ subproblems, one for each potential facility $i \in V_1$, as follows:

$$\text{Minimize} \sum_{j \in V_2} \left(c_{ij} + \lambda_j\right) x_{ij} + f_i y_i \tag{3.34}$$

subject to

$$\sum_{j \in V_2} d_j x_{ij} \leq q_i y_i \tag{3.35}$$

$$0 \leq x_{ij} \leq 1, \ j \in V_2 \tag{3.36}$$

$$y_i \in \{0, 1\}. \tag{3.37}$$

Let $\mathrm{LB}^{(i)}_{\mathrm{CPL}}(\lambda)$, $i \in V_1$, be the optimal objective function value of ith subproblem (3.34)–(3.37). We will have

$$\mathrm{LB}_{\mathrm{CPL}}(\lambda) = \sum_{i \in V_1} \mathrm{LB}^{(i)}_{\mathrm{CPL}}(\lambda) - \sum_{j \in V_2} \lambda_j.$$

The optimal solution of subproblem (3.34)–(3.37) can be determined easily by inspection, by observing that

- for $y_i = 0$, constraint (3.35) implies $x_{ij} = 0$, for each $j \in V_2$, and therefore $\mathrm{LB}^{(i)}_{\mathrm{CPL}}(\lambda)|_{y_i=0} = 0$;
- for $y_i = 1$, subproblem (3.34)–(3.37) is a continuous knapsack problem; it is well known that this problem can be solved in polynomial time by means of a greedy procedure, by sorting the decision variables x_{ij}, $i \in V_1$, $j \in V_2$, according to nondecreasing values of the ratios $(c_{ij} + \lambda_j)/d_j$, $j \in V_2$, and assigning, one by one, following the order in which the decision variables are sorted, the maximum possible value to each of them (compatibly with constraints satisfaction), whenever the corresponding Lagrangian cost $c_{ij} + \lambda_j$ is negative, and zero when the corresponding Lagrangian cost is non-negative (see the following example). It is worth noting that this case has to be taken into account only if there is at least one negative Lagrangian cost coefficient in the objective function (3.34).

Consequently, we will have

$$\mathrm{LB}_{\mathrm{CPL}}^{(i)}(\lambda) = \min\left\{0,\ \mathrm{LB}_{\mathrm{CPL}}^{(i)}(\lambda)|_{y_i=1}\right\},$$

that is,

$$\mathrm{LB}_{\mathrm{CPL}}^{(i)}(\lambda) \leq 0, \lambda \in \Re^{|V_2|}.$$

With reference to the relaxed Lagrangian problem reported in the previous box, we solve the subproblem $i = 1$, that is,

$$\begin{aligned} \text{Minimize } & -102.84x_{11} - 47.78x_{12} - 57.38x_{13} + 193.50x_{14} + \\ & -33.19x_{15} - 100.30x_{16} - 59.11x_{17} + 220y_1 \end{aligned}$$

subject to

$$36x_{11} + 42x_{12} + 34x_{13} + 50x_{14} + 27x_{15} + 30x_{16} + 43x_{17} \leq 80y_1$$

$$0 \leq x_{1j} \leq 1, j = 1, \dots, 7$$

$$y_1 \in \{0, 1\}.$$

By setting $y_1 = 1$, the following continuous knapsack problem is obtained:

$$\begin{aligned} \text{Minimize } & -102.84x_{11} - 47.78x_{12} - 57.38x_{13} + 193.50x_{14} - 33.19x_{15} + \\ & -100.30x_{16} - 59.11x_{17} + 220 \end{aligned}$$

subject to

$$36x_{11} + 42x_{12} + 34x_{13} + 50x_{14} + 27x_{15} + 30x_{16} + 43x_{17} \leq 80$$

$$0 \leq x_{1j} \leq 1, j = 1, \dots, 7,$$

which can be solved in the following manner. Decision variables $x_{1j} = 1, j = 1, \dots, 7$, are sorted by nondecreasing values of the ratios $\{-102.84/36, -47.78/42, -57.38/34, 193.50/50, -33.19/27, -100.30/30, -59.11/43\}$, that is,

$$\{x_{16}, x_{11}, x_{13}, x_{17}, x_{15}, x_{12}, x_{14}\}.$$

Hence, we set

$$\bar{x}_{16} = \min\{1, 80/30\} = 1.$$

The maximum throughput constraint becomes

$$36x_{11} + 42x_{12} + 34x_{13} + 50x_{14} + 27x_{15} + 43x_{17} \leq 50.$$

Then, we set

$$\bar{x}_{11} = \min\{1, 50/36\} = 1.$$

The maximum throughput constraint becomes

$$42x_{12} + 34x_{13} + 50x_{14} + 27x_{15} + 43x_{17} \leq 14.$$

Then, we set

$$\bar{x}_{13} = \min\{1, 14/34\} = 7/17.$$

Since the residual throughput is reduced to 0, we will have

$$\bar{x}_{12} = \bar{x}_{14} = \bar{x}_{15} = \bar{x}_{17} = 0;$$

$$\text{LB}_{\text{CPL}}^{(i)}(\lambda)|_{y_i=1} = -102.84 - 57.38 \times (7/17) - 100.30 + 220 = -6.77.$$

Since

$$\text{LB}_{\text{CPL}}^{(i)}(\lambda)|_{y_i=1} < 0,$$

it follows that

$$\text{LB}_{\text{CPL}}^{(i)}(\lambda) = -6.77,$$

and, therefore,

$$y_1^* = 1, x_{11}^* = 1, x_{12}^* = 0, x_{13}^* = 7/17, x_{14}^* = 0, x_{15}^* = 0, x_{16}^* = 1, x_{17}^* = 0.$$

In a similar way, the other subproblems $i = 2, \ldots, 6$, can be solved, obtaining

$$\text{LB}_{\text{CPL}}^{(2)}(\lambda) = -11.70;$$

$$y_2^* = 1; x_{21}^* = 11/12; x_{22}^* = 0; x_{23}^* = 0; x_{24}^* = 0; x_{25}^* = 1; x_{26}^* = 1; x_{27}^* = 0;$$

$$\text{LB}_{\text{CPL}}^{(3)}(\lambda) = -10.48;$$

$$y_3^* = 1; x_{31}^* = 1; x_{32}^* = 5/21; x_{33}^* = 1; x_{34}^* = 0; x_{35}^* = 0; x_{36}^* = 1; x_{37}^* = 0;$$

$$\text{LB}_{\text{CPL}}^{(4)}(\lambda) = -41.55;$$

$$y_4^* = 1; x_{41}^* = 1; x_{42}^* = 0; x_{43}^* = 1; x_{44}^* = 0; x_{45}^* = 20/27; x_{46}^* = 1; x_{47}^* = 0;$$

$$\text{LB}_{\text{CPL}}^{(5)}(\lambda) = -13.79;$$

$$y_5^* = 1; x_{51}^* = 0; x_{52}^* = 3/14; x_{53}^* = 1; x_{54}^* = 0; x_{55}^* = 1; x_{56}^* = 1; x_{57}^* = 0;$$

$$\text{LB}_{\text{CPL}}^{(6)}(\lambda) = -63.10;$$

$$y_6^* = 1; x_{61}^* = 1; x_{62}^* = 1; x_{63}^* = 0; x_{64}^* = 0; x_{65}^* = 0; x_{66}^* = 1; x_{67}^* = 12/43.$$

The optimal cost of the Lagrangian relaxed problem corresponds to

$$\text{LB}_{\text{CPL}}(\lambda) = -147.39 + 829 = 681.61.$$

Starting from the optimal solution of the Lagrangian relaxed problem, it is possible to construct a CPL feasible solution as follows.

Step 1. *Finding the facilities to be activated.* Let L be the list of potential facilities $i \in V_1$ sorted by nondecreasing values of $\mathrm{LB}_{\mathrm{CPL}}^{(i)}(\lambda)$, $i \in V_1$ (note that $\mathrm{LB}_{\mathrm{CPL}}^{(i)}(\lambda) \leq 0$, $i \in V_1, \lambda \in \Re^{|V_2|}$). Extract from L the minimum number of facilities capable of satisfying the total demand $\sum_{j \in V_2} d_j$. Let $\bar{V}_1$ be the set of facilities selected. Then $\bar{V}_1$ satisfies the relation

$$\sum_{i \in \bar{V}} q_i \geq \sum_{j \in V_2} d_j.$$

Step 2. *Customer allocation to the selected facilities.* Solve the demand allocation problem (3.26)–(3.29) considering $\bar{V}_1$ as the set of facilities to be opened. Let $\mathrm{UB}_{\mathrm{CPL}}(\lambda)$ be the cost (3.26) associated to the optimal allocation.

The heuristic first selects the facilities characterized by the smallest $\mathrm{LB}_{\mathrm{CPL}}^{(i)}(\lambda)$ values and then allocates optimally the demand to them.

With reference to the relaxed Lagrangian problem of the Milatog CPL problem, the list L of facilities is the following:

$$L = \{6, 4, 5, 2, 3, 1\}.$$

The set $\bar{V}_1$ of open facilities will be

$$\bar{V}_1 = \{4, 5, 6\},$$

since $\sum_{i \in \bar{V}_1} q_i = 340 > \sum_{j \in V_2} d_j = 262$. The demand allocation problem is therefore the following:

$$\begin{aligned}
\text{Minimize } & 75.60x_{41} + 123.48x_{42} + 42.84x_{43} + 195.00x_{44} + 72.90x_{45} + \\
& 88.20x_{46} + 59.34x_{47} + 139.32x_{51} + 107.10x_{52} + 78.54x_{53} + \\
& 121.50x_{54} + 36.45x_{55} + 65.70x_{56} + 99.33x_{57} + 83.88x_{61} + \\
& 92.82x_{62} + 124.10x_{63} + 134.50x_{64} + 75.87x_{65} + 69.90x_{66} + \\
& 48.59x_{67} + 745
\end{aligned}$$

subject to

$$x_{41} + x_{51} + x_{61} = 1$$

$$x_{42} + x_{52} + x_{62} = 1$$

$$x_{43} + x_{53} + x_{63} = 1$$

$$x_{44} + x_{54} + x_{64} = 1$$

$$x_{45} + x_{55} + x_{65} = 1$$

$$x_{46} + x_{56} + x_{66} = 1$$

$$x_{47} + x_{57} + x_{67} = 1$$

$$36x_{41} + 42x_{42} + 34x_{43} + 50x_{44} + 27x_{45} + 30x_{46} + 43x_{47} \leq 120$$

$$36x_{51} + 42x_{52} + 34x_{53} + 50x_{54} + 27x_{55} + 30x_{56} + 43x_{57} \leq 100$$

$$36x_{61} + 42x_{62} + 34x_{63} + 50x_{64} + 27x_{65} + 30x_{66} + 43x_{67} \leq 120$$

$$x_{ij} \geq 0, i = 4, 5, 6, j = 1, \ldots, 7,$$

with the following optimal solution:

$$\mathrm{UB_{CPL}}(\lambda) = 1\,229.48.$$

The daily demand of the seven farms is satisfied as can be deduced from Table 3.10.

Table 3.10 Fraction of the daily forage demand of each farm satisfied by the purchased silos 4, 5 and 6 for the Milatog problem.

	Farm						
Purchased silo	1	2	3	4	5	6	7
4	1	0	1	0	0	0	0
5	0	0	0	1	1	23/30	0
6	0	1	0	0	0	7/30	1

Thus, for each set of Lagrangian multipliers $\lambda \in \Re^{|V_2|}$, the above procedure computes both a lower and upper bound ($\mathrm{LB_{CPL}}(\lambda)$ and $\mathrm{UB_{CPL}}(\lambda)$, respectively). If these bounds coincide, an optimal solution has been found. Otherwise, in order to determine the Lagrangian multipliers corresponding to the maximum possible lower bound $\mathrm{LB_{CPL}}(\lambda)$ (or at least a satisfactory bound), the classical *subgradient algorithm* can be used. This algorithm also generates, in many cases, better upper bounds, since the feasible solutions generated from improved lower bounds are generally less costly. Here is a schematic description of the subgradient algorithm:

Step 0. *Initialization.* Select a tolerance value $\varepsilon \geq 0$. Set $\mathrm{LB} = -\infty$, $\mathrm{UB} = \infty$, $h = 1$ and $\lambda_j^{(h)} = 0$, $j \in V_2$.

Step 1. *Computation of a new lower bound.* Solve the Lagrangian relaxed problem (3.30)–(3.33) using $\lambda^{(h)} \in \Re^{|V_2|}$ as a vector of Lagrangian multipliers. If $\text{LB}_{\text{CPL}}(\lambda^{(h)}) > \text{LB}$, set $\text{LB} = \text{LB}_{\text{CPL}}(\lambda^{(h)})$.

Step 2. *Computation of a new upper bound.* Determine the corresponding feasible solution. Let $\text{UB}_{\text{CPL}}(\lambda^{(h)})$ be its cost. If $\text{UB}_{\text{CPL}}(\lambda^{(h)}) < \text{UB}$, set $\text{UB} = \text{UB}_{\text{CPL}}(\lambda^{(h)})$.

Step 3. *Check of the stopping criterion.* If $(\text{UB} - \text{LB})/\text{LB} \leq \varepsilon$, $STOP$. LB and UB represent the best upper and lower bounds available for z^*_{CPL}, respectively.

Step 4. *Updating of the Lagrangian multipliers.* Determine the subgradient of the jth relaxed constraint

$$s_j^{(h)} = \sum_{i \in V_1} x_{ij}^{(h)} - 1, \; j \in V_2,$$

where $x_{ij}^{(h)}$ is the solution of the Lagrangian relaxed problem (3.30)–(3.33) using $\lambda^{(h)} \in \Re^{|V_2|}$ as Lagrangian multipliers. Then set

$$\lambda_j^{(h+1)} = \lambda_j^{(h)} + \beta^{(h)} s_j^{(h)}, \; j \in V_2, \tag{3.38}$$

where $\beta^{(h)}$ is a suitable scalar coefficient. Let $h = h + 1$ and go back to Step 1.

This algorithm attempts to determine for a ε-optimal solution, that is, a feasible solution with a maximum user-defined deviation ε from the optimal solution.

Computational experiments have shown that the initial values of the Lagrangian multipliers do not significantly affect the behaviour of the heuristic. Hence, the Lagrangian multipliers are set equal to zero in Step 0. Formula (3.38) can be explained in the following way. If, at the hth iteration, the left-hand side of constraint (3.19) is higher than the right-hand side ($\sum_{i \in V_1} x_{ij}^{(h)} > 1$) for a certain $j \in V_2$, the subgradient $s_j^{(h)}$ is positive and the corresponding Lagrangian multiplier has to be increased in order to heavily penalize the constraint violation. Vice versa, if the left-hand side of constraint (3.19) is lower than the right-hand side ($\sum_{i \in V_1} x_{ij}^{(h)} < 1$) for a certain $j \in V_2$, the associated subgradient $s_j^{(h)}$ is negative and the value of the associated Lagrangian multiplier must be decreased to make the service of the unsatisfied demand fraction $1 - \sum_{i \in V_1} x_{ij}^{(h)}$ more attractive. Finally, if the jth constraint (3.19) is satisfied ($\sum_{i \in V_1} x_{ij}^{(h)} = 1$), the corresponding Lagrangian multiplier is unchanged.

The term $\beta^{(h)}$ in Equation (3.38) is a proportionality coefficient defined as

$$\beta^{(h)} = \frac{\alpha(\mathrm{UB} - \mathrm{LB_{CPL}}(\lambda^{(h)}))}{\sum_{j=1}^{|V_2|} (s_j^{(h)})^2}, \tag{3.39}$$

where α is a scalar arbitrarily chosen in the interval $(0, 2]$. The use of parameter $\beta^{(h)}$ in Equation (3.38) limits the variations of the Lagrangian multipliers when the lower bound $\mathrm{LB_{CPL}}(\lambda^{(h)})$ approaches the current upper bound UB.

Finally, note that the $\{\mathrm{LB_{CPL}}(\lambda^{(h)})\}$ sequence produced by the subgradient algorithm does not decrease monotonically. Therefore, there could exist iterations h for which $\mathrm{LB_{CPL}}(\lambda^{(h)}) < \mathrm{LB_{CPL}}(\lambda^{(h-1)})$ (this explains the lower bound update in Step 1). In practice, $\mathrm{LB_{CPL}}(\lambda^{(h)})$ values exhibit a zigzagging pattern.

The procedure is particularly efficient. Indeed, it generally requires, for $\varepsilon \approx 0.01$, only a few thousand iterations for problems with hundreds of vertices in V_1 and in V_2.

The Lagrangian heuristic is used to determine a suboptimal solution of the Milatog CPL problem as follows:

- the proportionality coefficient $\beta^{(h)}$ is determined at each iteration h according to (3.39) with $\alpha = 1/2$;
- a random choice of facilities to insert into the list $L^{(h)}$ in case of identical values of the optimal cost of the corresponding Lagrangian subproblems at each iteration h (Step 2).

In what follows, the results of the first three iterations are reported:

$$\lambda^{(1)} = [0, 0, 0, 0, 0, 0, 0]^{\mathrm{T}};$$

$$\mathrm{LB_{CPL}}(\lambda^{(1)}) = 0;$$

$$\mathrm{LB} = 0;$$

$$y^{(1)} = [1, 0, 1, 0, 1, 0]^{\mathrm{T}};$$

$$\mathrm{UB_{CPL}}(\lambda^{(1)}) = 1\,227.680;$$

$$\mathrm{UB} = 1\,227.680;$$

$$s^{(1)} = [-1, -1, -1, -1, -1, -1, -1]^{\mathrm{T}};$$

$$\beta^{(1)} = 87.691;$$

$$\lambda^{(2)} = [-87.691, -87.691, \ldots, -87.691]^{\mathrm{T}};$$

$$\mathrm{LB_{CPL}}(\lambda^{(2)}) = 613.840;$$

$$\text{LB} = 613.840;$$

$$y^{(2)} = [1, 0, 1, 0, 0, 1]^{\text{T}};$$

$$\text{UB}_{\text{CPL}}(\lambda^{(2)}) = 1\,255.280;$$

$$\text{UB} = 1\,227.680;$$

$$s^{(2)} = [-1, -1, -1, -1, -1, -1, -1]^{\text{T}};$$

$$\beta^{(2)} = 43.846;$$

$$\lambda^{(3)} = [-131.537, -131.537, \ldots, -131.537]^{\text{T}};$$

$$\text{LB}_{\text{CPL}}(\lambda^{(3)}) = 920.760;$$

$$\text{LB} = 920.760;$$

$$y^{(3)} = [0, 0, 1, 1, 0, 1]^{\text{T}};$$

$$\text{UB}_{\text{CPL}}(\lambda^{(3)}) = 1\,266.150;$$

$$\text{UB} = 1\,227.680;$$

$$s^{(3)} = [-1, -1, -1, -1, -1, -1, -1]^{\text{T}};$$

$$\beta^{(3)} = 21.923.$$

At the 23rd iteration, the procedure determines a LB value equal to 1 079.143, whereas UB = 1 218.080 (i.e. the optimal cost), with a value of (UB − LB)/LB = 0.129.

3.3.3 Single-commodity two-echelon discrete location problems

When the transport costs from the predecessor nodes to the facilities to be located and from these to the successor nodes are both relevant, it is necessary to use a two-echelon formulation.

In this section, a *single-commodity two-echelon* (SCTE) discrete location model will be presented, based on the following assumptions. Let $G(V_1 \cup V_2 \cup V_3, A_1 \cup A_2)$ be a complete directed tripartite graph in which the vertices in V_1 represent the predecessor nodes, the vertices in V_2 represent the potential facilities to be located and the vertices in V_3 represent the successor nodes; the arcs in $A_1 = V_1 \times V_2$ are associated with the commodity flows between the predecessor nodes and the potential facilities, while the arcs in $A_2 = V_2 \times V_3$ correspond to the commodity flows between the potential facilities and the successor nodes.

Let $t = 1, \ldots, T$, be the time periods composing the planning horizon. Denote by o_{it}, $i \in V_1$, the maximum quantity of commodity available at the predecessor node i at time period t, and q_{jt}, $j \in V_2$, the maximum throughput of potential facility j, at time period t, and d_{rt}, $r \in V_3$, the demand estimate of the successor node r at time period t. Similarly to the SCSE formulation illustrated in the previous Section 3.3.2, we limit our investigation to the case in which o_{it}, $i \in V_1$, q_{jt}, $j \in V_2$ and d_{rt}, $r \in V_3$, are constant for each time period t, $t = 1, \ldots, T$, of the planning horizon, that is,

$$o_{it} = o_i, i \in V_1, t = 1, \ldots, T;$$

$$q_{jt} = q_j, j \in V_2, t = 1, \ldots, T;$$

$$d_{rt} = d_r, r \in V_3, t = 1, \ldots, T.$$

Under these assumptions, the SCTE location problem can be referred to a generic time period, instead of considering the whole planning horizon. Let a_{ijr}, $i \in V_1$, $j \in V_2$, $r \in V_3$, be the unit transport cost of the commodity from the predecessor node i to successor node r through the facility j.

Finally, it is assumed that the facility cost of each potential facility j, $j \in V_2$, in a time period can be expressed in terms of a fixed cost f_j and a constant marginal cost per unit of time g_j.

Supposing that the commodity flows are divisible, the SCTE problem can be formulated using the following decision variables: y_j, $j \in V_2$, is binary, equal to 1 if potential facility j is activated, 0 otherwise; $s_{ijr} \geq 0$, $i \in V_1$, $j \in V_2$, $r \in V_3$, which represents the average quantity of commodity transported, in each time period $t = 1, \ldots, T$, from the predecessor node i to the successor node r through the facility j:

$$\text{Minimize} \sum_{i \in V_1} \sum_{j \in V_2} \sum_{r \in V_3} a_{ijr} s_{ijr} + \sum_{j \in V_2} (f_j y_j + g_j \sum_{i \in V_1} \sum_{r \in V_3} s_{ijr}) \tag{3.40}$$

subject to

$$\sum_{j \in V_2} \sum_{r \in V_3} s_{ijr} \leq o_i, i \in V_1 \tag{3.41}$$

$$\sum_{i \in V_1} \sum_{j \in V_2} s_{ijr} = d_r, r \in V_3 \tag{3.42}$$

$$\sum_{i \in V_1} \sum_{r \in V_3} s_{ijr} \leq q_j y_j, j \in V_2 \tag{3.43}$$

$$y_j \in \{0, 1\}, j \in V_2 \tag{3.44}$$

$$s_{ijr} \geq 0, i \in V_1, j \in V_2, r \in V_3. \tag{3.45}$$

The objective function (3.40) includes both the transport cost and the cost of operating the facilities. Constraints (3.41) impose an upper bound on the commodity flow from each predecessor node; equations (3.42) impose the demand satisfaction for every successor node; relations (3.43) represent the maximum throughput constraints on the facilities to be activated. They also impose that no flow can traverse a facility if it is not activated.

With reference to the Milatog location problem, we assume now the presence of two plants for the daily supply of the silos to be purchased.

These plants have a maximum daily throughput (in quintals of forage) equal to 120 and 150, respectively. The kilometric distances from the manufacturing plants to the potential silos are shown in Table 3.11.

Table 3.11 Distances, in kilometres, between the manufacturing plants and the potential silos in the Milatog problem.

	Silo					
Manufacturing plant	1	2	3	4	5	6
1	20	25	18	22	15	25
2	19	22	25	28	24	21

Model (3.40)–(3.45) becomes

$$V_1 = \{1, 2\} \text{ is the set of the manufacturing plants;}$$

$$V_2 = \{1, 2, 3, 4, 5, 6\} \text{ is the set of the potential silos;}$$

$$V_3 = \{1, 2, 3, 4, 5, 6, 7\} \text{ is the set of farms.}$$

The fixed costs associated to the silos and the average daily storage costs are those shown in the previous section; regarding the unit transport costs, we have: $a_{1111} = 0.12 \times (18 + 20) = €\ 4.56$ (the other costs a_{ijr}, $i \in V_1$, $j \in V_2$, $r \in V_3$, can be calculated in a similar way).

Indicating with y_j, $j = 1, \ldots, 6$, the binary decision variable associated with each potential site j, having value 1 if the silo j is purchased by Milatog, 0 otherwise, and with s_{ijr}, $i = 1, 2$, $j = 1, \ldots, 6$, $r = 1, \ldots, 7$, the decision variable expressing the average daily quantity requested by farm r satisfied by silo j and coming from manufacturing plant i, we will have the following formulation of the SCTE problem:

$$\begin{aligned}\text{Minimize } & 4.56 s_{111} + 5.16 s_{112} + \cdots + 3.48 s_{117} + 5.52 s_{121} + 5.16 s_{122} + \\ & \cdots + 5.40 s_{127} + \cdots + 4.68 s_{261} + 4.56 s_{262} + \cdots + 3.48 s_{267} + \\ & + 220 y_1 + 240 y_2 + 260 y_3 + 275 y_4 + 240 y_5 + 230 y_6 + \\ & + 0.15 \times (s_{111} + s_{112} + \cdots + s_{117} + s_{211} + s_{212} + \cdots + s_{217}) + \\ & + 0.18 \times (s_{121} + s_{122} + \cdots + s_{127} + s_{221} + s_{222} + \cdots + s_{227}) +\end{aligned}$$

$$\cdots$$

$$+0.17 \times (s_{161} + s_{162} + \cdots + s_{167} + s_{261} + s_{262} + \cdots + s_{267})$$

subject to

$$s_{111} + s_{112} + \cdots + s_{117} + s_{121} + s_{122} + \cdots + s_{127} + \cdots +$$

$$+ s_{161} + s_{162} + \cdots + s_{167} \leq 120$$

$$s_{211} + s_{212} + \cdots + s_{217} + s_{221} + s_{222} + \cdots + s_{227} + \cdots +$$

$$+ s_{261} + s_{262} + \cdots + s_{267} \leq 150$$

$$s_{111} + s_{121} + \cdots + s_{161} + s_{211} + s_{221} + \cdots + s_{261} = 36$$

$$s_{112} + s_{122} + \cdots + s_{162} + s_{212} + s_{222} + \cdots + s_{262} = 42$$

$$\cdots$$

$$s_{117} + s_{127} + \cdots + s_{167} + s_{217} + s_{227} + \cdots + s_{267} = 43$$

$$s_{111} + s_{112} + \cdots + s_{117} + s_{211} + s_{212} + \cdots + s_{217} \leq 80y_1$$

$$s_{121} + s_{122} + \cdots + s_{127} + s_{221} + s_{222} + \cdots + s_{227} \leq 90y_2$$

$$\cdots$$

$$s_{161} + s_{162} + \cdots + s_{167} + s_{261} + s_{262} + \cdots + s_{267} \leq 120y_6$$

$$y_j \in \{0, 1\}, j = 1, \ldots, 7$$

$$s_{ijr} \geq 0, i = 1, 2, j = 1, \ldots, 6, r = 1, \ldots, 7.$$

The optimal solution amounts to setting set up silos 1, 5 and 6, and corresponds to an overall daily cost equal to € 1 788.86. Table 3.12 shows the average quantities transported to every farm.

Table 3.12 Average daily quantity (in quintals) of cereals transported from the manufacturing plants to the farms by means of the three silos purchased by Milatog.

Manufacturing plant	Activated silo	Farm						
		1	2	3	4	5	6	7
	1	12	0	0	0	0	0	0
1	5	0	0	23	50	27	0	0
	6	0	0	0	00	0	0	0
2	1	24	0	11	0	0	30	3
	5	0	0	0	0	0	0	0
	6	0	42	0	0	0	0	40

3.3.4 The multicommodity case

The vast majority of practical cases involve more heterogeneous commodities, which leads to the definition of location-allocation problems that are much more complex than those previously examined.

In the sequel, we will restrict the discussion to the multicommodity version of model (3.40)–(3.45). The problem can be formulated on the complete directed tripartite graph G used for the SCTE problem. Let H be the set of commodities, whose flows are assumed to be expressed in the same units of measurement (i.e. conventional load units). Denoted by: $o_{ih}, i \in V_1, h \in H$, the average quantity of commodity h available at the predecessor node i in a time period; q_j, $j \in V_2$, the maximum throughput of the potential facility j, assumed to be constant for every time period of the planning horizon; $d_{rh}, r \in V_3, h \in H$, the demand estimate of commodity h requested by the successor node r in a time period, and $a_{ijrh}, i \in V_1, j \in V_2, r \in V_3, h \in H$, the unit transport cost of commodity h from the predecessor node i to the successor node r through the facility j. As in the SCTE model, we assume that the cost of each facility $j, j \in V_2$, in a time period can be expressed in terms of fixed costs f_j and marginal costs g_j.

The *multicommodity two-echelon* (MCTE) location problem can be formulated using the following decision variables: $y_j, j \in V_2$, is binary, equal to 1 if potential facility j is selected, 0 otherwise; $s_{ijrh}, i \in V_1, j \in V_2, r \in V_3, h \in H$, representing the average quantity of commodity h transported in a time period from the predecessor node i to the successor node r through the facility j. The model is as follows:

$$\text{Minimize} \sum_{i \in V_1} \sum_{j \in V_2} \sum_{r \in V_3} \sum_{h \in H} a_{ijrh} s_{ijrh} + \sum_{j \in V_2} (f_j y_j + g_j) \sum_{i \in V_1} \sum_{r \in V_3} \sum_{h \in H} s_{ijrh}$$

subject to

$$\sum_{j \in V_2} \sum_{r \in V_3} s_{ijrh} \leq o_{ih}, i \in V_1, h \in H$$

$$\sum_{i \in V_1} \sum_{j \in V_2} s_{ijrh} = d_{rh}, r \in V_3, h \in H$$

$$\sum_{i \in V_1} \sum_{r \in V_3} \sum_{h \in H} s_{ijrh} \leq q_j y_j, j \in V2$$

$$y_j \in \{0, 1\}, j \in V_2$$
$$s_{ijrh} \geq 0, i \in V_1, j \in V_2, r \in V_3, h \in H.$$

K9 is a German petrochemical company. The firm's management intends to renovate its production and distribution network, which is presently composed of two refining plants, two DCs and hundreds of sales points (gas pumps and liquefied gas retailers). After a series of meetings, it was decided to relocate the DCs, leaving the position and features of the two production plants unchanged. The products of K9 are subdivided into two homogeneous commodities (represented by the indices $h = 1, 2$): fuel for motor transport and liquefied gas (the latter sold in cylinders). There are four potential sites suited to receive a DC and their maximum daily throughput q_j (expressed in hectolitres) are, respectively, 1 500, 1 200, 2 300 and 2 500.

The sales points have been grouped into three districts ($r = 1, 2, 3$) characterized by the daily demands shown in Table 3.13. The annual fixed costs (in €) of the distribution centres $f_j, j \in V_2$, are the following: 960 000, 880 000, 1 540 000, 1610 000. The daily storage facility costs are, respectively, 0.15, 0.14, 0.20 and € 0.25/hl. The transport costs $a_{ijrh}, i \in V_1$, $j \in V_2, r \in V_3, h = 1, 2$, are obtained by multiplying the cost per kilometre and per hectolitre (equal to € 0.0067 for $h = 1$ and to € 0.0082 for $h = 2$) by the return trip distances between the manufacturing plants $i \in V_1$ and the centroids of the sales districts $r \in V_3$ through DC $j \in V_2$ (see Table 3.14).

Finally, Table 3.15 shows the daily average quantities of the two commodities available at the two manufacturing plants.

Table 3.13 Average daily sales districts demand (in hl) of the two commodities in the K9 problem.

	Commodity	
District	1	2
1	800	300
2	600	400
3	700	500

The transport costs $a_{ijrh}, i = 1, 2, j = 1, \ldots, 4, r = 1, 2, 3, h = 1, 2$, can be deduced from Tables 3.16 and 3.17. For example, the cost a_{1111} is obtained as

$$a_{1111} = 0.0067 \times 2 \times 423 = € \ 5.6682/\text{hl}.$$

The fixed annual costs of the DCs are transformed in daily fixed costs by assuming the annual horizon is composed of 220 working days during which the distribution operations of the two commodities take place. Therefore, the following fixed daily costs (in €) are obtained for each of the four potential DCs: 4 363.64, 4 000.00, 7 000.00 and 7 318.18. Indicating with $y_j, j \in V_2$, the binary decision variable associated with each potential DC j, having value 1 if the site j is used to host a K9 DC, 0 otherwise, and with $s_{ijrh}, i \in V_1, j \in V_2, r \in V_3$ and $h = 1, 2$, the decision variable expressing the average daily quantity of commodity h requested by the sales district r satisfied by the DC

Table 3.14 Kilometric distances between the refining plants and the centroids of the sales districts through the potential DC in the K9 problem.

Refining plant	DC	Sales district 1	Sales district 2	Sales district 3
1	1	423	612	1 108
	2	613	434	927
	3	1 031	631	918
	4	1 628	1 236	954
2	1	826	1 028	1 531
	2	864	638	1 158
	3	838	464	782
	4	1 227	871	544

Table 3.15 Average daily quantity (in hl) of the two commodities available at the refining plants in the K9 problem.

Refining plant	Commodity 1	Commodity 2
1	1 200	500
2	1 500	800

Table 3.16 Transport costs (in €/hl) for the first commodity in the K9 problem.

Refining plant	DC	Sales district 1	Sales district 2	Sales district 3
1	1	5.6682	8.2008	14.8472
	2	8.2142	5.8156	12.4218
	3	13.8154	8.4554	12.3012
	4	21.8152	16.5624	12.7836
2	1	11.0684	13.7752	20.5154
	2	11.5776	8.5492	15.5172
	3	11.2292	6.2176	10.4788
	4	16.4418	11.6714	7.2896

Table 3.17 Transport costs (in €/hl) for the second commodity in the K9 problem.

Refining plant	DC	Sales district 1	2	3
1	1	6.9372	10.0368	18.1712
	2	10.0532	7.1176	15.2028
	3	16.9084	10.3484	15.0552
	4	26.6992	20.2704	15.6456
2	1	13.5464	16.8592	25.1084
	2	14.1696	10.4632	18.9912
	3	13.7432	7.6096	12.8248
	4	20.1228	14.2844	8.9216

j and coming from the refining plant *i*, the following formulation MCTE problem is obtained:

$$\begin{aligned}
\text{Minimize } & 5.6682s_{1111} + 8.2008s_{1112} + 14.8472s_{1121} + 6.9372s_{1122} + \\
& + \cdots + 14.2844s_{2431} + 8.9216s_{2432} + \\
& + 4\,363.64y_1 + 0.15(s_{1111} + s_{1112} + \cdots + s_{2231} + s_{2232}) + \\
& + 4\,000.00y_2 + 0.14(s_{1211} + s_{1212} + \cdots + s_{2231} + s_{2232}) + \\
& \ldots \\
& + 7\,318.18y_4 + 0.25(s_{1411} + s_{1412} + \cdots + s_{2431} + s_{2432})
\end{aligned}$$

subject to

$$\begin{aligned}
& s_{1111} + s_{1121} + s_{1131} + \cdots + s_{1411} + s_{1421} + s_{1431} \le 1\,200 \\
& s_{1112} + s_{1122} + s_{1132} + \cdots + s_{1412} + s_{1422} + s_{1432} \le 500 \\
& \ldots \\
& s_{2112} + s_{2122} + s_{2132} + \cdots + s_{2412} + s_{2422} + s_{2432} \le 800 \\
& s_{1111} + s_{1211} + \cdots + s_{2311} + s_{2411} = 800 \\
& s_{1112} + s_{1212} + \cdots + s_{2312} + s_{2412} = 300 \\
& \ldots \\
& s_{1132} + s_{1232} + \cdots + s_{2332} + s_{2432} = 500 \\
& s_{1111} + s_{1112} + \cdots + s_{2131} + s_{2132} \le 1\,500y_1 \\
& s_{1211} + s_{1212} + \cdots + s_{2231} + s_{2232} \le 1\,200y_2
\end{aligned}$$

$$\cdots$$

$$s_{1411} + s_{1412} + \cdots + s_{2431} + s_{2432} \leq 2\,500 y_4$$

$$y_j \in \{0, 1\}, j \in V_2$$

$$s_{ijrh} \geq 0, i \in V_1, j \in V_2, r \in V_3, h = 1, 2.$$

The optimal solution has a daily cost of € 39 330.80, corresponding to the set-up of the first and the third DCs. The optimal average daily quantities (in hl) of each commodity requested by every district and satisfied by the selected DC and coming from each refining plant (for the sake of conciseness, only the quantities different from zero are shown) are

$$s^*_{1111} = 800;\ s^*_{1112} = 300;\ s^*_{1332} = 100;\ s^*_{2321} = 600;$$

$$s^*_{2322} = 400;\ s^*_{2331} = 700;\ s^*_{2332} = 400.$$

3.3.5 Location-covering problems

In location-covering problems, the aim is to locate a least-cost set of service facilities in such a way that each user can be reached within a limited travel time from the closest facility.

The simplest problem can be modelled on a graph $G = (V_1 \cup V_2, E)$, where vertices in V_1 represent the potential facilities, vertices in V_2 describe the users to be reached and each edge $(i, j) \in E$ corresponds to a least-duration path between i and j. Let $f_i, i \in V_1$, be the fixed cost of potential facility i; $a_{ij}, i \in V_1, j \in V_2$, a binary constant equal to 1 if potential facility i is able to serve customer j, 0 otherwise (given a user-defined limit time T, $a_{ij} = 1$, if $t_{ij} \leq T_{ij}, i \in V_1, j \in V_2$, otherwise $a_{ij} = 0$). The decision variables are binary: $y_i, i \in V_1$, is equal to 1 if facility i is opened, 0 otherwise. The problem is modelled as follows:

$$\text{Minimize} \sum_{i \in V_1} f_i y_i \tag{3.46}$$

subject to

$$\sum_{i \in V_1} a_{ij} y_i \geq 1, j \in V_2 \tag{3.47}$$

$$y_i \in \{0, 1\}, i \in V_1. \tag{3.48}$$

The model (3.46)–(3.48) is a classic *set-covering* (SC) problem and falls into the class of NP-hard problems. A good feasible solution can be determined by using the following simple heuristic.

Step 1. If $f_i = 0$, for some $i \in V_1$, set $\bar{y}_i = 1$ and remove all the constraints $j \in V_2$ in which y_i appears with coefficient a_{ij} equal to 1.

Step 2. If $f_i > 0$, for some $i \in V_1$ and y_i does not appear with coefficient a_{ij} equal to 1 in each of the remaining constraints $j \in V_2$, then set $\bar{y}_i = 0$.

Step 3. For all the remaining decision variables, determine the ratio f_i/n_i, where n_i is the number of constraints in which y_i appears with a unit coefficient. Choose the decision variable k for which the ratio f_k/n_k is minimum; set $\bar{y}_k = 1$ and remove all the constraints in which y_k appears with a unit coefficient.

Step 4. If no constraints remain, *STOP*, and set the remaining decision variables equal to zero, otherwise go back to Step 1.

In Portugal, the municipal administration of Coimbra needs to locate emergency fire stations to cover all the seven residential districts of the city within a maximum time of 16 minutes. The minimum distances between the centroids of the seven urban areas are reported in Table 3.18. The stations' annual costs (in tens of thousands of €) are: 200, 160, 240, 220, 180, 180 and 220. The average travelling speed is assumed to be 65 km/h. The SC model is as follows:

$$\text{Minimize } 200y_1 + 160y_2 + 240y_3 + 220y_4 + 180y_5 + 180y_6 + 220y_7$$

subject to

$$y_1 + y_2 + y_7 \geq 1$$

$$y_1 + y_2 + y_5 + y_6 + y_7 \geq 1$$

$$y_3 + y_4 + y_5 + y_6 \geq 1$$

$$y_3 + y_4 + y_7 \geq 1$$

$$y_2 + y_3 + y_5 + y_6 \geq 1$$

$$y_2 + y_3 + y_5 + y_6 \geq 1$$

$$y_1 + y_2 + y_4 + y_7 \geq 1$$

$$y_1, y_2, y_3, y_4, y_5, y_6, y_7 \in \{0, 1\}.$$

We apply the above heuristic. Since $f_i > 0, i = 1, \ldots, 7$, and the binary decision variables $y_i, i = 1, \ldots, 7$, appear in all the constraints with unit coefficient, we calculate the ratios

$$f_1/n_1 = 200/3 = 66.66;$$

$$f_2/n_2 = 160/5 = 32;$$

$$f_3/n_3 = 240/4 = 60;$$
$$f_4/n_4 = 220/3 = 73.33;$$
$$f_5/n_5 = 180/4 = 45;$$
$$f_6/n_6 = 180/4 = 45;$$
$$f_7/n_7 = 220/4 = 55.$$

The lowest ratio is obtained for $k = 2$; for this reason, $\bar{y}_2$ is set equal to 1 and the first two and the last three constraints are removed. The problem becomes

$$\text{Minimize } 200y_1 + 240y_3 + 220y_4 + 180y_5 + 180y_6 + 220y_7$$

subject to

$$y_3 + y_4 + y_5 + y_6 \geq 1$$
$$y_3 + y_4 + y_7 \geq 1$$
$$y_1, y_3, y_4, y_5, y_6, y_7 \in \{0, 1\}.$$

The coefficient $f_1 > 0$ and the decision variable y_1 does not appear in any constraint, so that $\bar{y}_1$ can be set equal to 0. We calculate the ratios

$$f_3/n_3 = 240/2 = 120;$$
$$f_4/n_4 = 220/2 = 110;$$
$$f_5/n_5 = 180/1 = 180;$$
$$f_6/n_6 = 180/1 = 180;$$
$$f_7/n_7 = 220/1 = 220.$$

The lowest ratio is obtained for $k = 4$, so that $\bar{y}_4$ is set equal to 1 and the two constraints are removed because in both y_4 appears with coefficient equal to 1. The problem appears now as

$$\text{Minimize } 240y_3 + 180y_5 + 180y_6 + 220y_7$$

subject to

$$y_3, y_5, y_6, y_7 \in \{0, 1\}.$$

This means that $\bar{y}_3 = \bar{y}_5 = \bar{y}_6 = \bar{y}_7 = 0$ are set. Hence, the feasible solution involves the location of the emergency stations in areas 2 and 4. The corresponding total annual cost amounts to € 3 800 000.

Table 3.18 Distances (in km) between the centroids of the seven districts of Coimbra.

	Area						
Area	1	2	3	4	5	6	7
1	0	8	24	18	30	20	16
2		0	30	120	14	4	6
3			0	16	12	10	18
4				0	18	20	6
5					0	4	100
6						0	54
7							0

Several variants of the SC model can be used in practice. For example, if the fixed costs f_i are identical for all potential facilities $i \in V_1$, it can be convenient to discriminate among all the solutions with the least number of open facilities the one corresponding to the least total travelling time, or to the most equitable demand distribution among the facilities. In the former case, let $x_{ij}, i \in V_1, j \in V_2$, be a binary decision variable equal to 1 if customer j is served by facility i, 0 otherwise. The problem can be modelled as follows:

$$\text{Minimize} \sum_{i \in V_1} M y_i + \sum_{i \in V_1} \sum_{j \in V_2} t_{ij} x_{ij} \tag{3.49}$$

subject to

$$\sum_{i \in V_1} a_{ij} x_{ij} \geq 1, j \in V_2 \tag{3.50}$$

$$\sum_{j \in V_2} x_{ij} \leq |V_2| \, y_i, i \in V_1 \tag{3.51}$$

$$y_i \in \{0, 1\}, i \in V_1 \tag{3.52}$$

$$x_{ij} \in \{0, 1\}, i \in V_1, j \in V_2. \tag{3.53}$$

In the objective function (3.49), M is an arbitrarily large positive constant chosen so that the number of facilities to be activated is always as small as possible; constraints (3.50) guarantee that all customers $j \in V_2$ are serviced, while constraints (3.51) ensure that if facility $i \in V_1$ is not set up ($y_i = 0$), then no customer $j \in V_2$ can be served by it.

In the English county of Cornwall, a consortium of 10 municipalities (Sennen Cove, Porth Curno, Trevilley, Botallack, Morvah, Treen, Zennor, St. Ives, St. Erth and Hayle) has decided to improve its fire-fighting service. The person responsible for the project has established that each centre of the community must be reached within 10 minutes from the nearest fire station. Since the main aim is just to provide a first aid in case of fire, the decision maker has decided to assign a single vehicle to each station. The annual cost of a station inclusive of the expenses of the personnel is £ 123 000. To determine the number and the optimal location of the fire stations, since all the municipalities of the consortium should be served, the location-covering model (3.49)–(3.53) can be used, for which $V_1 = V_2 =$ {Sennen Cove, Porth Curno, Trevilley, Botallack, Morvah, Treen, Zennor, St. Ives, St. Erth, Hayle}, while the fastest travelling times $t_{ij}, i \in V_1, j \in V_2$, are reported in Tables 3.19 and 3.20.

Table 3.19 Part I–Travel times (in minutes) between the municipalities of the consortium in Cornwall.

	Sennen Cove	Porth Curno	Trevilley	Botallack	Morvah
Sennen Cove	0.00	3.07	1.80	6.33	9.53
Porth Curno		0.00	4.67	9.20	12.33
Trevilley			0.00	5.07	8.20
Botallack				0.00	3.73
Morvah					0.00

Table 3.20 Part II–Travel times (in minutes) between the municipalities of the consortium in Cornwall.

	Treen	Zennor	St. Ives	St. Erth	Hayle
Sennen Cove	12.20	14.07	17.07	16.27	16.80
Porth Curno	15.00	16.33	16.93	16.13	16.67
Trevilley	10.87	12.07	15.80	15.00	15.53
Botallack	6.40	7.67	12.33	15.20	15.73
Morvah	4.00	5.27	9.87	13.80	14.40
Treen	0.00	2.00	6.60	14.73	15.27
Zennor		0.00	5.13	10.13	10.73
St. Ives			0.00	5.07	5.60
St. Erth				0.00	2.27
Hayle					0.00

The coefficients a_{ij}, $i \in V_1$, $j \in V_2$, were obtained from Tables 3.19 and 3.20, by imposing that $a_{ij} = 1$ if $t_{ij} \leq 10$ minutes, $a_{ij} = 0$ otherwise, $i \in V_1, j \in V_2$. The minimum number of fire stations turns out to be two.

The facilities are located in Trevilley and St. Ives. The fire station located in Trevilley serves Sennen Cove, Porth Curno, Botallack, Morvah and Trevilley itself, and the remaining ones are served by the fire station located in St. Ives.

The presence of coefficients in the objective function (3.49) with different orders of magnitude can cause serious numerical difficulties for any solver of the problem (3.49)–(3.53). To overcome these difficulties, we can adopt a different solution method. First, we solve the problem (3.46)–(3.48) with $f = f_i$, $i \in V$, in such a way to determine the number p^* of facilities to activate. Then, we can solve a modified version of problem (3.49)–(3.53), in which the objective function (3.49) is replaced by

$$\text{Minimize} \sum_{i \in V_1} \sum_{j \in V_2} t_{ij} x_{ij}$$

and we have the following additional constraint:

$$\sum_{i \in V_1} y_i = p^*.$$

3.3.6 *p*-centre problems

In p-centre problems, the aim is to locate p facilities in such a way that the maximum travel time from a user to the closest facility is minimized. The p-centre model finds its application when it is necessary to ensure equity in servicing users spread over a wide geographical area. The problem can be modelled on a directed, undirected or mixed graph $G = (V, A, E)$, where V is a set of vertices representing both user sites and road intersections, while A and E (the set of arcs and edges, respectively) describe the road connections among the vertices. Exactly p facilities have to be located either on a vertex, or on an arc or edge. For $p \geq 2$, the p-centre model is NP-hard.

If G is a directed graph, it can be easily shown that an optimal solution of the p-centre problem exists such that every facility location is a vertex (the *vertex location* property).

If G is undirected or mixed, the optimal locations of the facilities could be vertices or internal points of edges. In what follows, an algorithm is described for the 1-centre problem. The reader is referred to the literature for a discussion of the more general case.

If G is directed, due to the vertex location property, the 1-centre can be easily determined in the following way. Let t_{ij}, $i, j \in V$, be the shortest travel time from i to j.

1. For each $i \in V$, determine the maximum travel time from i to every other vertex $j \in V$, that is, $T_i = \max_{j \in V}\{t_{ij}\}$.
2. Locate the facility in the vertex i^* such that $T_{i^*} = \min_{i \in V}\{T_i\}$.

To locate a first aid mobile unit in the town of Krosno in Poland, the town centre was subdivided into 13 areas and the minimum distances between the relative centroids was computed (see Table 3.21).

Table 3.21 Minimum distances (in m) between the centroids of the 13 urban areas of the town of Krosno.

	1	2	3	4	5	6	7	8	9	10	11	12	13
1	0	682	188	428	548	621	653	723	738	377	404	149	143
2	655	0	841	754	804	438	506	257	352	524	283	762	755
3	139	748	0	301	579	572	665	639	611	452	381	102	160
4	356	658	270	0	547	565	563	603	363	707	462	337	316
5	600	839	575	523	0	701	707	738	428	458	612	675	698
6	441	779	604	482	393	0	482	630	562	436	643	575	582
7	625	314	594	587	790	457	0	269	309	624	437	696	737
8	406	369	585	705	710	323	555	0	497	732	483	520	512
9	786	609	745	529	827	568	364	505	0	663	622	823	657
10	652	847	701	486	391	746	747	822	709	0	259	638	528
11	437	616	613	733	793	582	692	810	828	644	0	519	489
12	108	720	156	318	480	576	706	742	637	478	285	0	185
13	141	596	191	294	435	544	769	727	650	471	408	208	0

We assume that the mobile first aid unit can be located in any centroid. Hence, we determine the farthest area from each centroid: the maximum value is determined for each of the 13 rows of the matrix shown in Table 3.21, thus, obtaining the following array: $T = [738, 841, 748, 707, 839, 779, 790, 732, 827, 847, 828, 742, 769]^T$. Consequently, the 1-centre is located in the fourth centroid, corresponding to the minimum value of $T_i, i = 1, \ldots, 13$, equal to 707 m. On the basis of this choice, the most disadvantaged area of the town is the tenth.

In the case of an undirected or mixed graph, as already observed above, the 1-centre could be a vertex or an internal point of an edge.

To simplify the discussion, we will refer only to the case of an undirected graph ($A = \emptyset$), although the procedure can be easily applied to a mixed graph. For each $(i, j) \in E$, let a_{ij} be the traversal time of edge (i, j). Furthermore, for each pair of vertices $i, j \in V$, denote by t_{ij} the shortest travel time between i and j, corresponding to the sum of the travel times of the edges of the shortest path between i and j. Note that, on the basis of the definition of travel time, we have

$$t_{ij} \leq a_{ij}, (i, j) \in E.$$

Finally, denote by $\tau_h(p_{hk})$ the travel time along edge $(h, k) \in E$ between vertex $h \in V$ and a point p_{hk} of the edge. In this way, the travel time $\tau_h(p_{hk})$

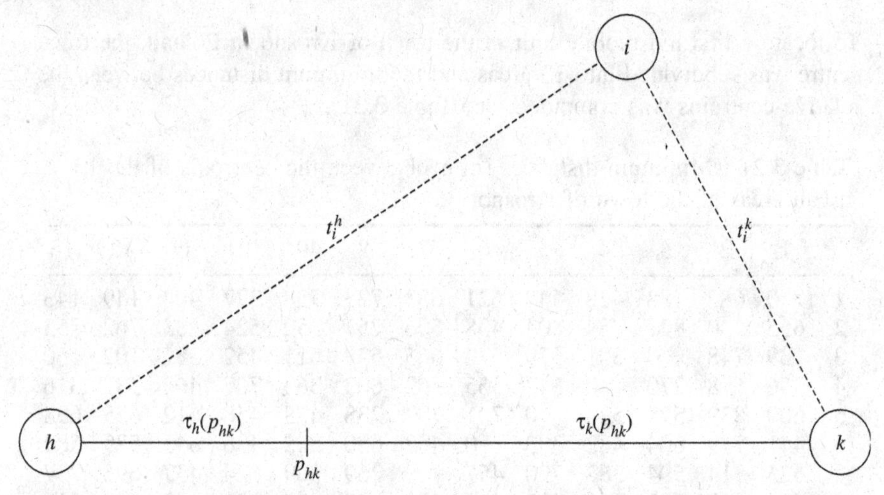

Figure 3.6 Computation of the travel time $T_i(p_{hk})$ from a user $i \in V$ to a facility in p_{hk}.

along the edge (h, k) between vertex $k \in V$ and p_{hk} is (see Figure 3.6):

$$\tau_k(p_{hk}) = a_{hk} - \tau_h(p_{hk}).$$

The 1-centre problem can be solved by the following algorithm proposed by Hakimi.

Step 1. *Computation of the travel time.* For each edge $(h, k) \in E$ and for each vertex $\in V$, determine the travel time $T_i(p_{hk})$ from $i \in V$ to a point p_{hk} of the edge (h, k) (see Figure 3.7):

$$T_i(p_{hk}) = \min\{t_{ih} + \tau_h(p_{hk}), t_{ik} + \tau_k(p_{hk})\}. \tag{3.54}$$

Step 2. *Finding the local centre.* For each edge $(h, k) \in E$, determine the local centre p^*_{hk} as the point on (h, k) minimizing the travel time from the most disadvantaged vertex:

$$p_{hk} = \arg\min \max_{i \in V} \{T_i(p_{hk})\},$$

where $\max_{i \in V}\{T_i(p_{hk})\}$ corresponds to the superior envelope of the functions $T_i(p_{hk})$, $i \in V$ (see Figure 3.8).

Step 3. *Determination of the* 1*-centre.* The 1-centre p^* is the best local centre p^*_{hk}, $(h, k) \in E$, that is,

$$p^* = \arg \min_{(h,k) \in E} \left\{ \min \max_{i \in V} \{T_i(p_{hk})\} \right\}.$$

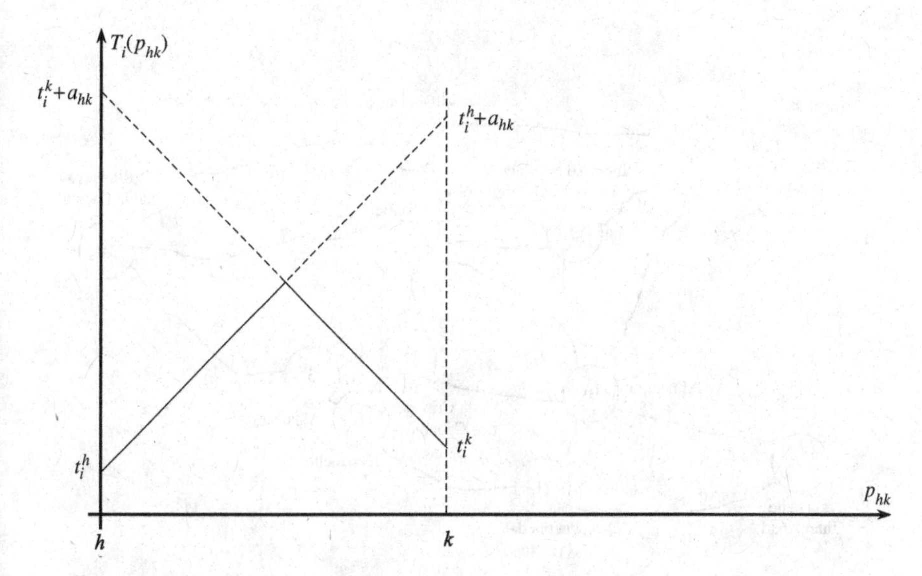

Figure 3.7 Travel time $T_i(p_{hk})$ versus the position of point p_{hk} along edge (h, k).

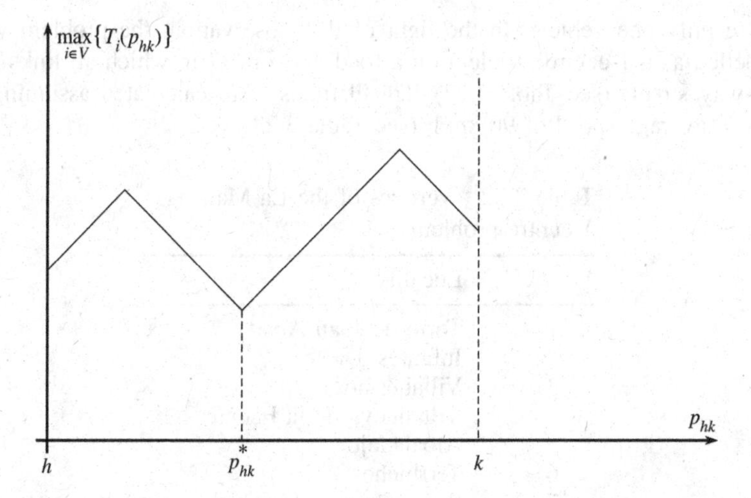

Figure 3.8 Determination of the local centre of edge $(h, k) \in E$.

In the La Mancha region of Spain (see Figure 3.9), a consortium of town councils, located in a rural area, decided to locate a parking place for ambulances. A preliminary examination of the problem revealed that the probability of receiving a service request during the completion of a previous call was extremely low because of the small number of the inhabitants of the zone. For this reason the team responsible for the service decided

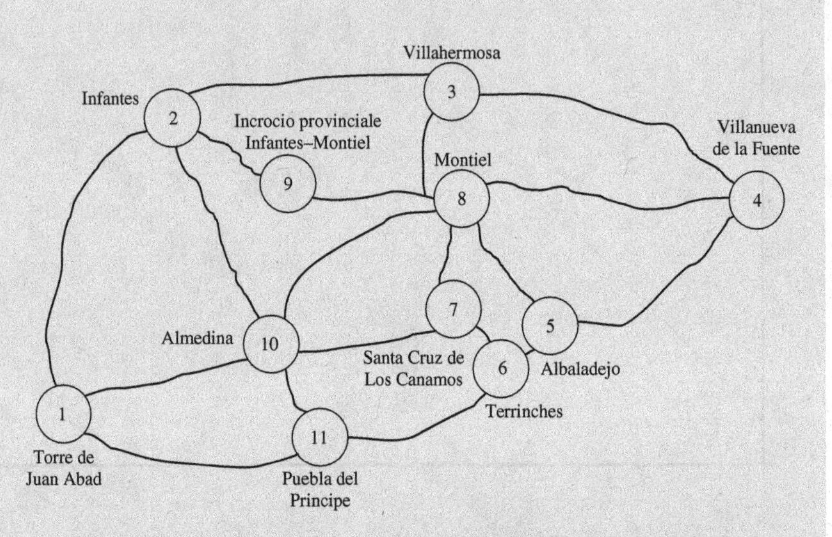

Figure 3.9 Location problem in the La Mancha region.

to use only one vehicle. In the light of this observation, the problem was modelled as a 1-centre problem on a road network G in which all links are two-way streets (see Table 3.22). Travel times were calculated assuming a vehicle average speed of 90 km/h (see Table 3.23).

Table 3.22 Vertices of the La Mancha 1-centre problem.

Vertex	Locality
1	Torre de Juan Abad
2	Infantes
3	Villahermosa
4	Villanueva de la Fuente
5	Albaladejo
6	Terrinches
7	Santa Cruz de los Canamos
8	Montiel
9	Infantes–Montiel crossing
10	Almedina
11	Puebla del Principe

Travel times t_{ij}, $i, j \in V$, are reported in Table 3.24. For each edge $(h, k) \in E$ and for each vertex $i \in V$, the travel time $T_i(p_{hk})$ from vertex i to a point p_{hk} of the edge (h, k) can be defined through Equation (3.54).

Table 3.23 Travel time (in minutes) on the road network edges in the La Mancha problem.

(i, j)	a_{ij}	(i, j)	a_{ij}	(i, j)	a_{ij}
(1,2)	12	(3,9)	4	(6,11)	5
(1,10)	6	(4,5)	9	(7,8)	4
(1,11)	8	(4,8)	10	(7,10)	5
(2,3)	9	(5,6)	2	(8,9)	1
(2,9)	8	(5,8)	6	(8,10)	7
(2,10)	9	(6,7)	3	(10,11)	4
(3, 4)	11				

Table 3.24 Travel times (in minutes) t_{ij} $i, j \in V$, in the La Mancha problem.

	i										
j	1	2	3	4	5	6	7	8	9	10	11
1	0	12	18	23	15	13	11	13	14	6	8
2		0	9	19	15	16	13	9	8	9	13
3			0	11	11	12	9	5	4	12	16
4				0	9	11	14	10	11	17	16
5					0	2	5	6	7	10	7
6						0	3	7	8	8	5
7							0	4	5	5	8
8								0	1	7	11
9									0	8	12
10										0	4
11											0

This enables the construction for each edge $(h, k) \in E$ of the function $\max_{i \in V} \{T_i(p_{hk})\}$, whose minimum corresponds to the local centre p^*_{hk}. For example, Figure 3.10 depicts the function $\max_{i \in V} \{T_i(p_{23})\}$, and Table 3.25 gives, for each edge $(h, k) \in E$, both the position of p^*_{hk} and the value $\max_{i \in V} \left\{T_i(p^*_{hk})\right\}$.

Consequently the 1-centre corresponds to the point p^* on the edge (8, 10). Therefore, the optimal location of the parking place for the ambulance should be on the road between Montiel and Almedina, at 2.25 km from the centre of Montiel. The cities least advantaged by this location decision are Villanueva de la Fuente and Torre de Juan Abad, since the ambulance would take an average time of 11.5 minutes to reach these villages.

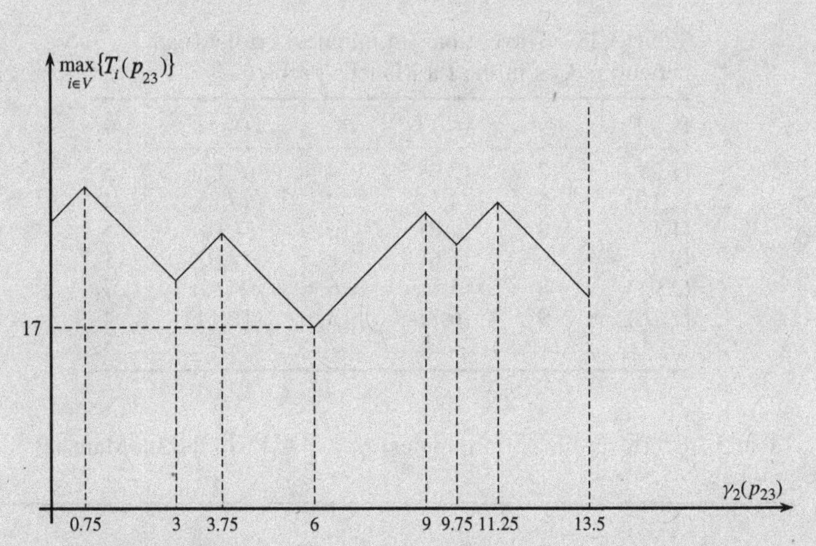

Figure 3.10 *Time* $\max_{i\in V}\{T_i(p_{23})\}$ *versus position* $\gamma_2(p_{23})$ *of* p_{23} *in the La Mancha problem.*

Table 3.25 Position of local centres p^*_{hk} and $\max_{i\in V}\{T_i(p^*_{hk})\}$ (in minutes) in the La Mancha problem.

(h,k)	$\gamma_h(p^*_{hk})$	$\max_{i\in V}\{T_i(p^*_{hk})\}$	(h,k)	$\gamma_h(p^*_{hk})$	$\max_{i\in V}\{T_i(p^*_{hk})\}$
(1,2)	18.00	19.0	(1,11)	12.00	16.0
(2,3)	6.00	17.0	(2,9)	12.00	14.0
(3,4)	0.00	18.0	(2,10)	13.50	17.0
(4,5)	13.50	15.0	(3,9)	6.00	14.0
(5,6)	0.00	15.0	(4,8)	15.00	13.0
(6,7)	3.75	13.5	(5,8)	9.00	13.0
(7,8)	2.25	12.5	(6,11)	4.50	15.0
(8,9)	0.00	13.0	(7,10)	0.00	14.0
(8,10)	2.25	11.5	(1,10)	9.00	17.0
(10,11)	6.00	16.0			

3.4 Hybrid methods

Hybrid methods share some of the distinguished features of qualitative and quantitative location methods. They can be used when the location of the facilities is affected by qualitative factors but the alternative solutions are quite numerous and cannot be enumerated.

Let V be the set of potential facilities. We also assume that the location decision is affected by m qualitative location factors. Applying the weighted scoring method (see Section 3.2), a score r_i can be assigned to the potential site $i \in V$, while a location cost c_i can be determined by using a quantitative method. Denoting by $\alpha \in [0, 1]$ the weight assigned to the quantitative factor for each site $i \in V$, the following hybrid measure of location is obtained:

$$m_i = \alpha \frac{c_i}{c_{max}} + (1 - \alpha)\left(1 - \frac{r_i}{r_{max}}\right), i \in V,$$

where

$$c_{max} = \max_{i \in V}\{c_i\}$$

and

$$r_{max} = \max_{i \in V}\{r_i\}.$$

The preferred site i^* is the one that achieves the lowest hybrid measure, that is, $i^* = \arg\min_{i \in V}\{m_i\}$.

We go back to the Jet Market problem (see Section 3.2), and now assume that the fixed costs (in €) for the three sites are: 2 350 000, 2 250 000 and 2 430 000. Choosing $\alpha = 0.8$, the following values of m_1, m_2 and m_3 are obtained:

$$m_1 = 0.8 \times \frac{2\,350\,000}{2\,430\,000} + (1 - 0.8) \times \left(1 - \frac{4.48}{4.65}\right) = 0.78;$$

$$m_2 = 0.8 \times \frac{2\,250\,000}{2\,430\,000} + (1 - 0.8) \times \left(1 - \frac{4.14}{4.65}\right) = 0.76;$$

$$m_3 = 0.8 \times \frac{2\,430\,000}{2\,430\,000} + (1 - 0.8) \times \left(1 - \frac{4.65}{4.65}\right) = 0.80.$$

On the basis of these measures, the optimal location corresponds to the second site.

3.5 Stochastic location models

The facility location models discussed so far are deterministic, i.e., they are based on the hypothesis that the relevant parameters (customer demands, location and transport costs etc.) are known with certainty. However, very often this assumption proves not to be very realistic. A location decision taken on the basis of point estimates or expected parameter values can result to be very poor.

As previously mentioned, the facility location problem is very often not separable from that of the commodity flow allocation, so that the overall decision-making process has an intrinsic *two-stage* nature, corresponding to two distinct decision levels. At the strategic level, the location of the facilities is established before knowing exactly the demand and cost values. Subsequently, once the values of the uncertain parameters are disclosed, the decisions at a tactical level are taken, that is, the commodity flows are allocated.

Regarding the representation of parameter uncertainty, the simplest way is to use discrete random variables, whose possible realizations constitute the so-called *scenarios*, each characterized by a specific probability of occurrence. On the basis of these scenarios, the recourse decisions are defined, which represent 'reactions' to the observation of the evolution of the uncertain parameters.

In the sequel, a model is described for the SCSE problem with uncertain demands and costs and represented by means of a set of scenarios S. For each scenario $s \in S$, characterized by a probability of occurrence p_s, an estimate of the demand d_{js} for every successor node $j \in V_2$, and of transport costs c_{ijs}, $i \in V_1, j \in V_2$, are available. The first-stage decision amounts to locating the facilities (i.e., to set the binary variables $y_i, i \in V_1$), whereas the second-stage decision is to allocate the commodity flows to the successor nodes for each scenario $s \in S$. The SCSE model with uncertain parameters can then be formulated as follows:

$$\text{Minimize} \sum_{s \in S} p_s \sum_{i \in V_1} \sum_{j \in V_2} c_{ijs} x_{ijs} + \sum_{i \in V_1} f_i y_i \tag{3.55}$$

subject to

$$\sum_{i \in V_1} x_{ijs} = 1, \; j \in V_2, s \in S \tag{3.56}$$

$$\sum_{j \in V_2} d_{js} x_{ijs} \leq q_i y_i, \; i \in V_1, s \in S \tag{3.57}$$

$$x_{ijs} \geq 0, \; i \in V_1, j \in V_2, s \in S \tag{3.58}$$

$$y_i \in \{0, 1\}, \; i \in V_1. \tag{3.59}$$

The objective function (3.55) represents the sum of the location costs and of the expected values of the transport costs over all scenarios, while constraints (3.56) and (3.57) extend the constraints (3.19) and (3.20) of the deterministic model to the stochastic case.

The model (3.55)–(3.59) belongs to the family of two-stage stochastic programming problems with integer variables and is known to be NP-hard. As a rule, the need to use a large number of scenarios makes the size of this model fairly large.

With reference to the Milatog location problem, we suppose that the company would like to reorganize its distribution logistics and to relocate its current silos. Relocating should take about six months and increase the maximum daily throughput of the silos by 10%. The relocating costs (in €), on a four-yearly basis, are estimated to be 428 000, 456 900, 526 400, 558 000, 496 000 and 542 000 for the six potential silos, respectively. They affect forage distribution from the silos to the farms for the three and a half years successive to the restructuring, that is, for $365 \times 3 + 183 = 1\,268$ days.

The forage demand of the farms cannot be estimated with certainty. However, the logistician thinks it plausible to imagine five ($|S| = 5$) equiprobable scenarios ($p_s = 0.2, s \in S$), identified by using a forecasting method. Each scenario is characterized by a daily average demand for each farm to be supplied, as shown in Table 3.26.

Table 3.26 Average daily demand scenarios (in quintals) for the Milatog problem.

	Scenario				
Farm	1	2	3	4	5
1	36	38	36	34	36
2	42	44	40	42	44
3	34	36	30	32	32
4	50	50	54	54	48
5	27	25	25	27	29
6	30	28	30	34	27
7	43	41	51	39	46

The costs $c_{ijs}, i \in V_1, j \in V_2, s \in S$, are: $c_{ijs} = c_{ij}, i \in V_1, j \in V_2, s \in S$. The model (3.55)–(3.59) is as follows:

$V_1 = \{1, 2, 3, 4, 5, 6\}$ is the set of the potential sites;

$V_2 = \{1, 2, 3, 4, 5, 6, 7\}$ is the set of farms;

$S = \{1, 2, 3, 4, 5\}$ is the set of scenarios;

$f_1 = 428\,000/1\,268 =$ € 334.90 (similarly, we can easily obtain the other costs $f_i, i = 2, \dots, 6$);

the daily average demand $d_{ijs}, i \in V_1, j \in V_2, s \in S$, varies for each scenario and the corresponding values can be obtained from Table 3.26;

$q_1 = 80 \times (1 + 0.1) = 88$ (an identical calculation is applied for computing the other capacities of the silos $q_i, i = 2, \dots, 6$, which are all increased by 10%).

We denote by $y_i, i = 2, \ldots, 6$, the binary decision variable associated with each potential site i (equal to 1 if the silo i is purchased and renovated by Milatog, 0 otherwise) and with $x_{ijs}, i = 2, \ldots, 6, j = 1, \ldots, 7, s = 1, \ldots, 5$, the decision variable corresponding to the fraction of the average daily quantity requested by farm j satisfied by silo i whenever scenario s occurs.

The problem can hence be formulated as:

$$\text{Minimize } 0.2(83.16x_{111} + \cdots + 48.59x_{671}) + \cdots +$$
$$+ 0.2(83.16x_{113} + \cdots + 48.59x_{673}) +$$
$$+ 0.2(83.16x_{114} + \cdots + 48.59x_{674}) +$$
$$+ 0.2(83.16x_{115} + \cdots + 48.59x_{675}) +$$
$$+ 334.90y_1 + 357.51y_2 + 411.89y_3 + 436.62y_4 +$$
$$+ 388.11y_5 + 424.10y_6$$

subject to

$$x_{111} + x_{211} + x_{311} + x_{411} + x_{511} + x_{611} = 1$$
$$\ldots$$
$$x_{115} + x_{215} + x_{315} + x_{415} + x_{515} + x_{615} = 1$$
$$\ldots$$
$$x_{171} + x_{271} + x_{371} + x_{471} + x_{571} + x_{671} = 1$$
$$\ldots$$
$$x_{175} + x_{275} + x_{375} + x_{475} + x_{575} + x_{675} = 1$$
$$36x_{111} + 42x_{121} + 34x_{131} + 50x_{141} +$$
$$+ 27x_{151} + 30x_{161} + 43x_{171} = 88y_1$$
$$\ldots$$
$$36x_{115} + 44x_{125} + 32x_{135} + 48x_{145} +$$
$$+ 29x_{155} + 27x_{165} + 46x_{175} = 88y_1$$
$$\ldots$$
$$36x_{611} + 42x_{621} + 34x_{631} + 50x_{641} +$$
$$+ 27x_{651} + 30x_{661} + 43x_{671} = 99y_6$$
$$\ldots$$
$$36x_{615} + 44x_{625} + 32x_{635} + 48x_{645} +$$
$$+ 29x_{655} + 27x_{665} + 46x_{675} = 99y_6$$

$$x_{ijs} \geq 0, i = 1, \ldots 6, j = 1, \ldots, 7, s = 1, \ldots, 5$$

$$y_i \in \{0, 1\}, i = 1, \ldots, 6.$$

The optimal solution amounts to renovate silos 1, 2 and 5, with an expected overall cost equal to € 1 613.98. The forage demand allocation of the farms to the silos depends on the scenario that will occur in the subsequent six months after the renovation of the facilities, as shown in Table 3.27.

Table 3.27 Demand allocation scenarios to the silos in the Milatog problem.

		Farm						
Scenario	Activated silo	1	2	3	4	5	6	7
	1	1	0	0	0	0	9/30	1
1	2	0	1	1	0	0	0	0
	5	0	0	0	1	1	21/30	0
2	1	1	0	0	0	0	9/28	1
	2	0	1	1	0	0	0	0
	5	0	0	0	1	1	19/28	0
3	1	1	0	0	0	0	1/30	1
	2	0	1	1	0	0	0	0
	5	0	0	0	1	1	29/30	0
4	1	1	0	0	0	0	15/34	1
	2	0	1	1	0	0	0	0
	5	0	0	0	1	1	19/34	0
5	1	1	0	0	0	0	6/27	1
	2	0	1	1	0	0	0	0
	5	0	0	0	1	1	21/27	0

As can be seen, changing the scenario leads to a different allocation of the forage flow between silos and farms, in particular for farm 6. For example, if scenario 3 occurred (average daily demand of 30 quintals of forage), only one quintal would be supplied by silo 1, whereas, if scenario 4 occurred, 15 out of the 34 required quintals would be supplied by silo 1.

3.6 Case study: Container warehouse location at Hardcastle

Hardcastle is a North European leader in intermodal transport. In the last year, the company operated nearly 240 000 containers, with an annual transport cost of about 50 million euros. Like other intermodal transport companies, Hardcastle manages both full and empty containers. When a customer places an order for freight transport, Hardcastle sends one or several empty containers of the appropriate type in terms of size and characteristics (refrigeration etc.). The containers are then loaded and sent empty. At the destination point, the containers are emptied and sent back to the company unless there is an outgoing load requiring the same kind of container (corresponding to compensation between the demand and the supply of empty containers). Compensation allows reducing the operating costs but, in practice, this is not always the case because

- the origin-destination demand matrix is strongly asymmetrical (some locations are mainly sources of materials while some others are mainly points of consumption);
- at a given location, the demand and supply for empty containers do not usually occur at the same time;
- containers may have a large number of sizes and features; as a result, it is unlikely that the containers incoming at a customer facility are suitable for outgoing goods.

To avoid surpluses or shortages of containers at the customer locations, the company has to provide a periodic distribution of the empty containers. Their operational modes are affected by the particular costs structure. The fixed costs are usually low (because the warehouses are often managed by a public administration); consequently, the location decision is typically reversible in the medium term. The transport costs along a route are an increasing and concave function of the number of containers (empty and full); therefore, the company prefers to send in one single route the loads whose origins and destinations are adjacent (consolidation), instead of sending single loads (see Figure 3.11).

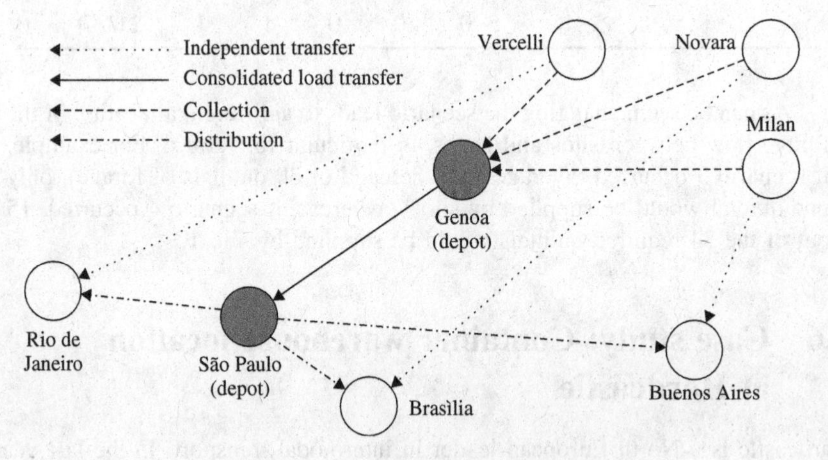

Figure 3.11 Freight transport at Hardcastle.

This last characteristic prohibits the direct distribution of the empty containers among the customers. The empty containers are stored in regional warehouses, usually sited next to ports and railways. In Figure 3.12, this distribution logistics is described in reference to the case of a manufacturing company of Pavia, operating in sending a load to an American customer located in Boston.

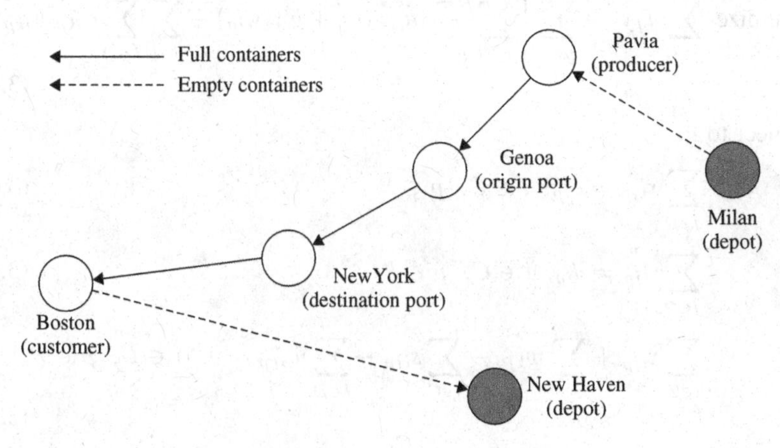

Figure 3.12 Transfer of a container at Hardcastle.

The management of the empty containers is a complex decision-making process made up of two stages:

- at a tactical level, one has to determine, on the basis of forecasted origin-destination transport demands, the number and locations of warehouses, as well as the expected container flows among warehouses;
- at an operational level, shipments are scheduled and vehicles are dispatched on the basis of the orders collected and of short-term forecasts.

Before its redesign, the company operated 87 depots (23 close to a sea terminal) and the empty containers' movements accounted for nearly 40% of the total freight traffic. In order to redesign its logistics system, Hardcastle aggregated its customers into 300 demand points (defined in the following as customers). Let C be the set of customers, D the set of potential depots, P the set of different types of containers, f_j, $j \in D$, the fixed cost of depot j, a_{ijp}, $i \in C$, $j \in D$, $p \in P$, the transport cost of a container of type p from customer i to depot j; b_{ijp}, $i \in C$, $j \in D$, $p \in P$, the transport cost of a container of type p from depot j to customer i; c_{jkp}, $j \in D$, $k \in D$, $p \in P$, the transport cost of an empty container of type p from depot j to depot k; d_{ip}, $i \in C$, $p \in P$, the number of containers of type p requested by the customer i, and o_{ip}, $i \in C$, $p \in P$, the supply of containers of type p from customer i. Furthermore, let y_j, $j \in D$, be a binary decision variable equal to 1 if the depot j is selected, and 0 otherwise; x_{ijp}, $i \in C$, $j \in D$, $p \in P$, the flow of empty containers of type p from customer i

to depot j; s_{ijp}, $i \in C$, $j \in D$, $p \in P$, the flow of empty containers of type p from depot j to customer i, and w_{jkp}, $j \in D$, $k \in D$, $p \in P$, the flow of empty containers of type p from depot j to depot k. The problem was formulated as follows:

$$\text{Minimize} \sum_{j \in D} f_j y_j + \sum_{p \in P} \left[\sum_{i \in C} \sum_{j \in D} (a_{ijp} x_{ijp} + b_{ijp} s_{ijp}) + \sum_{j \in D} \sum_{k \in D} c_{jkp} w_{jkp} \right] \tag{3.60}$$

subject to

$$\sum_{j \in D} x_{ijp} = o_{ip}, \; i \in C, \; p \in P \tag{3.61}$$

$$\sum_{j \in D} s_{ijp} = d_{ip}, \; i \in C, \; p \in P \tag{3.62}$$

$$\sum_{i \in C} x_{ijp} + \sum_{k \in D} w_{kjp} - \sum_{i \in C} s_{ijp} - \sum_{k \in D} w_{jkp} = 0, \; j \in D, \; p \in P \tag{3.63}$$

$$\sum_{p \in P} \sum_{i \in C} (x_{ijp} + s_{ijp}) + \sum_{p \in P} \sum_{k \in D} (w_{jkp} + w_{kjp}) \leq y_j \sum_{p \in P} \sum_{i \in C} (o_{ip} + d_{ip} + 2M), \; j \in D \tag{3.64}$$

$$x_{ijp} \geq 0, \; i \in C, \; j \in D, \; p \in P \tag{3.65}$$

$$s_{ijp} \geq 0, \; i \in C, \; j \in D, \; p \in P \tag{3.66}$$

$$w_{jkp} \geq 0, \; j \in D, \; k \in D, \; p \in P \tag{3.67}$$

$$y_j \in \{0, 1\}, \; j \in D, \tag{3.68}$$

where M in constraints (3.64) is an upper bound on the w_{jkp}, $j \in D$, $k \in D$, $p \in P$. The objective function (3.60) is the sum of warehouse fixed costs and empty container variable transport costs (between customers and warehouses, and between pairs of warehouses). Constraints (3.61)–(3.63) impose empty container flow conservation. Constraints (3.64) state that if $y_j = 0$, $j \in D$, then the incoming and outgoing flows from site j are equal to zero. Otherwise, constraints (3.64) are not binding since

$$x_{ijp} \leq o_{ip}, \; i \in C, \; j \in D, \; p \in P,$$

$$s_{ijp} \leq d_{ip}, \; i \in C, \; j \in D, \; p \in P,$$

$$w_{jkp} \leq M, \; j \in D, \; k \in D, \; p \in P.$$

The implementation of the optimal solution of model (3.60)–(3.67) yielded a reduction in the number of warehouses to 48 and a 47% reduction in transport cost.

3.7 Case study: The organ transplantation location-allocation policy of the Italian National Transplant Centre

In Italy the institution of reference for the organ-sharing system is the National Transplant Centre (CNT), managed by the Ministry of the Health. CNT ensures the coordination of all the structures in which the organs are explanted and transplanted and, in particular, three inter-regional centres (CIR): the North Italy Transplantation (NITp), which includes Friuli-Venezia Giulia, Liguria, Lombardy, Marche, Veneto and the independent Province of Trento, the Inter-Regional Association of Transplantation (AIRT), which includes Piedmont, Aosta Valley, Tuscany, Emilia-Romagna, Apulia and the independent Province of Bolzano, and the Centre-South Organization of Transplantation (OCST), which includes Abruzzo, Basilicata, Umbria, Campania, Lazio, Calabria, Molise, Sardinia and Sicily.

The transplant waiting list management uses a prioritization scheme, on the basis of the degree of histological compatibility. When an organ becomes available within a region, it will be offered first to the regional waiting list, then to the corresponding CIR waiting lists, and finally to the nation at large. Under this perspective, an organ is a 'local resource', because it is considered as 'produced' in the region in which becomes available: the concept of local primacy is respected. For example, if an organ becomes available in Apulia (a region belonging to the AIRT) and there are no compatible patients in this region, it is then offered to the patient having the highest priority on the AIRT waiting list. An eligible patient living in Piedmont will have to sustain a high travelling cost and, above all, a higher risk, since the likelihood of a successful transplant strongly depends on the organ cold-ischaemia time. A patient waiting in Calabria (a region belonging to OCST) would have benefited more from the transplant because of the better condition of the harvested organ. The current CIR distribution affects the organs allocation policy and produces unexpected and unwanted consequences on the patient selection.

Recently, a more efficient organization of the Italian organ-sharing system for three different organs (heart, liver and kidney) has been proposed by the resolution, for each organ, of a mathematical model for the transplant centre's location and the allocation of the organs to the patients in the waiting lists.

Let V_1 and V_2 be, respectively, the set of explantation centres and the set of potential transplant centres, p of which have to be opened. Each explantation centre will refer to one transplant centre. Moreover, let K be the set of 'demand points', which aggregate potential patients of a specific geographic area. For each demand point $k \in K$ the annual demand d_k, that is, the number of patients on the waiting list, is assumed known in advance.

Both the explantation centres and the demand points are organized for each province (105), while 52 potential locations for transplant centres have been selected. Each transplant centre is assumed to have a circular covering area, with radius r which depends on the cold-ischaemia time of the organ.

Moreover, let a_{ij}, $i \in V_1$, $j \in V_2$, be the aerial distance between the explantation centre i and the transplant potential centre j (transport of explanted organ is usually performed by means of an emergency helicopter); let b_{kj}, $k \in K$, $j \in V_2$, be the terrestrial distance between the demand point k and the transplant potential centre j. We can define T_k, $k \in K$, as the set of all those candidates that are within an acceptable distance of the demand point k, that is, $T_k = \{j \in V_2 : b_{kj} \leq r\}$.

The binary decision variables are the following: x_{ij}, $i \in V_1$, $j, j \in V_2$, that is 1 if explantation centre i is assigned to the transplant centre j, 0 otherwise; y_{kj}, $k \in K$, $j \in V_2$, that is 1 if the demand point k is assigned to the transplant centre j, 0 otherwise; z_j, $j \in V_1$, that is 1 if transplant potential centre j is activated, 0 otherwise.

The location-allocation model has been formulated as a multiobjective model as follows:

$$\text{Minimize} \sum_{i \in V_1} \sum_{j \in V_2} a_{ij} x_{ij} \tag{3.69}$$

$$\text{Minimize} \sum_{k \in K} \sum_{j \in T_k} d_k b_{kj} y_{kj} \tag{3.70}$$

$$\text{Minimize } M \tag{3.71}$$

subject to

$$\sum_{j \in V_2} x_{ij} = 1, i \in V_1 \tag{3.72}$$

$$\sum_{j \in T_k} y_{kj} = 1, k \in K \tag{3.73}$$

$$x_{ij} \leq z_j, i \in V_1, j \in V_2 \tag{3.74}$$

$$y_{kj} \leq z_j, k \in K, j \in V_2 \tag{3.75}$$

$$\sum_{j \in V_2} z_j = p \tag{3.76}$$

$$M \geq \sum_{k \in K} d_k y_{kj}, j \in V_2 \tag{3.77}$$

$$x_{ij} \in \{0, 1\}, i \in V_1, j \in V_2 \tag{3.78}$$

$$y_{kj} \in \{0, 1\}, k \in K, j \in V_2 \tag{3.79}$$

$$z_j \in \{0, 1\}, j \in V_2. \tag{3.80}$$

The objective function (3.69) represents the total distance between explantation centres and transplant centres that have been activated; objective function (3.70) is the total distance between demand points and transplant centres, weighted with the annual demand levels; objective function (3.71) is the number of patients in the longest waiting list. The minimization of M aims at balancing the allocation of demand points to the transplant centres (as observed in Section 3.1, for logistic facilities in the public utility sector the presence of equal conditions for all the users is mandatory).

The three objectives can be conflicting. Thus, for the solution of model (3.69)–(3.80), a reasonable trade-off has been imposed, by identifying a priority criterion among the objectives and defining a unique objective function as their weighted sum, using weights that are proportional to the assigned priority levels.

Constraints (3.72) guarantee that each explantation centre $i \in V_1$ can be assigned to just one transplant centre $j \in V_2$, while constraints (3.73) impose that each demand point $k \in K$ can be associated to just one transplant centre $j \in V_2$. Conditions (3.74) and (3.75) are linking constraints among decision variables and, therefore, restrict the assignment of explantation centres and demand points to transplant centres that are really activated; equation (3.76) imposes the activation of exactly p transplant centres. Constraints (3.77), together with the minimization of the objective function (3.71), ensure that the term M is equal to the number of patients on the longest waiting list.

The solution of model (3.69)–(3.80) yields for the Italian Ministry of Health a more efficient and effective organ-sharing system, for heart, liver and kidney. For example, in the case of heart transplant, a new configuration has been proposed by the model, keeping the number of activated transplant centres constant at 24: nine existing transplant centres should be closed and nine new centres should be opened in different locations. The new configuration allows a reduction of about the 18% of the overall distance between the explantation centres and the transplant centres, as well as a 24% reduction of the distance between the demand points and the transplant centres. It also yields a better balance of the size of the waiting lists of the different transplant centres.

3.8 Questions and problems

3.1 Illustrate why crude oil refineries are customarily located near home-heating and automotive fuel markets.

3.2 Conforge is one of the world's leading casual shoes producers. Its authorized dealership operating in Switzerland has recently decided to open a retail outlet in Zurich. Four existing potential commercial areas have been selected. The location factors considered are: potential market, size of the area, variety of the products sold, estimated position in the area, presence of competing stores, proximity of complementary shops, presence of residential area and workplaces within one km, pedestrian and vehicular traffic and presence of free parking. The potential market is expressed as the quantity of shoes sold to the resident population in the territory, during one year. The position in the area has been evaluated both qualitatively and quantitatively by considering whether the retailer outlet is located in: (a) a primary area (i.e. the area with an amount of customers between 55% and 70% of the total), (b) a secondary area (near the primary area but with a greater distance and amount of customers approximately equal to 15–20%) or (c) a marginal area (the farthest from the store and such that the amount of customer is approximately 5–10% of the total). A group of seven experts has assigned a weight to each of the eight location factors (see Table 3.28) and the scores to the four potential sites for each location factor (see Table 3.29). By applying the weighted scoring method, determine the commercial area to be selected.

Table 3.28 Weights associated to each location factor in the Conforge problem.

ID	Location factor	Weight
1	Potential market	0.20
2	Size of the area	0.20
3	Variety of the products sold	0.15
4	Estimated position in the area	0.14
5	Presence of competing stores	0.11
6	Proximity of complementary shops	0.07
7	Presence of houses and workplaces within 1 km	0.07
8	Pedestrian and vehicular traffic	0.03
9	Presence of free parking	0.03
		1.00

Table 3.29 Performance scores of the four potential commercial areas in the Conforge problem.

	Score			
Location factor	Site 1	Site 2	Site 3	Site 4
1	6	6	7	8
2	7	8	7	9
3	8	7	8	6
4	9	5	6	7
5	6	6	7	7
6	5	8	7	6
7	7	7	8	7
8	7	8	5	8
9	8	7	8	8

3.3 The Ring Offshore company, in the North Sea, has deployed 13 drilling platforms to be connected through a network of pipelines with a hub assembly whose best position can be determined by solving a SCSE continuous location problem. To this end, the Cartesian coordinates and the average daily flow rate of each drilling platform are reported in Table 3.30. Solve the problem by using the Weiszfeld heuristic (with at least three iterations).

Table 3.30 Coordinates and average flow rates of the drilling platforms in the North Sea.

Drilling platform	Abscissa (km)	Ordinate (km)	Flow rate (quintals/day)
1	0.00	8.54	198
2	11.56	0.00	191
3	5.58	12.45	212
4	22.60	11.35	279
5	8.88	17.38	205
6	37.28	21.56	230
7	12.72	18.65	278
8	35.65	5.42	198
9	9.27	24.32	226
10	25.56	34.54	188
11	15.87	28.55	244
12	31.53	10.12	215
13	29.72	31.40	248

3.4 Describe how fixed costs f_i, $i \in V_1$, in the CPL model should be computed in order to take labour costs, property taxes and site developments costs into account.

3.5 Your company has to close 20 out of its 125 warehouses. Suppose the SCSE hypotheses hold. How would you define V_1? What is the value of p?

3.6 Modify the CPL model to take into account the fact that a subset of already existing facilities $V_1' \subseteq V_1$ cannot be closed (but can be upgraded). Indicate the current fixed cost and maximum throughput of facility $i \in V_1'$ as f_i' and q_i', respectively. Denote f_i'' and q_i'' by the fixed cost and the maximum throughput if facility $i \in V_1'$ is upgraded, respectively.

3.7 Borachera is a major Spanish wine wholesaler currently operating two CDCs in Salamanca and Albacete, and a number of RDCs all over the Iberian peninsula. In order to reduce its overall logistics cost, the company wishes to redesign its distribution logistics system by replacing its current RDCs with three (possibly new) RDCs. Based on a preliminary qualitative analysis, an RDC should be located in the Castilla-Leon region, in Valladolid, Burgos or Soria. A second RDC should be located in the

Extremadura region, in Badajoz, Plasencia or Caceres. Finally, the third RDC should be located in the Argon region, in Barbastro, Saragossa or Teruel. Transport costs from RDCs to retailers are charged to retailers. Formulate the Borachera problem as an SCSE discrete location model.

3.8 As illustrated in Section 3.3.2, when designing a distribution logistics system it is customary to remove excessively long transport links from the model in order to allow for a timely delivery to customers. How should the CPL Lagrangian heuristic be modified in this case?

3.9 Labro is a Portuguese producer of olive oil. It is interested in establishing two new tasting stands for its products in hypermarkets located in the Central Alentejo region. From a preliminary analysis, the hypermarkets available to host them are in the following locations (one tasting stand per hypermarket): Alandroal, Borba, Évora, Mourão, Portel, Redondo, Sousel, Vendas Novas, Viana do Alentejo and Vila Viçosa. The oil mill plant has three warehouses, which are located in the towns of Borba, Mourão and Redondowhich, respectively, and can be used for distributing the oil products to the two tasting stands. The following data are available:

- f_j costs (in €/year) for opening and managing tasting stand j;
- average daily amount of oil r_j (expressed as a number of packages of 12-litre bottles) needed to supply the tasting stand if it is located at hypermarket j;
- average daily availability d_i of oil (for 12-litre bottles) at warehouse i.

Formulate the location problem by considering that: (a) each tasting stand will be supplied with a daily door-to-door connection, (b) the average daily transport cost grows linearly with the number of cases of 12-litre oil bottles transported and (c) 220 working days should be considered per year.

3.10 Coffa is a Sicilian distributor of canned tuna. It is interested in opening a new warehouse in the province of Siracusa. Avola and Buccheri are the two potential locations of the warehouse, shown by a preliminary analysis. If the warehouse is located in Avola, the annual total facility cost is € 127 500, whereas if it is located in Buccheri, this cost is € 123 000. The goal is to supply, four days a week, all dealers located in the following cities: Syracuse, Augusta, Avola, Buccheri, Ferla, Lentini, Melilli, Noto, Pachino and Priolo. Due to the limited capacity of the vehicles, transport is carried out with a single trip from the warehouse to each of the cities listed above (door-to-door transport). The distances (in km) between the cities are listed in Table 3.31. Formulate and solve the location problem by assuming 200 working days in a year and a transport cost per km equal to € 0.95. Furthermore, assume that it is possible to determine the warehouse location in an alternative way, by using the centre of gravity. To this end, use the Cartesian coordinates reported in the last two

Table 3.31 Distances (in km) between the 10 cities and Avola and Buccheri, respectively, as well as their Cartesian coordinates, in the Coffa problem.

	Avola	Buccheri	Abscissa	Ordinate
Siracusa	25	55	7.8	7.7
Augusta	56	54	6.6	11.4
Avola	0	57	5.2	4.3
Buccheri	57	0	0.0	9.0
Ferla	52	12	1.7	8.9
Lentini	67	35	2.8	12.6
Melilli	45	36	5.0	10.2
Noto	9	45	4.0	3.9
Pachino	26	69	4.3	0.0
Priolo	38	59	6.1	9.6

columns of Table 3.31. According to this approach, the optimal warehouse location is that with the minimum Euclidean distance from the centre of gravity among the 10 cities (one can restrict the selection to the best three candidate cities by visually inspecting Figure 3.13).

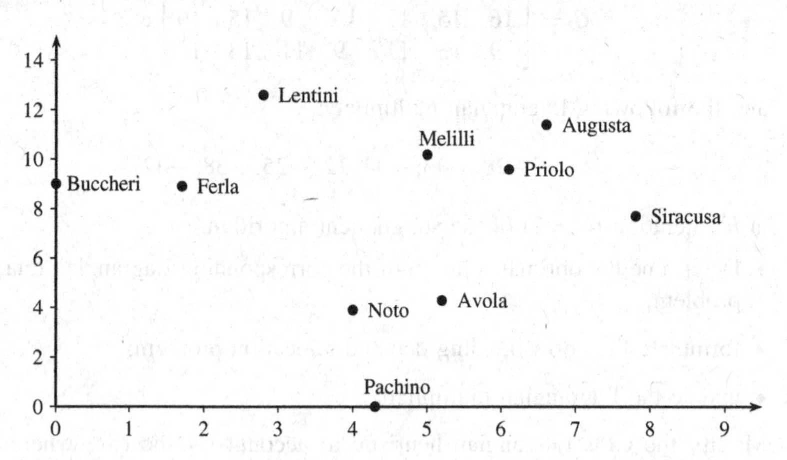

Figure 3.13 Location of the 10 cities in the Coffa problem.

3.11 Consider the following SCSE discrete location problem:

$$\text{Minimize} \sum_{i \in V_1} \sum_{j \in V_2} c_{ij} x_{ij} + \sum_{i \in V_1} f_i y_i$$

subject to

$$\sum_{i \in V_1} x_{ij} = 1, \; j \in V_2$$

$$\sum_{j \in V_2} d_j x_{ij} \leq q_i y_i, \; i \in V_1$$

$$\sum_{i \in V_1} y_i = 2$$

$$x_{ij} \geq 0, i \in V_1, \; j \in V_2$$

$$y_i \in \{0, 1\}, \; i \in V_1,$$

where $|V_1| = 3$, $|V_2| = 7$, $f = [132, 138, 147]^T$, $q = [1\,500, 1\,300, 2\,800]^T$, and $d = [52, 67, 88, 47, 91, 45, 68]^T$. Determine the optimal solution of this location problem (hint: observe that $q_i \geq \sum_{j \in V_2} d_j, i \in V_1$).

3.12 Extend the CPL model to the case of demand varying over the planning horizon. Assume that, once opened, a facility cannot be closed.

3.13 Consider the following CPL formulation: $|V_1| = 3$, $|V_2| = 7$, $f = [44, 46, 21]^T$, $q = [220, 100, 240]^T$ and $d = [72, 80, 68, 45, 58, 68, 60]^T$ and

$$C = \begin{bmatrix} 12 & 13 & 11 & 13 & 19 & 15 & 11 \\ 16 & 15 & 17 & 13 & 9 & 15 & 19 \\ 9 & 13 & 11 & 9 & 11 & 13 & 13 \end{bmatrix}.$$

and the following Lagrangian multipliers:

$$\lambda^{(h)} = [-28, -43, -34, 12, -25, -38, -42]^T$$

at h^{th} iteration ($h = 2$) of the subgradient algorithm.

- Determine the optimal solution of the corresponding Lagrangian relaxed problem;
- formulate the corresponding demand allocation problem;
- update the Lagrangian multipliers.

3.14 Modify the CPL Lagrangian heuristic to account for the case where the demand is indivisible (i.e. the demand of any successor node must be satisfied by a single facility). Is the modified heuristic still time-polynomial? How can be one determine whether an instance of the modified problem is feasible?

3.15 Formulate a polling station location problem, taking into account the following binding constraints: (a) the number of polling stations is fixed for each municipality and calculated according to the number of resident voters, (b) the number of voters assigned to each polling station may not exceed given lower and upper bounds and (c) the suitability of the potential sites is established by specific safety measures on the accessibility and typology of the buildings (e.g. in some countries, only public buildings, such as schools, are eligible). Other soft constraints impose the removal of any difficulty in going to the polls. This typically means that almost the same number of voters is assigned to each polling station, with the aim of limiting possible troubles and disservice (e.g. queues of voters or delays in communicating the vote-counting results). Another important requirement, generally handled as a goal to pursue rather than a constraint, is the minimization of the total distance covered by voters to reach their respective polling stations.

3.16 Modify the MCTE discrete location model to account for the case in which a successor node can be supplied by a single facility.

3.17 In the location problem of Coimbra (see Section 3.3.5), assume that the average travel speed is 55 km/h and that the annual cost of the seventh urban area is € 1 700 000 instead of € 2 200 000. Solve the related SC problem by the heuristic illustrated in Section 3.3.5. Compare the solution with the optimal solution obtained by using an integer programming (IP) solver. What can you say about the quality of the heuristic solution? How does the solution change if an emergency fire station has to be located in the seventh urban area?

3.18 A consortium of municipalities in Romania has decided to locate a first aid centre in a given area. Each intersection of the road network of the interested area is associated with a vertex of the set $V = \{1, 2, 3, 4, 5, 6, 7, 8, 9\}$ and each road segment is associated with an edge of the set $E = \{(1, 2), (1, 7), (2, 3), (2, 6), (3, 5), (3, 7), (3, 9), (4, 5), (4, 8), (4, 9), (5, 6), (6, 8), (7, 9), (8, 9)\}$. The traversal times a_{ij} for each edge $(i, j) \in E$ are shown in Table 3.32, while the shortest transfer travel time between i and j, for each pair of vertices $i, j \in V$, are reported in Table 3.33. The average speed on the road segments can be assumed equal to 60 km/h. Assuming that the first aid centre should be located on the edge $(3, 9)$, solve the 1-centre problem by using the Hakimi algorithm.

Table 3.32 Traversal time (in minutes) of the edges of the 1-centre problem in Romania.

(i, j)	a_{ij}	(i, j)	a_{ij}	(i, j)	a_{ij}
(1,2)	5	(3,7)	4	(5,6)	4
(1,7)	7	(3,9)	7	(6,8)	12
(2,3)	4	(4,5)	5	(7,9)	8
(2,6)	11	(4,8)	5	(8,9)	7
(3,5)	6	(4,9)	3		

3.19 Assume that in the Conforge problem (see Problem 3.2), the annual costs (in €) for renting the appropriate space in the four commercial areas are known and equal to 29 520, 39 312, 50 760 and 22 800, respectively. Select

Table 3.33 Shortest transfer travel time (in minutes) for any pair of vertices of the 1-centre problem in Romania.

	i								
j	1	2	3	4	5	6	7	8	9
1	0	5	9	18	15	16	7	22	15
2	5	0	4	14	10	11	8	18	11
3	9	4	0	10	6	10	4	14	7
4	18	14	10	0	5	9	11	5	3
5	15	10	6	5	0	4	10	10	8
6	16	11	10	9	4	0	14	12	12
7	7	8	4	11	10	14	0	15	8
8	22	18	14	5	10	12	15	0	7
9	15	11	7	3	8	12	8	7	0

the commercial area in which to locate the retail outlet by using the hybrid method with $\alpha = 0.2$. Does the solution change if we set $\alpha = 0.8$?

3.20 Consider the p-median problem of United Bank (see Section 3.3.2). Assume that the transport cost from each potential site to any district centroid is proportional to the product of the distance and the number of the customers in the district. Under the hypothesis that the number of customers in each district is not known in advance, but can assume three possible values, according to a given probability of occurrence, formulate a stochastic version of the p-median problem in which the decision variables $x_{ij}, i \in V_1, j \in V_2$, are scenario-dependent. (Hint: the total number of scenarios equal to 3^8.)

4

Selecting the suppliers

4.1 Introduction

In logistics systems planning, an increasing attention is paid to the provision of materials and services. Supplier management is a very important issue affecting the performances of a company, in terms of both profits and service levels offered to its customers. In order to guarantee an adequate provision of materials and services, companies usually evaluate different alternatives for each provision, not only on the basis of its type but also with respect to the relationship established with each supplier. It is indeed convenient to establish well-structured relationships with several suppliers (also with potential ones) even for the same product or service, with the aim of monitoring their efficiency and deciding periodically whether to renew existing contracts with them. Having a single supplier for a specific product or service (a *single-source*, SS, policy), or more than one (*multiple-source*, MS), yields different advantages and disadvantages. The first solution has the advantage of encouraging the supplier to carry on targeted investments, thus contributing to the development of a useful cooperation and simplifying the integration of the logistics processes between the company and the supplier (see Section 1.4). On the other hand, this solution increases the risk of dependence on the supplier and does not allow the company to make comparisons with alternative sources. The second solution reduces the risk of dependence and allows benefits resulting from competition. However, coordinating several suppliers can be rather complex.

Introduction to Logistics Systems Management, Second Edition. Gianpaolo Ghiani, Gilbert Laporte and Roberto Musmanno.

Supplier management procedures are made up of the following main steps:

1. definition of the set of potential suppliers;
2. definition of the selection criteria;
3. supplier selection.

These phases will be examined in detail in the following sections.

4.2 Definition of the set of potential suppliers

The search for suppliers is performed either when a company does not yet have a portfolio of suppliers (e.g. when entering a new market) or when it is about to renew its current pool of suppliers. Various sources of information can be used when searching for suppliers:

- *specialized journals*. Articles and ads published by specialized magazines can be used to collect useful information to compare alternative suppliers;
- *search engines*. The web contains plenty of information on alternative suppliers;
- *trade fairs*. These constitute a good opportunity to meet potential suppliers and analyze their offer of products and services;
- *organized meetings*. This amounts to meeting with suppliers' representatives, in order to negotiate prices, lead times and other relevant aspects.

Inspection visits to potential suppliers' plants can also be useful.

4.3 Definition of the selection criteria

Defining the selection criteria is certainly the most critical phase of the decision-making process, because these and their relative weights determine the selected suppliers. Once identified, the selection criteria are recorded on the supplier file. This file generally contains the classical information records about the supplier (company name, average business volume, kind of activity, certifications etc.). The choice of selection criteria should correspond to the company's strategies and also depends on the kind of market in which the company operates. Thus, it is not possible to define a generally applicable list of selection criteria. However, the specialized literature contains several lists of criteria which may be refined for the specific application context. A study of particular interest is that conducted by G.W. Dickson, which identifies 23 different evaluation criteria and weights (see Table 4.1). It is based on a survey carried out with 273 American managers belonging to the National Association of Purchasing Managers (NAPM). The factors that were considered the most critical are: (a) product quality, (b) contract terms for delivery, (c) performance history and (d) guarantee terms.

Table 4.1 Supplier selection criteria according to G.W. Dickson.

Classification	Criterion	Weight
1	Quality	3.508
2	Contract terms for delivery	3.147
3	Performance history	2.998
4	Guarantee terms	2.849
5	Structural and manufacturing capacity	2.775
6	Cost	2.758
7	Technical capacity	2.545
8	Financial position	2.514
9	Conformity to the procedures	2.488
10	Communication system	2.426
11	Reputation	2.412
12	Business attractiveness	2.256
13	Management and organization	2.216
14	Operative controls	2.211
15	Assistance service	2.187
16	Attitude	2.120
17	Impression	2.054
18	Packaging ability	2.009
19	Ended-works reports	2.003
20	Geographic position	1.872
21	Total of ended business	1.597
22	Training aids	1.537
23	Reciprocal agreements	0.610

More recently, T.Y. Choi and J.L. Hartley have studied supplier selection criteria in the auto industry. They have identified 23 criteria divided into eight main categories (see Table 4.2).

In conclusion, it can be stated that the most relevant selection criteria are: (a) purchase price and quality, specifically related to the product or service considered, and (b) delivery efficiency, flexibility, financial capabilities and reputation, related to an evaluation of the potential supplier rather than to the specific product or service to be purchased.

The presence of several selection criteria to be jointly considered forces the decision maker to define their relative weights, with the aim of establishing their impact on the supplier selection. Of course, not all the criteria have the same weight for all companies. A company focusing on the quality of its products will give a higher weight to the quality criterion; on the other hand, a company interested in reducing its inventory costs (e.g. using the just-in-time policy) will pay more attention to the geographic position of its suppliers.

To determine the weight of the different criteria, the following two-step procedure, derived from the *Analytical Hierarchic Process* (AHP), can be used.

Table 4.2 Selection criteria for suppliers (according to T.Y. Choi and J.L. Hartley) and their subdivision into categories.

Category	Criteria
Finances	Financial conditions Profitability Financial information availability Performance awards
Consistency	Product conformity Consistent delivery times Quality philosophy Response times
Relations capacity	Long-term relations Closeness in relations Openness in communications Reputation
Flexibility	Changes in production volumes Reduction of equipping times Reduction of delivery times Resolution of conflicts
Technological capabilities	Design capabilities Technical capabilities
Services	Post-sales assistance Sale representative competence
Reliability	Incremental improvements Product reliability
Price	Initial price

- Let m be the number of selection criteria, arbitrarily sorted, and let A be an $m \times m$ matrix whose entry a_{kj}, $k = 1, \dots, m$, $j = 1, \dots, m$, $k \neq j$, indicates the relative importance of criterion k with respect to criterion j and $a_{jk} = 1/a_{kj}$, $k = 1, \dots, m$, should be put in a formula, that is, $k = 1, \dots, m$, $j = 1, \dots, m$, $k \neq j$ (see Table 4.3). Entries a_{kk}, $k = 1, \dots, m$, of the diagonal are equal to 1.

- For each criterion k, determine the quantity

$$\hat{w}_k = \sqrt[m]{\prod_{j=1}^{m} a_{kj}}, k = 1, \dots, m \tag{4.1}$$

Table 4.3 Standard scale of values associated with a comparison between criteria k and j.

Comparison judgement between criteria k and j	a_{kj}	a_{jk}
k and j are equally important	1	1
k is moderately preferable to j	3	1/3
k is quite preferable to j	5	1/5
k is decidedly preferable to j	7	1/7
k is extremely preferable to j	9	1/9
j is moderately preferable to k	1/3	3
j is quite preferable to k	1/5	5
j is decidedly preferable to k	1/7	7
j is extremely preferable to k	1/9	9

representing the geometric mean of the elements of the kth row of A. The weight of each criterion k is then determined as

$$w_k = \frac{\hat{w}_k}{\sum_{i=1}^{m} \hat{w}_i}, k = 1, \ldots, m. \tag{4.2}$$

The resulting weights from (4.2) are normalized so that

$$\sum_{k=1}^{m} w_k = 1.$$

Suntech Solar is a Mexican company specialized in the assembly of monocrystalline photovoltaic panels. To evaluate its suppliers of monocrystalline photovoltaic cells, Suntech decided to use the following five selection criteria (numbered from 1 to 5): product quality, product price, terms of delivery, financial situation and geographic position. To determine the weight of each criterion, the manager of the purchasing office adopted the standard scale of values shown in Table 4.3, obtaining the judgements shown in the Table 4.4 from the pairwise comparison. The matrix A of dimension 5×5 is therefore

$$A = \begin{bmatrix} 1 & 1 & 5 & 5 & 7 \\ 1 & 1 & 7 & 7 & 7 \\ 1/5 & 1/7 & 1 & 7 & 9 \\ 1/5 & 1/7 & 1/7 & 1 & 1/7 \\ 1/7 & 1/7 & 1/9 & 7 & 1 \end{bmatrix}.$$

Using (4.1), one obtains

$$\hat{w} = [2.8094, 3.2141, 1.1247, 0.2255, 0.4366]^{\mathrm{T}}$$

and, by (4.2), the following normalized vector of weights for the five criteria is deducted:

$$w = [0.3597, 0.4115, 0.1440, 0.0289, 0.0559]^{\mathrm{T}}.$$

The criterion with the greatest weight (41.15%) is the product price, followed by the product quality (with a weight of 35.97%).

4.4 Supplier selection

A large variety of methods can be used for supplier selection, and an exhaustive presentation of them is outside the scope of this textbook. In the following, we will focus on two of the most widespread methods of practical interest.

The weighted scoring method attributes a total score to each supplier, on the basis of the chosen selection criteria. It is the same as the method used for solving discrete location problems (see Section 3.2). Let V be the set of potential suppliers. A weight $w_k \in (0, 1)$ is associated to each selection criterion k on the basis of its importance with respect to the other criteria (note that $\sum_{k=1}^{m} w_k = 1$). Let s_{ik}, $i \in V$, be the score (from 0 to 10) associated with selection criterion k for supplier i. The total score r_i of the supplier $i \in V$ is calculated as

$$r_i = \sum_{k=1}^{m} w_k s_{ik},$$

and it is therefore possible to sort the suppliers according to the non-increasing order of their total score. If only a supplier i^* has to be selected, the one with the highest total score is chosen, that is, $i^* = \arg\max_{i \in V}\{r_i\}$.

In the Suntech Solar problem, an evaluation grid of four suppliers of monocrystalline photovoltaic cells (Table 4.4) is assumed to be available. Using the weighted scoring method, the most preferable supplier is the second one, followed by the fourth (in case of the adoption of a MS policy with two suppliers).

Table 4.4 Evaluation grid of four suppliers of photovoltaic cells for the Suntech Solar plant.

Criterion	Weight	Score			
		Supplier 1	Supplier 2	Supplier 3	Supplier 4
Quality	0.3597	7	9	7	7
Price	0.4115	8	8	9	9
Delivery	0.1440	6	10	7	6
Financial situation	0.0289	5	8	5	9
Geographical position	0.0559	8	6	6	9
	Score	7.27	8.54	7.71	7.85

The supplier selection problem can be also formulated as an optimization model. The scientific literature reveals a large variety of such models. In the sequel, we illustrate one of the simplest and most widely used model in the case of a single commodity, corresponding to a multiobjective LP model.

Let V be the set of potential suppliers; m is the number of identified selection criteria, each of which has a weight $w_k, k = 1, \ldots, m$; $s_{ik}, i \in V, k = 1, \ldots, m$, is the evaluation of supplier i with respect to criterion k; $r_i = \sum_{k=1}^{m} w_k s_{ik}, i \in V$, is the total score of the supplier i; D is the quantity required from the suppliers (total demand) during the planning horizon; c_i, $i \in V$, is the unit purchase price from supplier i and q_i, $i \in V$, is the capacity of supplier i, that is, the maximum quantity that the supplier can provide in the planning horizon.

The decision variables are x_i, $i \in V$, representing the quantity of commodity purchased from supplier i in the planning horizon. The mathematical model is as follows:

$$\text{Minimize} \sum_{i \in V} c_i x_i \tag{4.3}$$

$$\text{Maximize} \sum_{i \in V} v_i x_i \tag{4.4}$$

subject to

$$\sum_{i \in V} x_i = D \tag{4.5}$$

$$0 \leq x_i \leq q_i, i \in V, \tag{4.6}$$

where $v_i = r_i / \max_{k \in V}\{r_k\}, i \in V$. The first objective (4.3) amounts to minimizing the total purchase cost. The second objective (4.4) is to maximize the overall score of the selected suppliers (due to the choice of v_i values, $i \in V$, the objective function (4.4) cannot exceed D). Constraint (4.5) imposes that the whole demand must be satisfied. Constraints (4.6) impose lower and upper bounds on the quantity provided by each supplier.

In order to determine a Pareto optimal solution of problem (4.3)–(4.6), we make use of the ϵ-*constraint method*: one of the two objective functions is transformed into a constraint by bounding its value above (in case of minimization) or below (in case of maximization).

Ilax is a leading Japanese company specialized in silicon products for the communication and electronics industry. The provision of quartz sand, an important raw material in this field, takes place every month. The portfolio of quartz sand suppliers is made up of five companies which were already evaluated using a weighted scoring method on the basis of 18 different criteria. The results of this evaluation are shown in the second column of Table 4.5. The demand of quartz sand for the next month is 6 500 tons.

During the negotiation phase of the provision agreements, each supplier has indicated to the company the maximum quantity of quartz sand available for replenishment during the planning horizon and the unit price offered. These values are shown in the third and fourth columns of Table 4.5, respectively.

Table 4.5 Score, capacity and unit selling price of Ilax's quartz sands suppliers.

Supplier	Score	Capacity (Mg)	Price ($)
1	8.23	3 500	792
2	8.01	4 000	767
3	7.57	4 000	758
4	8.18	4 000	780
5	8.54	2 500	803

The corresponding multiobjective supplier selection model (4.3)–(4.6) is the following:

$$\text{Minimize } 792x_1 + 767x_2 + 758x_3 + 780x_4 + 803x_5$$

$$\text{Maximize } (8.23/8.54)x_1 + (8.01/8.54)x_2 + (7.57/8.54)x_3 + + (8.18/8.54)x_4 + (8.54/8.54)x_5 \quad (4.7)$$

subject to

$$x_1 + x_2 + x_3 + x_4 + x_5 = 6\,500$$

$$0 \leq x_1 \leq 3\,500$$

$$0 \leq x_2 \leq 4\,000$$

$$0 \leq x_3 \leq 4\,000$$

$$0 \leq x_4 \leq 4\,000$$

$$0 \leq x_5 \leq 2\,500.$$

Ilax considers the cost minimization more important than the maximization of the supplier score. For this reason, it was decided to transform the objective function (4.7) into the following additional constraint:

$$0.9637x_1 + 0.9379x_2 + 0.8864x_3 + 0.9578x_4 + 1.0000x_5 \geq \epsilon,$$

with $\epsilon = 6\,500\alpha$, where $\alpha \in [0, 1]$ is set to 0.95. The following Pareto optimal solution is obtained:

$$\bar{x}_1 = 0; \bar{x}_2 = 4\,000; \bar{x}_3 = 0; \bar{x}_4 = 1820.8333; \bar{x}_5 = 679.1667,$$

with a provision cost of \$ 5 033.62.

Model (4.3)–(4.6) deals with a single commodity to be supplied from the suppliers (e.g. corn) in one 'shot' (the quantity ordered is delivered once) and need not be an integer (e.g. 345.76 kg of corn). The model can be extended to consider additional features:

- integer quantities to be supplied (e.g. if the load unit is the pallet and not expressed in kilograms, 450 pallets could be a feasible solution, whereas 450.56 pallets is not correct);
- multicommodity (e.g. corn and, simultaneously, other cereals);
- the quantity ordered can be delivered from the suppliers in different provision periods (e.g. I ordered 100 pallets of a product with a deadline of two weeks: the first 30 pallets arrive at the end of the first week, and the remaining part at the end of the second week.

These cases are proposed as exercises.

4.5 Case study: The system for the selection of suppliers at Baxter

Baxter Healthcare Corporation is a worldwide company operating in the healthcare sector in over 110 countries with over 48 000 employees. The company was founded in 1931 as a manufacturer of intravenous solutions. Baxter provides, through subsidiaries, products and services for the care of patients who are in critical condition: those affected by hemophilia, immune deficiencies, infectious diseases, kidney disease and trauma. In particular, BTT (Baxter Transfusion Therapy), a leader in transfusion medicine for more than 40 years, is a manufacturing company and provider of services for the collection, storage and distribution of blood products. In 2007, BTT's manufacturing facility in San German, Puerto Rico, decided to request quality certification according to US Food and Drug Administration (FDA) policy.

In particular, the FDA approval requires each certified company adopts specific rules for purchasing. These rules require that

- the company should adopt written procedures which specify the requirements that suppliers and consultants must meet;
- the company evaluates and selects potential suppliers and consultants on the basis of their ability to meet specified requirements;
- the necessary control of the product or service to be delivered is properly defined and is based on analytical data evaluation processes;
- there are adequate systems in the company for the registration of suppliers and consultants;
- the company has proper documentation defining the requirements and quality indicators for the goods or services to buy.

To meet these standards, BTT decided to adopt a structured system for the selection of suppliers. The system, briefly described in this section, was based on an SS policy for each product.

Every trimester, BTT prepares an internal report on the performance of each supplier. The report is sent to the vendors in order to let them review it and modify their actions if necessary. For each delivery, the incoming goods are subjected to quality controls, which can yield corrective actions to manage the relationship with the supplier in case of noncompliance, and preventive actions to reduce the risks of noncompliance.

To each qualified supplier $i \in V$ is assigned a quality index, denoted as the *Supplier Quality Index* (*SQI*), which is updated monthly. For each type of product to be purchased, the company maintains a list of suppliers, sorted by increasing values of *SQI*.

The definition of the quality index is made using the weighted scoring method, based on three criteria:

- quality (Q), with a weight equal to 50%;
- punctuality (P), weight 40%;
- company-supplier relationship (R), weight 10%.

The SQI calculation rule is

$$SQI_i = 0.5Q_i + 0.4P_i + 0.1R_i, i \in V.$$

However, to be selected a supplier must maintain its SQI value above nine. For each scheduled delivery, the BTT purchasing manager selects from the list the first supplier who is able to accept the terms (quantity and timing) of the provision itself.

The quality of supplier $i \in V$ is established for each supply based on the verification of compliance with respect to the requirements of the goods supplied and on the supplier's capability to give prompt and effective responses to the received requests for corrective and preventive actions. The quality indicator is therefore calculated by means of three parameters:

- responsiveness to the verification of incoming goods (QA);
- responsiveness to comments arising from the quality control procedure (QB);
- responsiveness to requests for corrective and preventive actions (QC).

The procedure for monitoring incoming goods is described in the following. One out of every five lots is tested if there were no problems with the last 10 consecutive deliveries received; otherwise, all the lots are tested and inspected. For each lot, 40% of the components are tested through a random check of the components specifications. If the test ends successfully, the unit is certified as 'component meeting the requirements'. The remaining 60% of components are directly certified or tested. The direct certification is granted if there was no problem with the vendor during the entire previous year. The chemical raw materials, however, cannot be directly certified. A lot is if no problem is detected, that is, if all its components meet the requirements.

For this reason, the parameter QA for the supplier $i \in V$ corresponds to the ratio between the number of accepted lots and the number of lots inspected, multiplied by 10 (this ensures that $QA_i \in [0, 10], i \in V$).

The calculation of QB is based on the number of days g necessary to receive from the supplier a formal response to comments arising from the quality control procedure. Responses must be received by BTT within at most 10 days; otherwise suppliers are charged a penalty of 5% on the cost of components supplied. For these reasons, QB is calculated as

$$QB_i = \max\{0, \ 11 - g\}, i \in V. \tag{4.8}$$

Equation (4.8) ensures that $QB_i \in [0, 10], i \in V$. The same rule is used for the calculation of QC, since BTT gives a maximum time of 10 days to each vendor to respond to requests for corrective and preventive actions on supplies.

The quality indicator for each supplier is obtained as follows:

$$Q_i = 0.6QA_i + 0.2QB_i + 0.2QC_i, i \in V.$$

The indicator of punctuality P for the supplier $i \in V$ corresponds to the ratio between the number of deliveries made on time and the total number of deliveries made in the month, multiplied by 10. This ensures that $P_i \in [0, 10], i \in V$. A delivery is made on time if it takes place at most one day after the planned date.

The indicator on the company-supplier relationship R is calculated using five parameters, to each of which the BTT purchasing manager assigns a score (between a minimum and a maximum value) representing the level of effectiveness and efficiency of communications between the company and the supplier. The parameters are

- proactivity (RA), which measures the ability to initiate communication about potential noncompliances and on issues that may affect the quality of a delivery; $RA_i \in [0, 10], i \in V$;
- reactivity (RB), which measures the ability to quickly, effectively and efficiently change delivery procedures and eventually also the goods delivered following specific requests coming from the company; $RB_i \in [0, 15], i \in V$;
- the ability to organize emergency and extraordinary events (RC), which measures the ability to accommodate requests of extraordinary quality visits of the plants, to organize special and extraordinary deliveries and so on; $RC_i \in [0, 5], i \in V$;
- accessibility (RD), which measures the ability to respond promptly, efficiently and courteously to company inquiries; $RD_i \in [0, 10], i \in V$;
- flexibility (RE), which measures the supplier's ability to adopt adequate methods and supply contents to specific needs declared by the company; $RE_i \in [0, 10], i \in V$.

Therefore,

$$R_i = \frac{RA_i + RB_i + RC_i + RD_i + RE_i}{5}, i \in V.$$

The described BTT system for the selection of suppliers obtained the FDA's approval for the San German facility.

4.6 Questions and problems

4.1 Your company needs electric energy for a warehouse. Consult the web pages of at least three potential providers and evaluate them in terms of their average price during the morning shift, payment conditions and reputation. Which other criteria could you take into account? Adopt a weighted scoring method to select the provider.

4.2 A company has identified the following criteria for selecting its suppliers: product cost (*PC*), transport cost (*TC*), performance awards (*PA*), product quality (*PQ*), supplier attitude (*SA*), supplier reliability (*SR*), supplier experience (*SE*) and lead times (*LT*). Apply the AHP procedure to determine the weights of each criterion, on the basis of the pairwise comparative values reported in Table 4.6.

Table 4.6 Values associated to a pairwise comparison of the selection criteria indicated in Problem 4.2.

	PC	*TC*	*PA*	*PQ*	*SA*	*SR*	*SE*	*LT*
PC	1	8	1/5	1/8	3	1/7	1/2	2
TC	1/8	1	1/7	1/8	1/5	1/7	1/5	1/3
PA	5	7	1	1/5	1/3	1/3	1/4	1/2
PQ	8	8	5	1	1	1	1	3
SA	1/3	5	3	1	1	1/2	1	1/6
SR	7	7	3	1	2	1	2	3
SE	2	5	4	1	1	1/2	1	2
LT	1/2	3	2	1/3	6	1/3	1/2	1

4.3 Alphen is an Austrian company producing yoghurt. In order to produce a new product for coeliacs, the company has to select a supplier of goat milk (which does not contain gluten). The company has decided to consider two selection criteria: the price and the reliability of the potential supplier. After a brief market analysis, the candidates have been identified and ratings for each of the two criteria have been computed (see Table 4.7). The supplier selection has been conducted by using two different methods, which are variants of the weighted scoring method: the *absolute* scoring method (ASM) and the *relative* scoring method (RSM). In ASM, the score of a supplier is given by considering the judgement value expressed on each criterion independently from the values given for the same judgement to the other suppliers. In RSM, the judgement value on a criterion depends on the worst value assigned the other suppliers for the same criterion. The company has decided to give a 40% ($\alpha = 0.4$) weight to the price criterion and a 60% (i.e. $1 - \alpha$) weight to the reliability.

Determine the supplier to be selected by using ASM and RSM. Discuss how the solution changes according to different values of the parameter α.

Table 4.7 Potential suppliers and absolute judgements on the two criteria adopted by Alphen.

Supplier	Price	Reliability
1	4.0	1.8
2	5.0	1.5
3	5.5	1.2
4	3.8	1.9
5	5.2	1.6
6	4.8	1.7

4.4 To select suppliers, some companies tend to adopt the so-called *value equation* generally expressed as the ratio between supplier performance and price. Typically, the performance represents all nonprice factors (i.e. quality of service, technical competence, experience and willingness to agree to the contractual terms proposed) and its value is selected on a scale from 1 to 100 (the higher the number, the better the performance). The price represents the score given to the supplier pricing proposal, on a scale of 1 to 10 (the lower price, the lower the score). On the basis of the list of potential suppliers (and corresponding performance and price) given in Table 4.8, select the best supplier by using the value equation. What can you conclude about this method for selecting suppliers?

Table 4.8 Performance and prices of the potential suppliers for the value equation method illustrated in Problem 4.4.

Supplier	Performance	Price
A	80	10
B	50	5
C	70	3
D	90	1
E	65	4
F	45	2
G	25	3
H	65	8
I	70	2

4.5 Assume there are n commodities and let V be the set of potential suppliers; let p_{ij}, $i \in V$, $j = \ldots, n$, be the price required by supplier i for commodity j and c_i, $i \in V$, the fixed cost which is applied only if the supplier i is selected (for the provision of one commodity at least). Define a mathematical model for the supplier selection problem.

4.6 Perform a sensitivity analysis on parameter $\alpha \in [0, 1]$ in the Ilax problem (see Section 4.4).

4.7 Modify the supplier selection model (4.3)–(4.6) for the multicommodity case.

4.8 In the Ilax problem, assume that the provision of quartz sand for the next month should be planned on a weekly basis. Furthermore, assume that the capacity and the unit selling cost vary on a weekly basis, according to the values reported in Table 4.9. Formulate the multiperiod version of the Ilax problem, and determine the optimal solution by using the ϵ-constraint method (with the same value of ϵ as the one used in the original Ilax problem; see Section 4.4). How does the new solution differ from the previous one?

Table 4.9 Weekly capacity and unit selling price for the Ilax suppliers.

	Week 1		Week 2		Week 3		Week 4	
Supplier	Capacity (t)	Price ($)	Capacity (t)	Price ($)	Capacity (t)	Price ($)	Capacity (t)	Price ($)
1	830	791	860	793	910	789	900	788
2	800	770	1 150	765	1 150	765	900	769
3	750	765	680	772	950	756	1 620	750
4	650	800	795	798	1 110	775	1 445	765
5	600	799	450	806	700	805	750	803

4.9 An optimization model, suitable for the selection of suppliers of a single commodity, can be defined if the following data are available. Let T be the number of time periods in the planning horizon considered for the provision of a commodity; $t = 1, \ldots, T$, indicates the time period in which a provision takes place. Let V be the set of potential suppliers, from which n $(1 \leq n \leq |V|)$ suppliers have to be selected; r_i, $i \in V$, is the score of the supplier i; D is the total quantity required to the suppliers (total demand) to be supplied during the planning horizon; o_{it}, $i \in V, t = 1, \ldots, T$, is the minimum quantity to be eventually ordered to the supplier i at time period t; O_{it}, $i \in V$, $t = 1, \ldots, T$, is the maximum quantity that can be eventually ordered to the supplier i at time period

t; $c_{it}, i \in V, t = 1, \ldots, T$, is the unit purchasing cost of supplier i at time period t; b is the available budget for purchasing the quantity D required during the planning horizon, and α $(0 \leq \alpha \leq 1)$ is a tolerance measure, corresponding to the percentage of the ordered quantity that can be delivered late, after the end of the planning horizon, and $\alpha_i, i \in V$, is the average percentage of ordered quantity delivered late by supplier i. Formulate the supplier selection model. (Hint: make use of the following decision variables: $x_{it}, i \in V, t = 1, \ldots, T$, representing the quantity of commodity replenished by supplier i at time period t; $y_i, i \in V$ equal to 1 if supplier i is selected, and 0 otherwise, and $z_{it}, i \in V, t = 1, \ldots, T$, equal to 1 if supplier i is chosen at time period t, and 0 otherwise.)

4.10 Use the supplier selection model formulated in Problem 4.9 to solve the following problem. Saint-Gold is a British company specialized in the production of ceiling fans. For the production of the Airmax model, the company needs a specific numerical controller. There are four potential suppliers, whose scores, determined using the weighted scoring method, are 6.97, 6.75, 6.31 and 6.92, respectively. During the negotiation phase, each supplier has imposed to the company a daily minimum and maximum quantity to be ordered. These values are shown in the second and third columns of Table 4.10.

Table 4.10 Minimum and maximum daily supply quantities and percentage of late delivery of the four suppliers in the Saint-Gold problem.

Supplier	Minimum supply	Maximum supply	Late delivery (%)
1	1140	3 000	3.50
2	900	3 600	2.80
3	300	3 000	4.30
4	720	4 800	2.50

Table 4.11 Prices (in £) of the numerical controller in the Saint-Gold problem.

	Day					
Supplier	1	2	3	4	5	6
1	0.23	0.22	0.23	0.22	0.19	0.24
2	0.21	0.24	0.22	0.18	0.21	0.23
3	0.19	0.16	0.20	0.17	0.18	0.19
4	0.18	0.14	0.17	0.19	0.20	0.24

4.11 The planning horizon for the supply of 5 000 controllers is equal to a six-day working week. Two suppliers have to be selected. The suppliers have presented an offer to the company which indicates the price (inclusive of the transport cost). These are indicated in Table 4.11. Analyzing the historical data on the suppliers' efficiency, the manager of Saint-Gold has computed the average rates of the delivery late for each potential supplier (see the fourth column of Table 4.10). The management of Saint-Gold does not tolerate late delivery for more than 3.5% of products during the planning period. The total budget available for the weekly purchase of the cards is £ 1 000.

5

Managing a warehouse

5.1 Introduction

Storage is a key logistics activity which has taken on an increasingly determinant role over the years in the organization of logistics systems. This activity concerns a specific facility of the logistics system, generically identified by the term *warehouse*, although there are different and more specific denominations according to their functionality and to their positioning within the logistics system.

Owned warehouse It is the least expensive solution in the long term, mainly in the case of continuous high-demand products, and it is the only possible solution in the case where storage operations need highly specialized equipment and staff. Moreover, it can be used as a parking area for company vehicles or as a base for the sales or purchases office.

Rented warehouse It provides the same services as an owned warehouse but does not need long-term investments and is therefore less expensive in the short term. It should be preferred whenever a reorganization of the distribution system is foreseen in the medium term.

Public warehouse It is frequently found in areas such as ports and airports. This kind of warehouse has standard equipment to fulfil general and common needs, and is therefore poorly suitable to highly specialized uses. Its costs are directly proportional to the area occupied in the warehouse, to the utilization time and to the services used. This solution allows a rapid adjustment of the logistics system structure to the spatial-temporal variations of the product demands.

Introduction to Logistics Systems Management, Second Edition. Gianpaolo Ghiani, Gilbert Laporte and Roberto Musmanno.

Depot It is an area for the storage of raw materials, semi-finished commodities and products used prevalently to guarantee the firm the correct carrying out of the manufacturing phases.

Central warehouse It is a warehouse which, apart from storage, ensures the products' distribution to customers, possibly through the use of peripheral storage zones.

Peripheral warehouse It is a warehouse served by a central warehouse and used for product distribution to customers. Its aim is to ensure a better quality of customer service.

Distribution centre It is mainly used in large-scale distribution for the storage of products coming from different manufacturers, so that their distribution to the sales points (and possibly to other locations belonging to the firm), located in a specific geographic area, can be realized.

CDC Also called primary warehouse. Often it is a synonym for central warehouse.

RDC It is a warehouse of secondary level (also named secondary warehouse). Typically it is a synonym for peripheral warehouse.

Automatic warehouse It is a warehouse where the load units receiving and picking operations in the storage zone are automated, typically through the use of an electromechanical machine moving on rails.

Spare parts warehouse It is used to stock consumables, tools and the spare parts necessary for getting a manufacturing plant effectively working.

Cool warehouse It is used for perishable items (typically foodstuffs or pharmaceutical items), for which a precise control of the storage temperature is necessary. Perishable food products can be divided into frozen, refrigerated and fresh products. The frozen products must be preserved at a temperature of −18 °C (large-scale distribution often requires delivery of products at a temperature of −20 °C). To respect these conditions, the majority of cool warehouses maintain a temperature of −25 °C to guarantee an adequate buffer of cold temperature. Fresh or refrigerated products are stored at a temperature between 0 °C and 5 °C.

Cross-docking warehouse It is a warehouse where the incoming goods stop in for a few hours at most, and are therefore not stored in the storage zone.

Quarantine warehouse It is a warehouse where one stores products in quarantine or goods whose customs duties are still unpaid.

Customs depot It is a depot managed by a public authority where the incoming goods are from foreign states and are stored while they await the completion of regular customs operations.

In what follows, reference will generally be made to a warehouse as a facility where the activities of receiving, storing and shipping of products take place (the latter, in turn, is subdivided into activities of picking from the storage zone, consolidation for forming the outgoing load units and loading the vehicles used for the deliveries).

Each of these activities can be more or less important, according to the warehouse typology, the number of items (which can vary from few units to several tens of thousands) and the demand for each product (which can vary from a few units a month to some hundreds units a day). In the simplest cases, the warehouse operations involve few large-sized load units of the same product and, therefore, the main activities carried out concern the storage phase. However, if the outgoing load units are made up of many items, the picking activity becomes much more complicated, and laborious assembling, packing and packaging operations of the outgoing load units become indispensable.

The advantages of warehouses in a logistics system are essentially related to the storage and shipping of products. In particular, thanks to warehouses, production is dissociated from the demand, which helps attenuate the effects of the unreliability of forecasting data on product demand, of the unreliability of manufacturing plants and transport services, of the seasonal nature of the products and so on. The warehouse enables companies to guarantee a regular production rate so as to minimize costs by maintaining sufficient stock level to satisfy the variability in demand.

Moreover, warehouses ensure the use of the best means of transport from an economical point of view both at the receiving phase (e.g. trucks) and for the subsequent delivery operations (e.g. vans). Furthermore, the transport of aggregated quantities of products allows for the reduction of transport volumes with respect to single dispatches since the aggregated forecast data on most products are more reliable than disaggregated data (see Section 2.4).

Finally, it can be noted that as the number of warehouses increases, the distance between them and the customers decreases, while the total distance from the previous facilities in the logistics system increases. Consequently, the incoming transport costs increase, but when the vehicles travel fully loaded, the outgoing transport costs decrease more rapidly than the incoming ones, which ensures a better service for the customer since the delivery times are reduced.

However, the opening and management of a warehouse inevitably involves costs which can be conveniently classified as follows.

- **Investment costs**. These are paid in the setting up phase of the logistics activity of a warehouse and are, for example, the purchase cost of the property, the planning cost, and the cost of equipment.

- **Operating costs**. These are the costs associated with the management of the load units stocked in the warehouse and to their handling. The financial burdens that weigh on the value of the stocks stored fall into this category. These are the costs tied to the activities of raw materials, semi-finished and finished products receiving and of order picking, the packaging costs, the administrative costs and the depreciation rates.
- **Costs of risk**. The total cost of risk can be calculated by considering both the operational and the commercial risks. The costs of operational risks concern, for example, the products that can become obsolete within a determined time span, the damaged products and shortages. The costs of commercial risks instead are relative to losses both of sales to customers and of image (because of unsuitable packages, late deliveries with respect to agreed terms etc.).
- **Running costs**. These refer to energy consumption, plant maintenance, insurance policies, safety measures and so on.

Codo is a food distributor operating in Argentina. In 2008, the company decided to open a new distribution centre in the town of Santa Fe, the capital of the eponymous province. Within two years, at parity of sales volumes, the annual transport costs for Codo decreased by 1.7%, or 97 000 pesos, due to an increase of 46 000 pesos in transport costs to the distribution centres, counterbalanced by a reduction of 143 000 pesos in the annual transport costs from the distribution centres to the supermarkets. The annual cost of the new distribution centre, including the depreciation rate of the investment costs, was 120 000 pesos.

Owing to a favourable economic trend facilitated by the improvement of the service offered to the supermarkets, an average increase in turnover of about 3.5% a year since 2009 was observed for the already existing supermarkets served by the Santa Fe distribution centre, with an increase of the yearly profits which were about 350 000 pesos in 2009.

5.1.1 Performance parameters

A number of indicators can be used to assess the performance of a warehouse. To evaluate the capacity of exploitation of the available storage space, the following space utilization parameters can be used:

- *surface utilization rate*, defined by the ratio of the surface effectively used for storing load units to the total surface of the warehouse;
- *volume utilization rate*, given by the ratio of the volume occupied by the stored load units and the total volume of the warehouse.

The potential capability of the warehouse of managing load units is assessed by means of the following indices:

- *potential receptivity*, given by the maximum number of storable load units; this is a static measure of the capacity of the warehouse;
- *throughput*, expressed by the maximum number of load units in transit in the warehouse within a given time unit; this parameter expresses a dynamic capacity of the warehouse.

Warehouse efficiency can be assessed by its capability of exploiting its capacity. This can be computed by means of the following parameters:

- *potential receptivity saturation coefficient*, which represents a percentage of the theoretical potential receptivity normally usable within a reference period (daily or monthly), and defined as the ratio of the number of load units present on average in the warehouse within the time period considered, to the potential receptivity;
- *selectivity index*, expressed by the ratio of the number of load units directly accessible in the picking phase to the potential receptivity. Low selectivity values usually indicate a high number of material-handling operations necessary to pick up the goods making up the outgoing load units. Low values therefore reflect the possibility of producing a lower product flow compared with warehouses having the same physical characteristics, but higher selectivity values;
- *access index*, expressed by the ratio of the number of material-handling operations within the reference period considered (day, month or year) to the potential receptivity of the warehouse. It indicates the frequency with which material-handling operations are carried out. Warehouses with high-access indices are therefore highly dynamic warehouses, where the stored goods are characterized by a reduced warehouse holding time and are therefore frequently handled; in contrast, warehouses with a low access index are more static warehouses, with less frequent material-handling operations;
- *inventory turnover index*, which is calculated as the ratio of the value of the outgoing goods in a specific time period to the value of the average inventory level in the warehouse in the same time horizon. The inventory turnover index is therefore a nondimensional number which, defined in this way, expresses the degree of mobility of the capitals immobilized in stock (how many times the capital invested in stock rotates in a planning period).

Inseko is a Ukrainian company producing building materials. During the last two years, the Kiev warehouse has had outgoing movements to a total value of € 28 980 000 and € 25 490 000, respectively. The average inventory level in the same period of time was equal to € 4 317 820 and € 4 174 530. The inventory turnover index during the two years was equal to 6.71 and 6.11, respectively. If the inventory turnover index is assumed to be a measure of the 'speed' of the warehouse, the difference between the inventory turnover indices in the two years (−0.70) represents the average acceleration of the warehouse. Therefore, in this case, it can be stated that there has been a deceleration in the rotation of the Inseko warehouse from one year to the next.

Typical values of the inventory turnover index are included between 5 and 10 in the case of an annual time horizon. However, the efficiency of the warehouse can also be obtained with values of the inventory turnover index falling outside the indicated interval, which depends on the characteristics of the items stored in the warehouse.

The calculation of the inventory turnover index can also be associated with a single product or with a single category of products in the warehouse. In this case, the inventory turnover index is no longer a measure of the efficiency of the whole warehouse, but remains useful to assess the adequacy of the inventory management policies (see Section 5.3.2) adopted for a determined product or for a determined category of products.

The calculation of the inventory turnover index can be achieved in two ways since, besides the value of the products, the quantities can also be considered.

Fri-X5-A is an additive in liquid form for concretes. The product, in iron drums of 200 kg, was among the goods present in the Inseko warehouse last year. Table 5.1 shows the information relative to the inventory levels of this product in the warehouse during the last year (composed of 10 time periods). Denoting by n_i and q_i, $i = 1, \ldots, 10$, respectively, the number of days and the inventory level of the Fri-X5-A additive for each time period i, the average inventory level of the product can be determined as

$$\bar{q} = \frac{\sum_{i=1}^{10} n_i q_i}{\sum_{i=1}^{10} n_i} = 1\,577.26.$$

The sales recorded during the last year were equal to 8 596 drums. Consequently, the year inventory turnover index of the Fri-X5-A additive, calculated on the quantitative product basis, is equal to 8 596/1 577.26 = 5.45.

Table 5.1 Inventory levels (in drums) of the Fri-X5-A additive in the 10 periods of last year for the Inseko problem.

Period	Time interval	Days	Inventory
1	01/01–20/01	20	1 000
2	21/01–07/03	46	2 200
3	08/03–08/04	32	1 400
4	09/04–05/07	88	800
5	06/07–10/07	5	600
6	11/07–01/09	53	2 100
7	02/09–20/10	49	1 600
8	21/10–22/10	2	600
9	23/10–29/12	68	2 100
10	30/12–31/12	2	1 300

Finally, to test the economic efficiency of the warehouse, the following cost indices can be used:

- *receptivity cost*, expressed by the ratio of the value of the annual warehouse cost to the potential receptivity;
- *handling cost*, which expresses the value of the cost for each load unit handled. It is calculated by summing up all annual cost items which are directly or indirectly imputable to material-handling activities, and dividing them by the annual throughput. The direct costs include, for example, the staff cost and the energy cost related to the material-handling operations, whereas the indirect cost is mainly composed of equipment maintenance costs directly used in the material-handling activities.

5.1.2 Decision-making problems

Decisions relative to the design, organization and management of the warehouse, from the point of view of logistics planning, can be classified as strategic, tactical and operational (see Table 1.8); the decision-making problems that derive from it will be presented in the following sections taking this classification into account.

5.2 Warehouse design

The design of a warehouse or the reorganization of an existing one is a strategic logistics planning problem whose objective is the minimization of the sum of the estimated costs of investment, management, risk and operation for a given set of incoming and outgoing materials flows, in the presence of constraints on the availability of space, capital and labour.

A complete design study provides for, among others, the following decision phases:

- site selection;
- choice of warehouse systems;
- choice of warehouse layout.

The study should also contemplate the design of the building (foundations, bearing structure, roofs, walls and floors), but this phase is beyond the scope of this volume and the reader should refer to more specialized texts. Site selection is a decision-making problem to be addressed by means of location techniques already discussed in Chapter 3. Here, we will therefore focus on the remaining phases.

5.2.1 Choice of warehouse systems

The choice of warehouse systems is made by considering a number of factors. One of the most important is given by the packaging characteristics (load units) of the products stored in the warehouse. In general, three possible packages are considered, on the basis of the type of product, its value, its size and the average quantities delivered:

- *primary package* (distribution units). This package comes into direct contact with the product. This is the sales unit destined to the customer. The determination of the characteristics of the primary package is associated to marketing considerations (quantity of product normally purchased by the consumer, manageability of the package, modality of use, conservation of the product in time, sales promotion, regulations of the industry sector, ecological problems involving the product or the package etc.). In the warehouses, some specific products can be stored in their primary package form in the different storage locations.
- *secondary package* (parcel). Packages containing one or more primary packages (see Figure 5.1). The number and layout of the distribution units need to be considered for the correct formation of the secondary packages. The constraints involve the ease of parcel handling within the warehouses and in the facilities to which the secondary packages are delivered.
- *tertiary packages* (pallets). The activity of receiving and storing load units in a warehouse typically involves the so-called palletized load unit, using suitable bases (pallets), which are also used to identify the tertiary packages (see Figure 5.1).

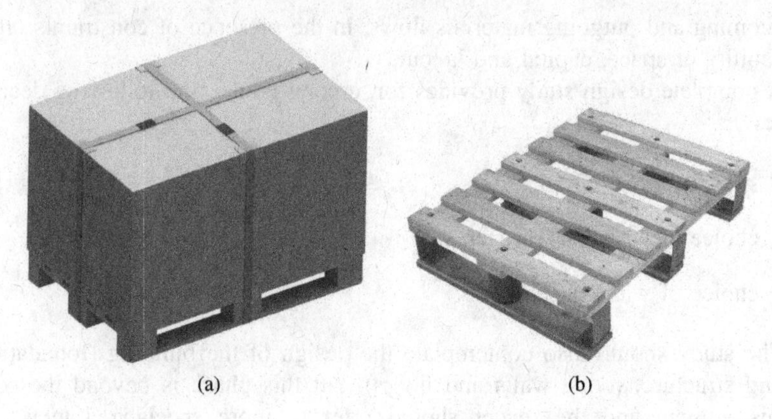

(a) (b)

Figure 5.1 (a) Palletized load unit; (b) pallet.

A pallet is a portable platform, usually made of wood, corrugated cardboard or plastic, on which the goods are stacked for transport and storage.

The advantages related to pallets are as follows:

- they help goods movement thanks to the use of automatic machineries;
- they allow an increase in loadable weight and volume;
- they enable a better use of warehouse spaces, because stacks of different heights can be formed.

Moreover, the use of pallets allows the isolation of goods from the underlying surface, therefore reducing damages caused, for example, by contact with a wet floor. The range of pallets used in the different industrial and commercial sectors is, as far as sizes are concerned, notably diversified: 80 cm × 120 cm, 100 cm × 120 cm, 120 cm × 120 cm, 112 cm × 112 cm etc. In Europe, the most widespread sizes used are 80 cm × 120 cm (the so-called standardized europallet, or ISO1), and 100 cm × 120 cm (the standardized Anglo-Saxon pallet, or ISO2). Both these sizes allow the combination of secondary packages of 60 cm × 40 cm, so that the available surface can be completely filled. Obviously combinations of submultiples of the units considered can be used (as shown in Table 5.2, where x and y indicate the sizes, in cm, of the two sides of the base of the secondary packages).

Table 5.2 Box size (in cm) compatible with filling the surface of the europallet and the Anglo-Saxon pallet.

	y	$y/2$	$y/3$	$y/4$	$y/5$
x	60 × 40	60 × 20	60 × 13.3	60 × 10	60 × 8
$x/2$	30 × 40	30 × 20	30 × 13.3	30 × 10	30 × 8
$x/3$	20 × 40	20 × 20	20 × 13.3	20 × 10	20 × 8
$x/4$	15 × 40	15 × 20	15 × 13.3	15 × 10	15 × 8
$x/5$	12 × 40	12 × 20	12 × 13.3	12 × 10	12 × 8

Once the packages are defined, the warehouse system is subdivided into the following three categories: storage systems, material-handling systems and identification systems. Other systems are used in the packaging activities (binders, packaging machines, wrapping machines, enveloping machines etc.) and in measuring (counters, rulers, scales etc.), but their description is beyond the scope of this volume.

5.2.1.1 Storage systems

There are two kinds of storage systems for packages:

- static, including stacks, racks (conventional, drive-in, drive-through and cantilevers), multilevel shelvings, lockers and drawer cabinets;
- dynamic, such as mobile racks, storage carousels, live pallet racks and push-back racks.

The main characteristics of the storage systems just listed will be briefly illustrated below.

Stacks Palletized load units are stacked in blocks separated by aisles, whose width depends on the kind of machinery used for movement. The stack (see Figure 5.2a) is useful to manage a limited number of products without a high rate of obsolescence and whose movement occurs in great quantities and not very frequently. This system is less expensive and the most flexible since it does not require investments in infrastructure, and the surface destined to the stack can be made rapidly available.

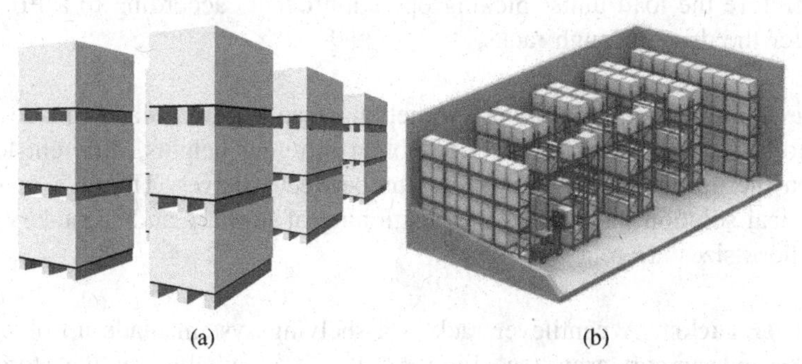

Figure 5.2 (a) Stacks; (b) conventional racks. Reproduced by permission of Mecalux.

Conventional racks Conventional racks (see Figure 5.2b) allow direct access to all the load units and therefore are particularly suitable for storing large volumes of single items. The height of the conventional racks, like the width of the aisles, is influenced by the choice of the means used to handle the load units. The picking of the load units can be carried out according to a FIFO (First-In, First-Out) or LIFO (Last-In, First-Out) logic.

Drive-in or drive-through racks These are systems similar to conventional stacks, with the difference that a suitable structure is used to support the palletized load units (so that problems of product superimposition are avoided). This system is useful for storing homogeneous products, in particular seasonal ones, and reduces potential damage to the packages.

With drive-in or drive-through racks, load units are stored several pallets deep and the lift equipment can enter the structure to store or retrieve pallets. The drive-in rack uses the same entry and exit point for each storage bay, providing LIFO access, while the drive-through rack is loaded on one side and unloaded from the other one (see Figure 5.3).

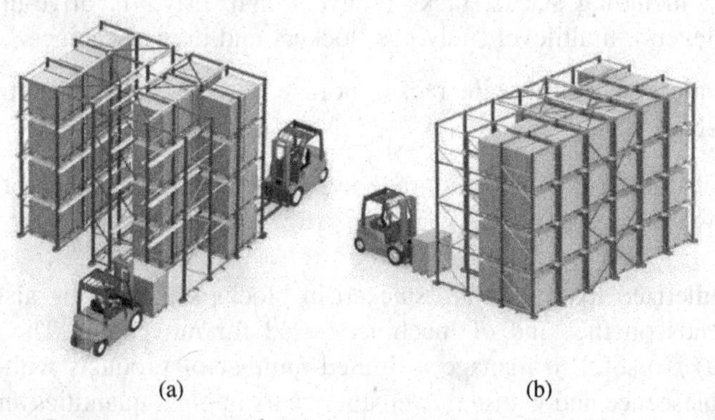

Figure 5.3 Loading and unloading operations for (a) drive-in or (b) drive-through racks. Reproduced by permission of Mecalux.

Therefore the load units' picking operation occurs according to a FIFO criterion for the drive-through racks.

Multilevel shelvings A multilevel shelving (see Figure 5.4a) is made up of standard-size components in order to obtain different heights, different lengths between the uprights and different depths between shelves. This is a relatively economical solution that allows the positioning of primary and secondary packages whose sizes are relatively small.

Cantilever racks A cantilever rack is a shelving system made up of upright elements and support arms (see Figure 5.4b). It is suitable for the storage of long, awkward products like bars, pipes, steel plates, beams and profiles, also at different heights. These are often found outdoors and in this case they are equipped with a gable roof for protection against atmospheric agents (sun, rain, snow etc.).

Lockers and drawer cabinets Lockers and drawer cabinets are used as stand-alone units prevalently for the storage of primary packages, or they can be inserted in multilevel shelvings.

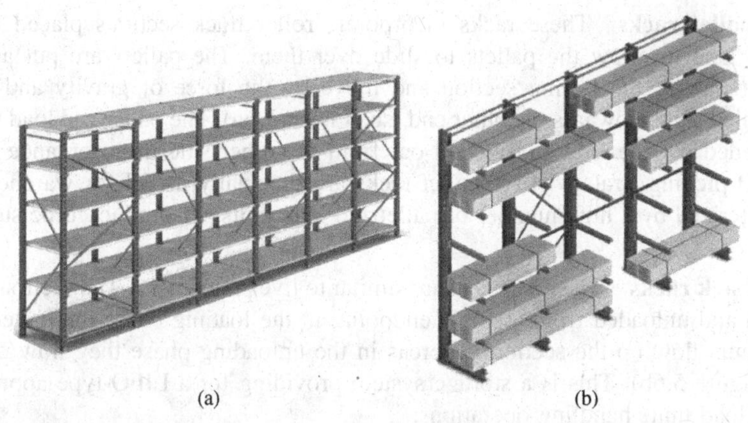

Figure 5.4 (a) Multilevel shelving; (b) cantilever rack. Reproduced by permission of Mecalux.

Mobile racks This is an expensive racking system capable of moving laterally to allow the storage aisle opening which is necessary every time the load units have to be picked (see Figure 5.5a). Mobile racks are used for load units which have to be handled with a low frequency, since the time necessary for the picking operations includes a waiting time to handle the racks.

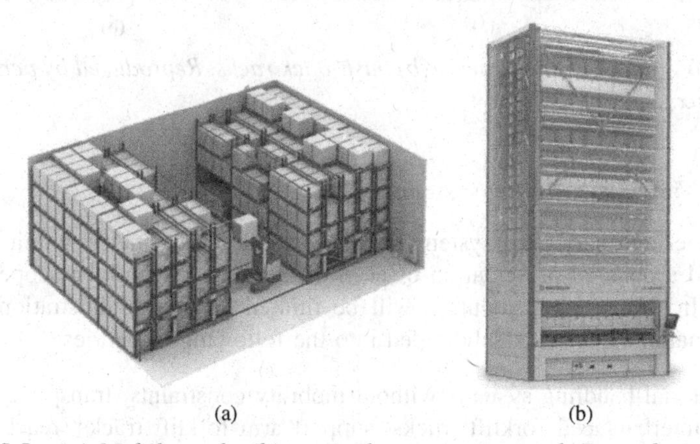

Figure 5.5 (a) Mobile rack; (b) vertical storage carousel. Reproduced by permission of Mecalux.

Storage carousels These are motorized systems where the bays move contemporaneously (vertically or horizontally) so as to present the storage bay concerned for storing and picking operations in correspondence to the fixed position of the operator (see Figure 5.5b). They enable the storage of many parcels in small volumes and make the storage and picking phases very quick. This solution has, however, two disadvantages: it needs high investments and it is not suitable for voluminous or very heavy load units.

Live pallet racks These racks incorporate roller track sections placed on a sloped lane to allow the pallets to slide over them. The pallets are put at the highest part of the rolling section and move by the force of gravity and at a controlled speed towards the other end (see Figure 5.6a). The palletized load units are loaded by one side and picked out from the other side in accordance with a FIFO picking strategy. Live pallet racks are especially useful for warehouses characterized by a high number of palletized load units per product to be stored.

Push-back racks These systems are similar to live pallet racks. The sections are loaded and unloaded from a single endpoint; in the loading phase the palletized load units flow up the section, whereas in the unloading phase they flow down (see Figure 5.6b). This is a storage system providing for a LIFO-type approach in the load units handling operations.

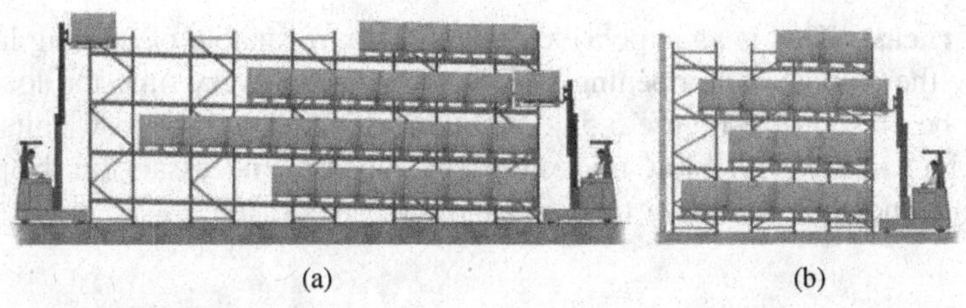

Figure 5.6 (a) Live pallet racks; (b) push-back racks. Reproduced by permission of Mecalux.

5.2.1.2 Material-handling systems

The number and variety of systems used to handle the packages within a warehouse and a detailed presentation of each of them are beyond the scope of this volume. In this section, attention will be limited to a brief illustration of the material-handling systems subdivided into the following categories:

- material-handling systems without mobility constraints (transpallet trucks, counterbalanced forklift trucks, support arm forklift trucks, reach trucks, order pickers and stacker cranes);
- material-handling systems with mobility constraints, which are systems where the transport of load units takes place following pre-established routes (fixed conveyors, floor-bound continuous conveyors and overhead conveyors);
- AGVS (automated guided vehicle systems).

Transpallet trucks These are trucks with forks, able to lift tertiary packages of relatively low weight in such a way that they can be horizontally handled.

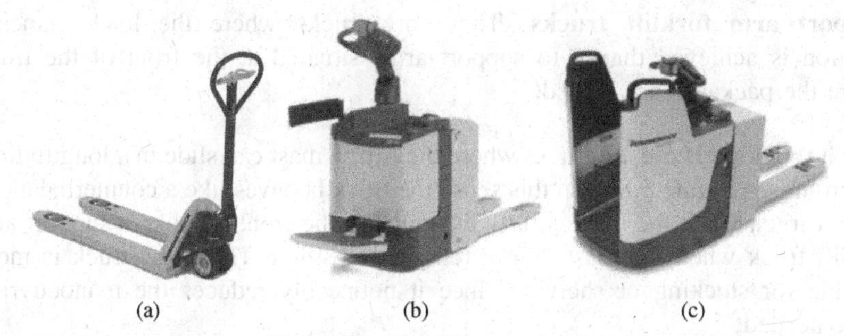

Figure 5.7 (a) Hand transpallet truck; (b) pedestrian electric transpallet truck; (c) stand-on electric transpallet truck. Reproduced by permission of Jungheinrich.

These trucks can be divided into hand transpallet trucks and electric transpallet trucks (see Figure 5.7).

The first ones are more efficient over short distances, whereas the electric transpallet trucks, in particular the stand-on ones, are more suitable for longer distances.

Counterbalanced forklift trucks They are characterized by the presence of a counterweight in the rear of the vehicle (see Figure 5.8a), in order to balance the load that will be sustained by the forks. Usually counterbalanced forklift trucks are identified as front-loading forklift trucks, where the forks lift the tertiary package frontally, or side-loading (bi-trilateral) forklift trucks, where the forks lift perpendicularly also to the longitudinal axis of the truck itself. Counterbalanced forklift trucks are used to load and unload vehicles from the ground and from a loading door; they can move on different kinds of floor surface, operate on whatever kind of shelving and deal with also considerable slopes.

Figure 5.8 (a) Counterbalanced forklift truck; (b) reach truck; (c) vertical order picker. Reproduced by permission of Jungheinrich.

Support arm forklift trucks These are trucks where the load-balancing function is achieved thanks to support arms situated at the front of the truck where the packages are loaded.

Reach trucks These are trucks where the lifting mast can slide in a longitudinal direction (see Figure 5.8b). In this sense the truck behaves like a counterbalanced forklift truck when the mast is fully extended to the front and like a support arm forklift truck when the mast is in a retracted position. The reach truck is more suitable for stocking or shelving since it noticeably reduces the manoeuvring space needed.

Order pickers With the horizontal order picker, the operator is in a fixed position and the forks lift to allow horizontal movements; with the vertical order picker (see Figure 5.8c), the operator can be lifted to enable the retrieval of pallets at different heights.

Stacker cranes These are systems integrated into automated storage and retrieval systems, also known as AS/RSs (see Figure 5.9), which are computer-controlled systems for storing and retrieving packages. They are made up of a platform (necessary to carry out the storage and retrieval operation of the load unit in the storage locations) fixed to a column crane that enables vertical transfer. The column is in turn fixed to the floor and to the ceiling by means of castors that allow the horizontal transfer movement of the stacker crane within the aisle. The advantages that this kind of system brings are that it provides the possibility of using narrow aisles, speed of movements and high precision in the load units' storage and picking activities. In a highly sophisticated system, multiple stacker cranes can be assigned to one aisle.

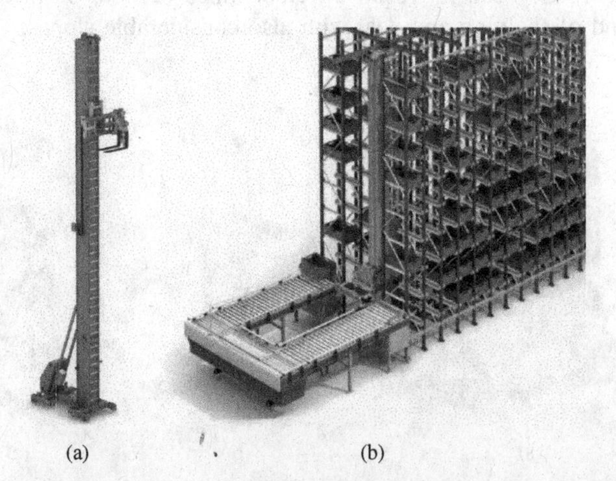

Figure 5.9 (a) Stacker crane for pallets; (b) AS/RS system served by stacker cranes for drawer cabinets. Reproduced by permission of Mecalux.

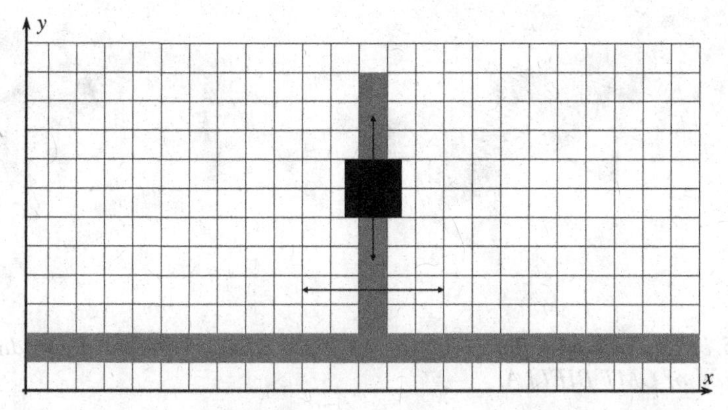

Figure 5.10 Representation of the movement of the stacker crane in a Cartesian plane.

Since the stacker crane has the possibility of moving simultaneously and independently along the horizontal axis x and along the vertical axis y (see Figure 5.10), if it is assumed that the velocities v_x and v_y along the two directions are constant, the movement time t satisfies the Chebyshev metric:

$$t = \max\left\{\frac{\Delta x}{v_x}, \frac{\Delta y}{v_y}\right\},$$

where Δx and Δy are the distances travelled in the directions of the sides of the shelves.

The Al Fokan warehouse in Bahrain provides for the use of a stacker crane in aisles for access to 20 m long shelves with a height of 12 m. The velocity of movement of the system along the horizontal guides is supposed constant at 40 cm/sec, whereas along the vertical guide the velocity is 25 cm/sec. The maximum waiting time for the completion of a retrieval operation is therefore

$$t = \max\left\{\frac{20}{0.40}, \frac{12}{0.25}\right\} = 50 \text{ sec.}$$

Some AS/RSs (called steering) are also able to negotiate curves on their route.

Fixed conveyors These are material-handling systems that remain fixed while the load units move. The movement of the load units takes place by means of motorized mechanisms that drive the load ahead by friction. Roller conveyors (see Figure 5.11a) and wheel conveyors (see Figure 5.11b) both belong to this family of material-handling systems.

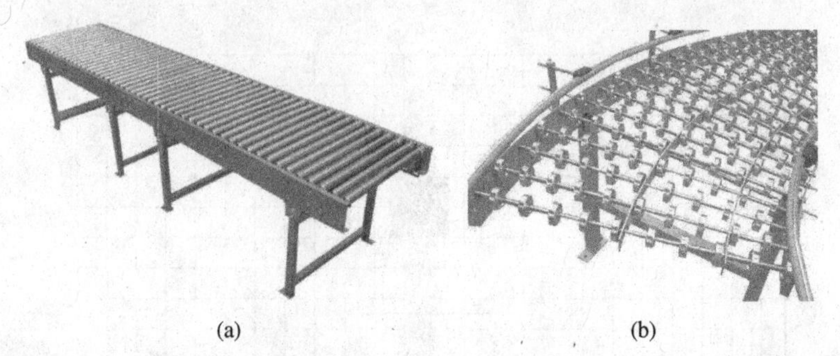

Figure 5.11 *(a) Fixed roller conveyor; (b) fixed wheel conveyor. Reproduced by permission of OMT BIELLA.*

Floor-bound continuous conveyors In this class of conveyors, the element in contact with the package (typically a belt, see Figure 5.12a or a slat) moves with the package itself, but cannot be detached from the transmission of the whole system.

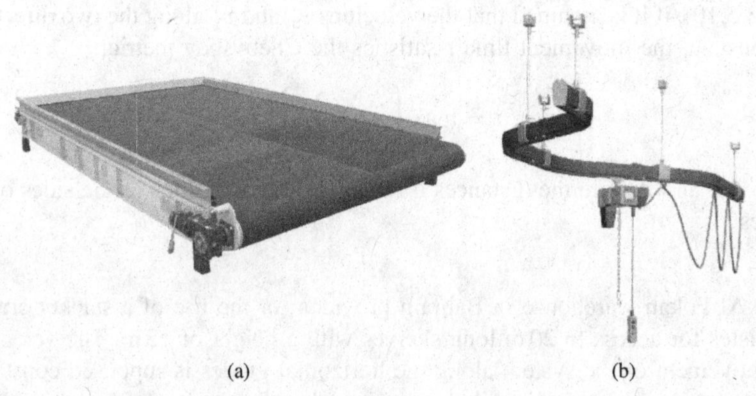

Figure 5.12 *(a) Floor-bound belt conveyor; (b) overhead conveyor. Reproduced by permission of OMT BIELLA.*

Overhead conveyors These material-handling systems, although accompanying the load units, however allow for temporary removal from the line, avoiding, among other things, any accumulation of packages for the successive warehouse operations (see Figure 5.12b).

Automated guided vehicle systems They are material-handling systems that use mobile robots (vehicles) with three or four wheels (see Figure 5.13), which move automatically, without a driver on board, inside a warehouse. These vehicles are fuelled by a system of batteries that supply energy to an electric motor and are driven by a numerically controlled machine that enables them to move along precise and easily programmable routes.

Figure 5.13 Automated guided vehicle. Reproduced by permission of Transtecnica.

5.2.1.3 Identification systems

These are systems for the codification of packages inside the warehouse by means of codes that can be scanned by automatic devices. The codification of the packages is of fundamental importance for the computerized management of the warehouse. The main identification systems are: bar codes, logistic labels and smart tags based on radiofrequency technology (radiofrequency identification, or RFID).

Bar codes The bar code is the optical conversion of a numerical or alphanumerical code which is used to identify a package. This optical conversion is represented by means of an alternating sequence of vertical bars and spaces (see Figure 5.14a).

There are several different bar code composition systems; these correspond to different codification requirements (amount of information, length of bar code etc.), determined by the kind of packages, and so by the customers' needs. The EAN (European Article Number) is among the most widespread codes and it is mostly used in large-scale distribution.

From the technological point of view two kinds of bar code scanning devices can be distinguished: optical scanners or laser scanners.

Optical scanners use a light source that illuminates the surface of the code enabling a suitable sensor to record the variations of the reflected ray. At every passage these perform a single scan of the code. In contrast, the laser scanner repeatedly explores the encoded surface at each passage, taking a series of pictures that allow a greater accuracy of scanning and therefore less sensitivity to the variations of the characteristics of the surface itself. This allows the scanning of compact codes, at high intensity, also on moving packages.

Logistic labels The logistic label records information, both in legible format (characters, numbers and graphic elements) and in the form of a bar code,

Figure 5.14 (a) Bar code; (b) logistic label.

also regarding the entities involved in material-handling operations as suppliers, shippers and customers (see Figure 5.14b).

Part of the logistic label is the SSCC (Serial Shipping Container Code), by means of which the physical path of the individual packages can be traced together with the information flow associated with it: thanks to the scanning of the SSCC, concerning the transport documents transmitted in electronic format, every moment of the shipping and delivery of the products can be checked, and the inventories can be updated as well, cataloguing the goods and following the various transport phases.

The label also contains information related to the package size, the EAN code of any container to be returned, the price per unit of measure, the reference code for any economic offer associated with the commodity etc.

Smart tags These are the main components of an automatic identification system that is based on RFID technology and is known as an RFID tag or transponder (see Figure 5.15a).

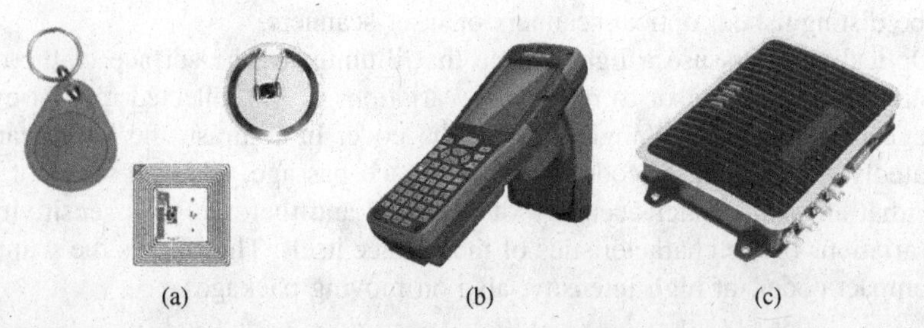

Figure 5.15 (a) Examples of RFID tags; (b) a portable RFID reader; (c) a fixed RFID reader.

An RFID tag can be active or passive. Passive tags, which are more economical and widespread, are made up of an aluminium or copper antenna, a memory microchip and a support for the protection of this chip. They do not have a battery and require no maintenance. The passive tags are made to transmit a unique serial code when they receive an appropriate electromagnetic stimulus. The active tags are more sophisticated electronic devices. Provided with an internal battery that powers them and that enables very great transmission distances to be achieved (over 400 m in the open for some models), they are equipped with a very complicated electronic system that allows the application to be customized on the basis of individual requirements. These are typically used in warehouses to monitor frequently transported pallets and, when necessary, to keep track of inventory data related to the manufacture history of the commodity or of information concerning the physical and organoleptic characteristics of the stored goods (temperature, humidity etc.).

The tag, as already pointed out, is activated by means of an electromagnetic field generated by the scanner (reader), which is the electronic device used for the exchange of information with the tag itself. The readers can be portable (see Figure 5.15b), and are used by operators or are installed on vehicles, integrated with an antenna. The fixed RFID readers (see Figure 5.15c) are integrated with groups of antennae such as gates, portals or tunnels.

5.2.1.4 Guidelines for the choice of warehouse systems

The choice of the most appropriate warehouse system should be made only after an in-depth analysis of the company logistics requirements in relation to the estimated economic investment. To help the readers understand how to effectively make this decision, we provide in this chapter a simplified scheme with a preliminary investigation of warehouse performances, on the basis of the selectivity index and of the access index. 'High' or 'low' values are assigned to these performance parameters, thus yielding four possible combinations (see Table 5.3), with specific solutions corresponding to each of them in terms of storage and material-handling systems.

Table 5.3 Scheme for the choice of storage and material-handling systems on the basis of the access index and the selectivity index.

		Selectivity index	
		High	Low
Access index	High	Shelvings and racks served by stacker	Push back racks
	Low	Shelvings and racks served by manual forklift trucks	Mobile racks

If high selectivity is requested and the access index is also very high, a solution based on shelving systems served by stacker cranes can be taken into consideration; in this case, shelving allows an almost unit selectivity, whereas the use of stacker cranes enables the operator to address the high number of access requests to the storage locations. On the other hand, when the access index is lower, but high selectivity is also desirable, more economical systems can be used. Therefore, systems based on shelving and racking are used to maintain a high selectivity, but are served by manual forklift trucks. Selectivity remains almost unit, since all the packages can be directly picked. A limited number of material-handling operations can also be achieved with manual systems, which are less high-performing but more economical compared with automated systems. When dealing with perishable goods and with the need to manage the warehouse with a FIFO policy, the selectivity requested can be lower. For the choice of typology of storage systems, the access index can be considered. In the case of a high value of the access index, storage systems such as push-back racks can be adopted. Finally, when both the selectivity and the access are low, the hypothesis of adopting mobile racks can be considered; this is particularly suitable in case limited space is available. As regards the choice of equipment for the identification of the incoming or outgoing load units to or from the warehouse, it can be useful to refer to what is illustrated in Table 5.4. This table shows the main differences to be found in the adoption of identification systems based on bar codes or on RFID technologies.

5.2.2 Choice of warehouse layout

The choice of warehouse layout involves sequentially solving two decision-making: it is first necessary to determine the arrangement of the warehouse zones where the warehouse activities take place (warehouse layout problem), and then size and organize each of the warehouse zones (internal layout problem).

The warehouse zones considered in this section are those where the three main warehouse activities (receiving, storaging and shipping) take place as seen in Section 5.1:

- the receiving zone is the area where the qualitative-quantitative control of checked-in packages, the preparation of the load units to be stored and the updating of the inventory are carried out;
- the storage zone is the area where the packages are stored. It is sometimes divided into a remote reserve zone (RZ), where the large load units coming from the receiving zone are stored, and a small-sized rapid access storage zone (pick zone (PZ)), from which the packages are retrieved for the formation of outgoing load units. This solution gives the possibility of keeping at a low level the workload (transfer time) of the staff involved in

material-handling operations, but it requires the periodic replenishment of the PZ with the packages stored in the RZ;

- the shipping zone is the area where the operations of the outgoing load units checking, assembly and packaging and of the shipping documents for the transport preparation take place.

Table 5.4 Comparison of bar code and RFID technology identification systems.

Bar code	RFID
The access mode is reading only	The access mode is reading or writing
The bar code must be directly visible to the scanner	The reader and tag do not need visual contact
The reading of the bar code is sequential, therefore it is possible to identify one product at a time	One reader is able also to communicate with hundreds of tags in a few seconds
The maximum reading distance is a few tens of centimetres	The maximum reading distance is in the order of metres with passive tags and kilometres with active tags
The maximum quantity of information storable is 100 bytes	Passive tags stores from 128 bytes to 8 Kbytes of information; active tags can arrive at 32 Kbytes
The scanners are extremely sensitive to lights, scratches and stains	The readers are totally insensitive to dirt and light
The reading phase requires predefined reading angles; the operation must be carried out at practically null speed	The tags can have any orientation during the reading or writing, and these operations can also take place with tags in movement
There are 26 different kinds of codification used according to the country or the field of application of the bar code	The tag has a single code at world the level. The uniqueness is guaranteed by the the manufacturers of the chip
There are no particular security systems	Information security access is guaranteed by cryptography systems
Duplication is extremely simple	Duplication is practically impossible
The cost is virtually zero	The cost is still prohibitive for some applications

One of the factors to be considered in the warehouse layout is the degree of complementarity between the pairs of activities carried out in them. The greater is the extent of the logistics flows (materials or information) between two activities, the greater is their degree of complementarity, and therefore the greater is the necessity to carry them out in zones that are close together. In relation to the logistics flow, the layout of the warehouse spaces can generally be designed in three different ways.

The first layout (see Figure 5.16), is called flow-through and provides for the arrangement of the receiving zone and the shipping zone at opposite ends of the warehouse. This kind of layout is based on the supposition that the majority of load units traversing the warehouse will need the same operations and they should therefore be processed in the same sequence: the packages traverse the warehouse from one part (receiving zone) to the other (shipping zone), passing through the storage zone. The flow-through layout is suitable for long, narrow spaces through which transit a high number of packages.

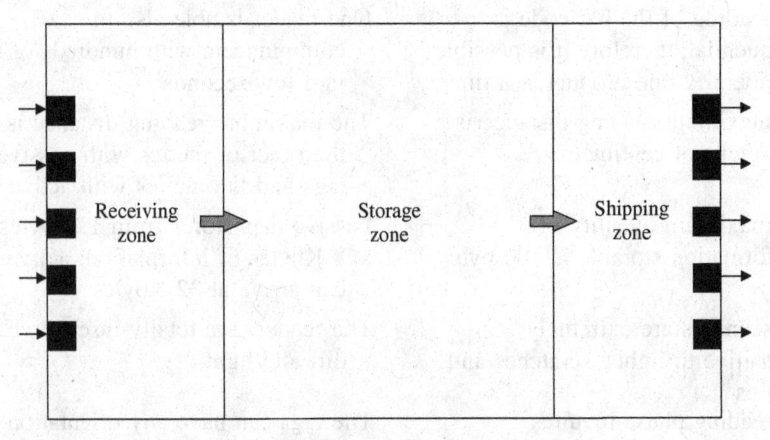

Figure 5.16 Flow-through warehouse layout.

The second layout (see Figure 5.17) is called a U-flow. It has a 'U' shape, which means that its receiving and shipping zones are located on the same side of the warehouse. The U-flow layout adds flexibility to the use of the spaces assigned to the receiving and shipping of the load units. The same docking doors can be used for receiving and shipping operations, and they can be suitably adapted to either activity according to the needs of the moment. This layout is particularly suitable for situations characterized by low material flows and enables the expansion of the warehouse along three sides (those sides without docking doors).

In the two warehouse layouts just illustrated, the orientation of the storage zone aisles is generally parallel to the direction followed by the load units flow. This yields a reduction of the transfer time of a load unit from the storage zone to the shipping zone or from the receiving zone to the storage zone.

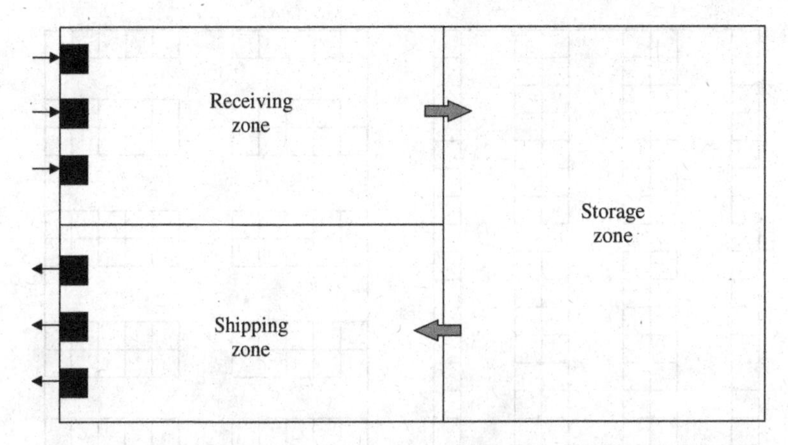

Figure 5.17 U-shaped warehouse layout.

Other possible warehouse layouts are classified as hybrid types and provide for the receiving zone and shipping zone on contiguous sides of the warehouse. This layout (Figure 5.18) is especially useful in situations characterized by low material flows and in warehouses located in square buildings. The hybrid layout does not constrain the orientation of the aisles in the storage zone. Its layout can be longitudinal, that is, with shelving laid out perpendicularly to the shipping zone, or transversal, with the shelving laid out parallel to the shipping zone (see Figure 5.19).

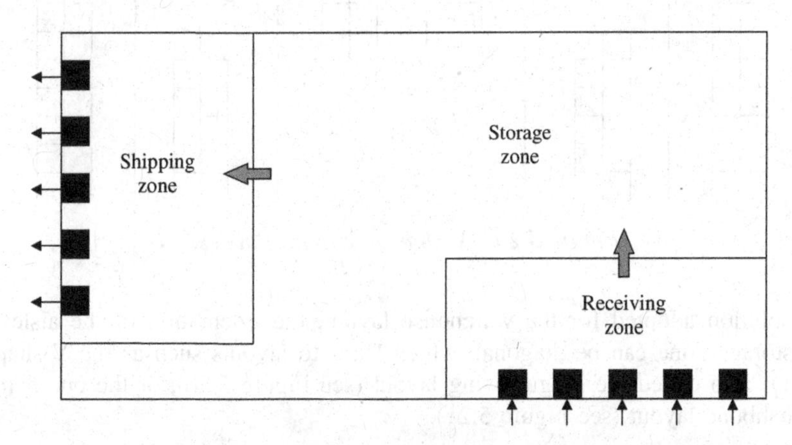

Figure 5.18 Hybrid warehouse layout.

The two different aisle layouts are substantially equivalent and the choice between them can be made by evaluating the potentiality of the warehouse in terms of incoming and outgoing operations. In some rare cases, in particular when storage systems based on stacks are used (see Section 5.2.1), independently of

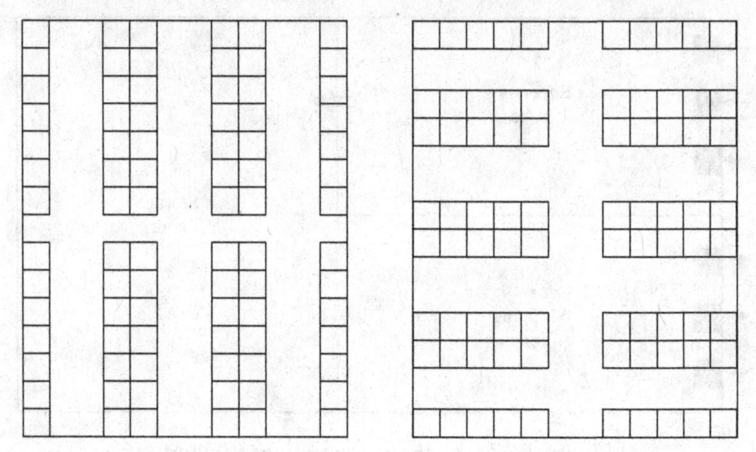

Figure 5.19 Longitudinal or transversal layout of aisles in the storage zone of Figure 5.18.

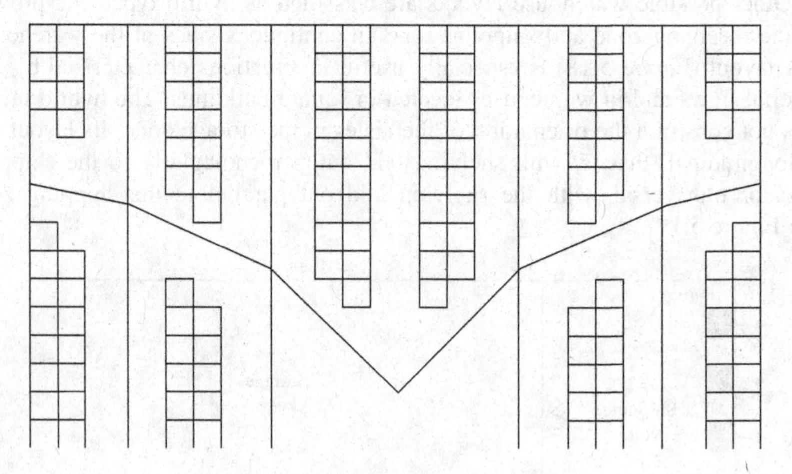

Figure 5.20 V-shaped layout of aisles.

the solution adopted for the warehouse layout, the orientation of the aisles in the storage zone can be diagonal, which leads to layouts such as the V-shaped layout, also called the seagull wing layout (see Figure 5.20), or the one named the fishbone layout (see Figure 5.21).

5.2.3 Sizing of the storage zone

Besides the material-handling systems and equipment, the capacity of the storage zone depends on the storage policy. In a *dedicated storage policy*, each product is assigned to a pre-established set of locations. This approach is easy to implement but results in an underutilization of the storage space. In fact, the space required

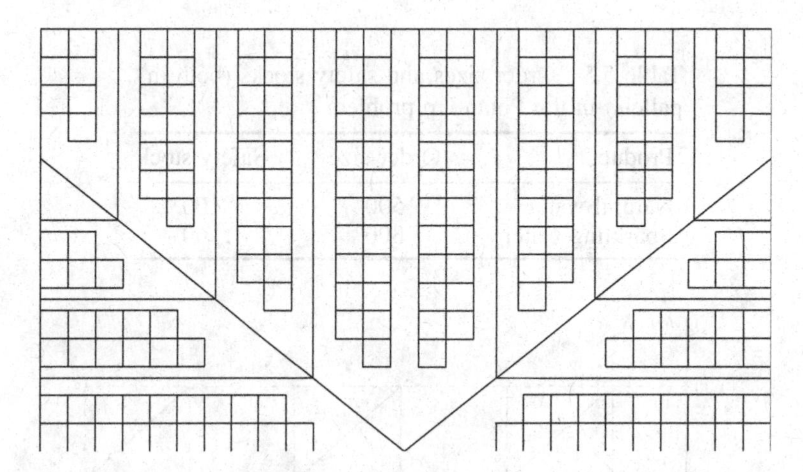

Figure 5.21 Fishbone layout of aisles in the storage zone.

is equal to the sum of the maximum inventory of each product in time. Let n be the number of products and let $I_j(t)$, $j = 1, \ldots, n$, be the inventory level (in terms of storage locations) of product j at time t. The number of required storage locations m_d in a dedicated storage policy is

$$m_d = \sum_{j=1}^{n} \max_t I_j(t). \tag{5.1}$$

In a *random storage policy*, the product allocation is decided dynamically on the basis of the current warehouse occupation and on future arrival and request forecasts. Therefore, the positions assigned to a product are variable in time. In this case, the number of storage locations m_r is

$$m_r = \max_t \sum_{j=1}^{n} I_j(t) \leq m_d. \tag{5.2}$$

The random storage policy allows for a higher utilization of the storage space, but requires that each package be automatically identified through a bar code (or a similar technique), and a database of the current position of all stocked packages is updated at every storage and retrieval.

Potan Up bottles two types of mineral water. In the warehouse located in Hangzhou, China, inventories are managed according to a reorder point policy (see Section 5.3.2.2). Order sized and safety stocks are reported in Table 5.5. Inventory levels as a function of time are illustrated in Figures 5.22 and 5.23. The company is currently using a dedicated storage

Table 5.5 Order sizes and safety stocks (both in pallets) in the Potan Up problem.

Product	Order size	Safety stock
Natural water	500	100
Sparkling water	300	60

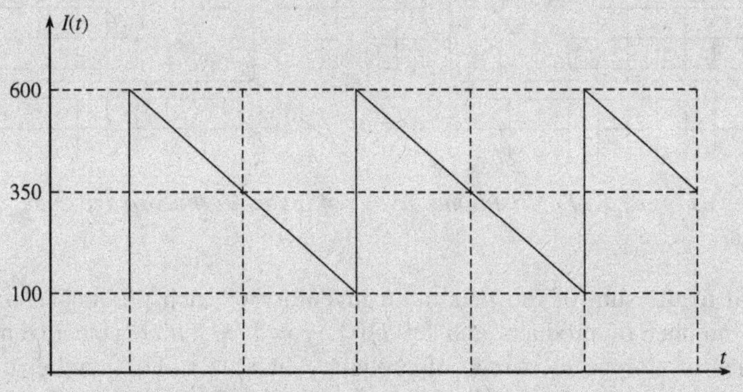

Figure 5.22 Inventory level of natural mineral water in the Potan Up problem.

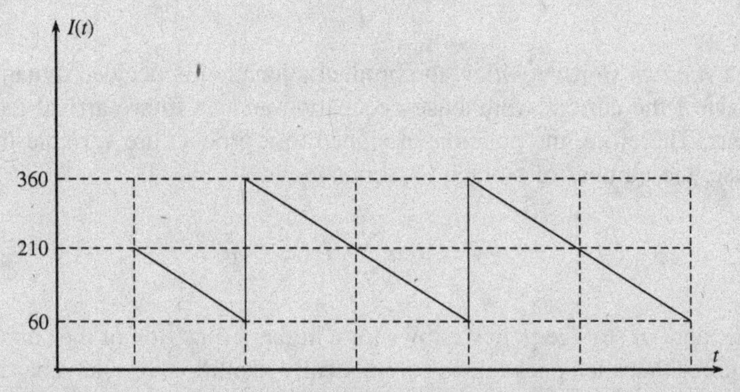

Figure 5.23 Inventory level of sparkling mineral water in the Potan Up problem.

policy. Therefore, the number of storage locations is given by Equation (5.1):

$$m_d = 600 + 360 = 960.$$

The firm is now considering the opportunity of using a random storage policy. The number of storage locations required by this policy would be (see Equation (5.2)):

$$m_r = 600 + 210 = 810.$$

In a *class-based storage policy*, the products are divided into a number of categories according to their demand, and each category is associated with a set of zones where the products are stored according to a random storage policy. The class-based storage policy reduces to the dedicated storage policy if the number of categories is equal to the number of products, and to the random storage policy if there is a single category.

Sisa adopted a class-based storage policy for its DC in Gioia Tauro, Italy. In particular, six product categories are provided for. The description, number of products per category and occupation of the storage zone are shown in Table 5.6.

Table 5.6 Product categories of the Sisa DC at Gioia Tauro.

Category	Description	Number of products	Occupation of storage zone (%)
1	Liquid food	1 105	17
2	Dry food	2 782	43
3	High-cost products	247	4
4	Domestic cleaning products	504	8
5	Personal hygiene	699	11
6	Other	1 163	18
	Total	6 500	100

Once the number of storage locations is known, the optimal size of the storage zone can be determined. It should be observed that a solution to this problem strongly depends on the selected warehouse layout, on the kind of storage and material-handling systems used, which affect the storage and retrieval times of the products in warehouse. The problem will therefore be solved for a specific warehouse layout, even though the approach followed remains valid in general, in the sense that it can be easily extended to other cases. In particular, it is supposed that the warehouse layout is of the flow-through type with a single entrance and a single exit for the storage zone, that the material-handling systems are without mobility constraints and that conventional racks are used in the storage zone with the layout shown in Figure 5.24. The case in which the storage zone is served by stacker cranes will be illustrated in Section 5.6.

The problem of the optimal sizing of the storage zone shown in Figure 5.24 consists in the determination of the measurement of length L_x and width L_y (the height of the warehouse is in fact determined by the height of the storage system). Let m be the required number of storage locations; α_x and α_y the occupation of a unit load along the directions x and y, respectively; w_x and w_y, the width of the side aisles and of the central aisle, respectively; n_z the number of storage

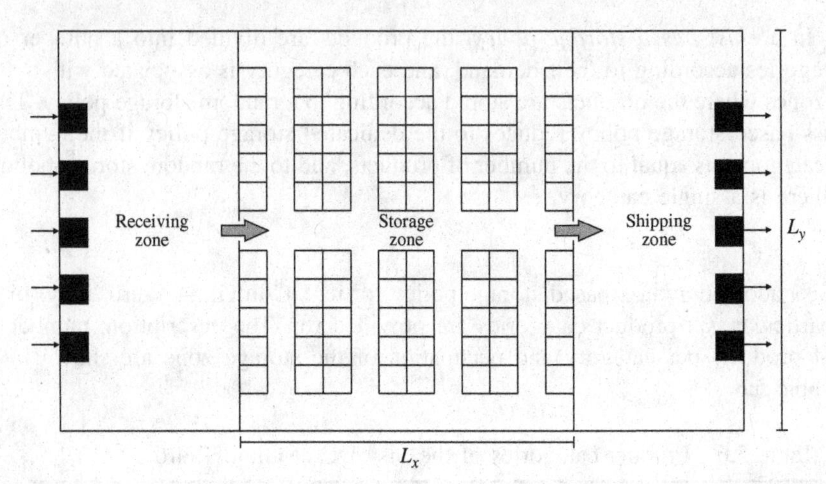

Figure 5.24 Warehouse layout chosen to illustrate the storage zone sizing problem.

locations along the z direction allowed by the storage system adopted and v the average speed of a picker. The decision variables are n_x, the number of storage locations along the x direction, and n_y, the number of storage locations along the y direction.

The extension L_x of the storage zone along the direction x is given by the following relation:

$$L_x = \left(\alpha_x + \frac{w_x}{2}\right) n_x$$

where, according to Figure 5.24, n_x is assumed to be an even number. Similarly, the extension L_y is

$$L_y = \alpha_y n_y + 3w_y.$$

Therefore, under the hypothesis that a material-handling operation consists of storing or retrieving a single load unit, and all storage locations have the same probability of being accessed, the average distance covered by a picker is $L_x + L_y/2$. Hence, the problem of sizing the storage zone can be formulated as follows:

$$\text{Minimize } \left(\alpha_x + \frac{w_x}{2}\right)\frac{n_x}{v} + \frac{\alpha_y n_y + 3w_y}{2v} \tag{5.3}$$

subject to

$$n_x n_y n_z \geq m \tag{5.4}$$

$$n_x,\ n_y \geq 0,\ \text{integer} \tag{5.5}$$

$$n_x \text{ even}, \tag{5.6}$$

where the objective function (5.3) is the average travel time of a picker, while inequality (5.4) states that the number of storage locations is at least equal to m.

Problem (5.3)–(5.6) can be easily solved in the following way. The integrity constraints on the variables n_x and n_y and constraint (5.6) are relaxed. Hence, inequality (5.4) will be satisfied at the optimum for the relaxed problem as an equality:

$$n_x = \frac{m}{n_y n_z}. \tag{5.7}$$

Therefore, n_x can be removed from the relaxed problem, whose formulation follows:

$$\text{Minimize } \left(\alpha_x + \frac{w_x}{2}\right)\frac{m}{n_y n_z v} + \frac{\alpha_y n_y + 3w_y}{2v} \tag{5.8}$$

subject to

$$n_y \geq 0.$$

Since the objective function (5.8) is convex, the minimizer n'_y can be found through the following relation:

$$\frac{d}{d\,n_y}\left(\left(\alpha_x + \frac{w_x}{2}\right)\frac{m}{n_y n_z v} + \frac{\alpha_y n_y + 3w_y}{2v}\right)\Bigg|_{n_y=n'_y} = 0.$$

Hence,

$$n'_y = \sqrt{\frac{2m\left(\alpha_x + \frac{w_x}{2}\right)}{\alpha_y n_z}}. \tag{5.9}$$

Finally, replacing n_y in Equation (5.7) by the n'_y value given by Equation (5.9), n'_x is determined:

$$n'_x = \sqrt{\frac{m\alpha_y}{2n_z(\alpha_x + \frac{w_x}{2})}}. \tag{5.10}$$

The optimal values n^*_x and n^*_y of the decision variables n_x and n_y are obtained by suitably integer rounding the values of n'_x and n'_y, in such a way to guarantee the satisfaction of inequality (5.4) and that n^*_x be even.

Wagner Bros plans to build a new warehouse near Sidney, Australia, in order to supply its sales points in New South Wales. On the basis of a preliminary analysis of the problem, it has been decided that the facility should accommodate at least 780 pallets of $90 \times 90\,\text{cm}^2$ dimension. The pallets will be stored onto conventional racks and transported by means of reach trucks. Each rack has four shelves, each of which can store a single pallet. Each pallet occupies a $1.05 \times 1.05\,\mathrm{m}^2$ area. Racks are arranged as in Figure 5.24, where side aisles are 3.5 m wide, while the central aisle is 4 m wide. The average speed of a reach truck is 5 km/h. Using Equations (5.9) and (5.10), we obtain

$$n'_x = \sqrt{\frac{780 \times 1.05}{2 \times 4 \times \left(1.05 + \frac{3.5}{2}\right)}} = 6.05;$$

$$n'_y = \sqrt{\frac{2 \times 780 \times \left(1.05 + \frac{3.5}{2}\right)}{1.05 \times 4}} = 32.25.$$

By rounding appropriately, we obtain $n^*_x = 6$ and $n^*_y = 33$; the total number of storage locations turns out to be 792, while $L_x = [1.05 + (3.5/2)] \times 6 = 16.8\,\mathrm{m}$ and $L_y = 1.05 \times 33 + 12 = 46.65\,\mathrm{m}$.

5.2.4 Sizing of the receiving zone

The size of the receiving zone is computed by the number p of docking doors which can be estimated by using the following relation:

$$p = \left\lceil \frac{dt}{qT} \right\rceil, \tag{5.11}$$

where d represents the average number of palletized load units entering the warehouse in a predefined time horizon (e.g. a week), t is the average time needed to unload a vehicle, q is the average capacity of the vehicles and T is the average time available in the time horizon for the vehicles' unloading operations.

Alternatively, p can be computed by considering the worst case, that is, the time period in which the maximum incoming movement to the warehouse is verified, but this leads to an oversizing of the receiving zone with a consequent significant cost increase of the warehouse.

The weekly average number of incoming palletized load units at the SISA DC at Gioia Tauro in Italy is 21 465. The average time needed to unload a vehicle is 18.3 minutes. The average vehicle capacity is 21.7 pallets. Considering that the daily average time available for the vehicle-loading operations is 6.5 hours, for six working days, for a total time of 2 340 minutes, using (5.11), the number p of docking doors for the unloading operations is

$$p = \left\lceil \frac{21\,465 \times 18.3}{21.7 \times 2\,340} \right\rceil = \lceil 7.8 \rceil = 8.$$

An estimate of the size of the receiving zone, including the surfaces assigned to the qualitative and quantitative check-in of load units and the updating of the corresponding inventory levels, is suggested by practical experience, which implies a space of about 50 m^2 for each docking door, sufficient enough to allow the easy unloading of vehicles for the movement of 50 palletized load units. The width of a docking door is 2.6 m on average, and the distance between two adjacent docking ports is about 1.5 m. This means that the dimensions L_x and L_y of the receiving zone can be computed as

$$L_y = 2.6p + 1.5(p + 1), \tag{5.12}$$

from which

$$L_x = \frac{50p}{L_y}. \tag{5.13}$$

The receiving zone of the SISA DC at Gioia Tauro measures about 400 m^2. The L_y dimension, on the basis of (5.12), is 34.3 m, whereas L_x, according to (5.13), is about 11.66 m.

5.2.5 Sizing of the shipping zone

The key factors that need to be taken into account in sizing the shipping zone are similar to those considered for the receiving zone, with the difference that it is sometimes necessary to consider the presence of equipments complementary to the shipping (packaging materials, material packaging lines, scales, labelling lies etc.). In general, it should be observed that the shipping zone can be smaller than the receiving zone since shipping is under the direct control of the warehouse manager, so that suitable planning can limit the traffic congestion of the outgoing packages. The number of docking doors of the shipping zone is determined by using formula (5.11), taking into account that the relevant parameters should be referred to the outgoing operations. However, it should be observed that, in the case of a warehouse with a flow-through layout, the number of docking doors for the incoming and outgoing operations of the warehouse is generally the same. Every docking door can be empirically limited to about 45 m^2 of shipping zone. However, the sizes L_x and L_y of the shipping zone can be determined also as a function of the selected warehouse layout. For example, the size L_y coincides with the corresponding size of the receiving zone in the flow-through-type warehouse; in contrast, the sizes L_x of the receiving and of the shipping zones will coincide in U-shaped warehouses.

There are seven docking doors for the shipping zone of the SISA DC at Gioia Tauro. These doors can be sometimes used also for the receiving zone in seasonal peak situations; the size L_y of the shipping zone is obtained by using (5.12) and results in 30.20 m. Since the warehouse is U-shaped, the size L_x coincides with that of the receiving zone and results in about 11.66 m. Consequently, the shipping zone measures about 352 m^2, that is, 50.30 m^2 for each docking door.

5.3 Tactical decisions for warehouse logistics planning

5.3.1 Product allocation to the storage locations

The allocation of products within the storage zone of a warehouse is based on the principle that fast moving products must be placed closer to the I/O ports in order to minimize the overall handling time. In the sequel, we examine the case of a dedicated storage policy. The allocation problem amounts to assigning each of the m_d storage locations available to a product. Let n be the number of products; m_j, $j = 1, \ldots, n$, the number of storage locations required for product j (in a dedicated storage policy, relation $\sum_{j=1}^{n} m_j \leq m_d$ holds); R the number of I/O ports of the storage zone; o_{jr}, $j = 1, \ldots, n$, $r = 1, \ldots, R$, the average number of handling operations on product j through I/O port r per time period; and t_{rk}, $r = 1, \ldots, R$, $k = 1, \ldots, m_d$, the travel time from I/O port r and storage location k.

Under the hypothesis that all storage locations have an identical utilization rate, it is possible to compute the cost c_{jk}, $j = 1, \ldots, n$, $k = 1, \ldots, m_d$, of assigning storage location k to product j:

$$c_{jk} = \sum_{r=1}^{R} \frac{o_{jr}}{m_j} t_{rk}, \tag{5.14}$$

where o_{jr}/m_j represents the average number of handling operations per time period on product j between I/O port r and any of the storage locations assigned to the product. Consequently, $(o_{jr}/m_j)t_{rk}$ is the average travel time due to storage location k if it is assigned to product j.

Let x_{jk}, $j = 1, \ldots, n$, $k = 1, \ldots, m_d$, be a binary decision variable, equal to 1 if storage location k is assigned to product j, 0 otherwise. The problem of seeking the optimal product allocation to the storage locations can be then modelled as follows:

$$\text{Minimize} \sum_{j=1}^{n} \sum_{k=1}^{m_d} c_{jk} x_{jk} \tag{5.15}$$

subject to

$$\sum_{k=1}^{m_d} x_{jk} = m_j, \quad j = 1, \ldots, n \tag{5.16}$$

$$\sum_{j=1}^{n} x_{jk} \leq 1, \; k = 1, \ldots, m_d \tag{5.17}$$

$$x_{jk} \in \{0, 1\}, \; j = 1, \ldots, n, \; k = 1, \ldots, m_d, \tag{5.18}$$

where constraints (5.16) state that all the packages must be allocated, while constraints (5.17) impose that each storage location k, $k = 1, \ldots, m_d$, can be assigned to at the most one product.

It is worth noting that because of the particular structure of constraints (5.16) and (5.17), relations (5.18) can be replaced with the simpler non-negativity conditions

$$x_{jk} \geq 0, \; j = 1, \ldots, n, \; k = 1, \ldots, m_d, \tag{5.19}$$

since it is known a priori that there exists an optimal solution of problem (5.15)–(5.17), (5.19) in which the decision variables take 0/1 values.

Malabar is an Indian company having a warehouse, whose storage zone has two I/O ports and 40 storage locations, arranged in four conventional racks (see Figure 5.25). The characteristics of the products at stock are reported in Table 5.7, while the distances between the two I/O ports and the storage locations are indicated in Tables 5.8 and 5.9. The optimal product allocation

Figure 5.25 Schematization of the storage zone of the Malabar warehouse.

can found through model (5.15)–(5.17), (5.19), in which $n = 5$, $m_d = 40$, while m_j, $j = 1, \ldots, 5$, are calculated on the basis of the second column of Table 5.7. The cost coefficients c_{jk}, $j = 1, \ldots, 5$, $k = 1, \ldots, 40$, are provided in Tables 5.10 and 5.11 and calculated using Equation (5.14), where it is assumed that travel time t_{rk} from I/O port $r = 1, 2$, to storage location k, $k = 1, \ldots, 40$, is directly proportional to the corresponding distance. The optimal solution is reported in Table 5.12. It is worth noting that two storage locations (locations 26 and 27) are not used since the positions available are 40, while $\sum_{j=1}^{5} m_j = 38$.

Table 5.7 Features of the products of the Malabar company.

Product	Number of storage locations required	Daily number of storages and retrievals	
		I/O port 1	I/O port 2
1	12	25	18
2	6	16	26
3	8	14	30
4	4	24	22
5	8	22	22

Table 5.8 Distance (in metres) between storage locations and I/O port 1 in the storage zone of the Malabar warehouse.

Storage location	Distance	Storage location	Distance	Storage location	Distance	Storage location	Distance
1	2	11	2	21	14	31	14
2	4	12	4	22	16	32	16
3	6	13	6	23	18	33	18
4	8	14	8	24	20	34	20
5	10	15	10	25	22	35	22
6	3	16	3	26	15	36	15
7	5	17	5	27	17	37	17
8	7	18	7	28	19	38	19
9	9	19	9	29	21	39	21
10	11	20	11	30	23	40	23

Table 5.9 Distance (in metres) between storage locations and I/O port 2 in the storage zone of the Malabar warehouse.

Storage location	Distance	Storage location	Distance	Storage location	Distance	Storage location	Distance
1	22	11	22	21	10	31	10
2	20	12	20	22	8	32	8
3	18	13	18	23	6	33	6
4	16	14	16	24	4	34	4
5	14	15	14	25	2	35	2
6	23	16	23	26	11	36	11
7	21	17	21	27	9	37	9
8	19	18	19	28	7	38	7
9	17	19	17	29	5	39	5
10	15	20	15	30	3	40	3

Table 5.10 Cost coefficients c_{jk}, $j = 1, \ldots, 5$, $k = 1, \ldots, 20$, in the Malabar problem.

	Assignment cost				
Storage location	Product 1	Product 2	Product 3	Product 4	Product 5
1	37.17	100.67	86.00	133.00	66.00
2	38.33	97.33	82.00	134.00	66.00
3	39.50	94.00	78.00	135.00	66.00
4	40.67	90.67	74.00	136.00	66.00
5	41.83	87.33	70.00	137.00	66.00
6	40.75	107.67	91.50	144.50	71.50
7	41.92	104.33	87.50	145.50	71.50
8	43.08	101.00	83.50	146.50	71.50
9	44.25	97.67	79.50	147.50	71.50
10	45.42	94.33	75.50	148.50	71.50
11	37.17	100.67	86.00	133.00	66.00
12	38.33	97.33	82.00	134.00	66.00
13	39.50	94.00	78.00	135.00	66.00
14	40.67	90.67	74.00	136.00	66.00
15	41.83	87.33	70.00	137.00	66.00
16	40.75	107.67	91.50	144.50	71.50
17	41.92	104.33	87.50	145.50	71.50
18	43.08	101.00	83.50	146.50	71.50
19	44.25	97.67	79.50	147.50	71.50
20	45.42	94.33	75.50	148.50	71.50

Table 5.11 Cost coefficients c_{jk}, $j = 1, \ldots, 5$, $k = 21, \ldots, 40$, in the Malabar problem.

Storage location	Assignment cost: Product 1	Product 2	Product 3	Product 4	Product 5
21	44.17	80.67	62.00	139.00	66.00
22	45.33	77.33	58.00	140.00	66.00
23	46.50	74.00	54.00	141.00	66.00
24	47.67	70.67	50.00	142.00	66.00
25	48.83	67.33	46.00	143.00	66.00
26	47.75	87.67	67.50	150.50	71.50
27	48.92	84.33	63.50	151.50	71.50
28	50.08	81.00	59.50	152.50	71.50
29	51.25	77.67	55.50	153.50	71.50
30	52.42	74.33	51.50	154.50	71.50
31	44.17	80.67	62.00	139.00	66.00
32	45.33	77.33	58.00	140.00	66.00
33	46.50	74.00	54.00	141.00	66.00
34	47.67	70.67	50.00	142.00	66.00
35	48.83	67.33	46.00	143.00	66.00
36	47.75	87.67	67.50	150.50	71.50
37	48.92	84.33	63.50	151.50	71.50
38	50.08	81.00	59.50	152.50	71.50
39	51.25	77.67	55.50	153.50	71.50
40	52.42	74.33	51.50	154.50	71.50

Table 5.12 Optimal allocation of products in the storage zone of the Malabar warehouse.

Storage location	Product	Storage location	Product	Storage location	Product	Storage location	Product
1	1	11	1	21	5	31	5
2	4	12	4	22	2	32	2
3	4	13	4	23	2	33	2
4	5	14	5	24	2	34	2
5	5	15	5	25	3	35	3
6	1	16	1	26	–	36	5
7	1	17	1	27	–	37	5
8	1	18	1	28	3	38	3
9	1	19	1	29	3	39	3
10	1	20	1	30	3	40	3

If the warehouse has a single I/O port ($R = 1$), the optimal solution of problem (5.15)–(5.18) can be easily found. In fact, under this hypothesis, cost coefficients c_{jk}, $j = 1, \ldots, n$, $k = 1, \ldots, m_d$, take the following form:

$$c_{jk} = \frac{o_{j1}}{m_j} t_{1k} = a_j b_k,$$

where $a_j = o_{j1}/m_j$ and $b_k = t_{1k}$ depend only on product j and on storage location k, respectively. Then, the optimal product allocation can be determined by using the following procedure.

Step 1. Construct a vector α of $\sum_{j=1}^{n} m_j$ components, in which there are m_j copies of each a_j, $j = 1, \ldots, n$. Sort the vector α by non-increasing values of its components. Define $\sigma_\alpha(i)$ in such a way that $\sigma_\alpha(i) = j$ if $\alpha_i = a_j$, $i = 1, \ldots, \sum_{h=1}^{n} m_h$.

Step 2. Let b be the vector of m_d components corresponding to values b_k, $k = 1, \ldots, m_d$. Sort the vector b by nondecreasing values of its components. Let β be the vector of $\sum_{j=1}^{n} m_j$ components, corresponding to the first $\sum_{j=1}^{n} m_j$ components of the sorted vector b. Define $\sigma_\beta(i)$ in such a way that $\sigma_\beta(i) = k$ if $\beta_i = b_k$, $i = 1, \ldots, \sum_{h=1}^{n} m_h$.

Step 3. Determine the optimal solution of problem (5.15)–(5.18) as:

$$x^*_{\sigma_\alpha(i), \sigma_\beta(i)} = 1, \; i = 1, \ldots, \sum_{j=1}^{n} m_j$$

and $x^*_{jk} = 0$, for all the remaining components.

This procedure is based on the fact that the minimization of the scalar product of two vectors α and β is achieved by ordering α by non-increasing values and β by non-decreasing values.

If the storage zone of the warehouse of Malabar company (see the previous problem) has a single I/O port (corresponding to port 1 in Figure 5.25), coefficients a_j, $j = 1, \ldots, 5$, are those reported in Table 5.13. For the sake of simplicity, travel times are assumed to be equal to distances (see Table 5.8). Values of α_i, $\sigma_\alpha(i)$, β_i, $\sigma_\beta(i)$, $i = 1, \ldots, 38$, are reported in Table 5.14. The optimal solution is reported in Table 5.15. It is worth noting that no package is allocated to the storage locations farthest from the I/O port (locations 30 and 40).

Table 5.13 Characteristics of products in the Malabar problem.

Product (j)	Number of storage locations required	Number of storages and retrievals per day	Coefficient a_j
1	12	43	3.58
2	6	42	7.00
3	8	44	5.50
4	4	46	11.50
5	8	44	5.50

Table 5.14 Values of α_i, $\sigma_\alpha(i)$, β_i, $\sigma_\beta(i)$, for $i = 1, \ldots, 38$, in the Malabar problem.

i	α_i	$\sigma_\alpha(i)$	β_i	$\sigma_\beta(i)$	i	α_i	$\sigma_\alpha(i)$	β_i	$\sigma_\beta(i)$
1	11.50	4	2	1	20	5.50	5	11	20
2	11.50	4	2	11	21	5.50	5	14	21
3	11.50	4	3	6	22	5.50	5	14	31
4	11.50	4	3	16	23	5.50	5	15	26
5	7.00	2	4	2	24	5.50	5	15	36
6	7.00	2	4	12	25	5.50	5	16	22
7	7.00	2	5	7	26	5.50	5	16	32
8	7.00	2	5	17	27	3.58	1	17	27
9	7.00	2	6	3	28	3.58	1	17	37
10	7.00	2	6	13	29	3.58	1	18	23
11	5.50	3	7	8	30	3.58	1	18	33
12	5.50	3	7	18	31	3.58	1	19	28
13	5.50	3	8	4	32	3.58	1	19	38
14	5.50	3	8	14	33	3.58	1	20	24
15	5.50	3	9	9	34	3.58	1	20	34
16	5.50	3	9	19	35	3.58	1	21	29
17	5.50	3	10	5	36	3.58	1	21	39
18	5.50	3	10	15	37	3.58	1	22	25
19	5.50	5	11	10	38	3.58	1	22	35

Table 5.15 Optimal product allocation to storage locations in the Malabar problem.

Storage location	Product	Storage location	Product	Storage location	Product	Storage location	Product
1	4	11	4	21	5	31	5
2	2	12	2	22	5	32	5
3	2	13	2	23	1	33	1
4	3	14	3	24	1	34	1
5	3	15	3	25	1	35	1
6	4	16	4	26	5	36	5
7	2	17	2	27	1	37	1
8	3	18	3	28	1	38	1
9	3	19	3	29	1	39	1
10	5	20	5	30	–	40	–

5.3.2 Inventory management

Inventory management in a warehouse requires decisions regarding the replenishment modality (how much to order of each product and when), aimed at providing a pre-established level of service at minimum cost. Such decision-making processes are substantially different if they deal with independent-demand products (that is, finished products), rather than dependent-demand products (components of finished products). For the latter situation, inventory management techniques are based on an material requirement planning (MRP) logic. For a deeper examination, readers should refer to specific texts, whereas attention in this chapter will be focused only to the inventory management policies of independent-demand products.

For these products the inventory management models will be classified as deterministic or stochastic, depending on when product demand and lead time (defined as the amount of time between placing an order and receiving the items ordered in the warehouse) are known with certainty or uncertainty, respectively.

5.3.2.1 Deterministic models

Deterministic models, besides having several applications (including warehouses of petrochemical companies supplied by means of pipelines, companies which use a large quantity of products in the warehouse, enough to justify stock replenishment at hourly or daily intervals etc.), present general features which can be used, as will be seen in this chapter, to derive other models equally widespread in practice.

Deterministic models with a constant demand rate In the case of a single product with constant demand, the replenishments are periodic and the objective pursued consists in the minimization of the average total cost in that period. The planning horizon is assumed to be infinite (for the finite horizon case, see Problem 5.14). Let d be the constant demand rate, T the time lapse between two consecutive orders and q the order size (i.e. the amount of product ordered at each replenishment). These quantities are such that

$$q = dT. \tag{5.20}$$

Let c be the value of an item (supposed to be independent of the order size), k the fixed reorder cost, h the holding cost per item and per time unit, u the shortage cost per item, independent of the duration of the shortage, and v the shortage cost per item and per time unit. The holding cost h can be expressed as a fraction p of c:

$$h = pc. \tag{5.21}$$

Parameter p is a banking interest rate (measuring capital cost) increased to take into account warehousing costs. Moreover, let $I(t)$ be the inventory level at time t, $A(t)$ the shortage level al time t, M the maximum inventory level, m the maximum shortage, l the lead time. The problem is to determine q (or, equivalently, T), and m in such a way that the overall average cost per time unit is minimum. We distinguish two different cases, according to whether the stock reorder time is instantaneous or not.

Case of non-instantaneous resupply. Let $T_r \geq 0$ be the *replenishment time*, that is, the time to make a replenishment, and r the *replenishment rate*, that is, the number of items per unit of time received during T_r. The following relation holds:

$$q = rT_r. \tag{5.22}$$

The inventory level $I(t)$ as a function of time t is shown in Figure 5.26. Dashed lines represent the cumulative number of items arriving at the warehouse during a replenishment (their slope is r). Since items are picked up at a rate d while a replenishment takes place, the net number of items stocked per time unit during T_r is $r - d$. Finally, after a replenishment, the stock level decreases at a rate d. Let T_1, T_2, T_3 and T_4 be the time the inventory level takes to go from $-m$ to 0, from 0 to M, from M to 0 and from 0 to $-m$, respectively.

The maximum inventory level M is given by

$$m + M = (r - d)T_r.$$

Therefore, from Equation (5.22),

$$M = (r - d)T_r - m = q(1 - d/r) - m.$$

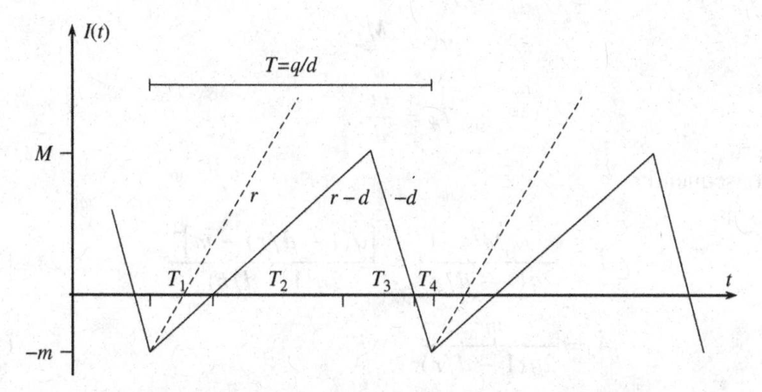

Figure 5.26 Inventory level as a function of time.

The total average cost per time unit $\mu(q, m)$ is

$$\mu(q, m) = \frac{1}{T}(k + cq + h\bar{I}T + um + v\bar{A}T). \qquad (5.23)$$

The quantity in parentheses in the right hand side is the *average cost per period*, given by the fixed and variable costs of a resupply (k and cq, respectively), and the holding cost $h\bar{I}T$, plus the shortage costs (um and $v\bar{A}T$). The holding and shortage costs depend on the average inventory level $\bar{I}$, and on the average shortage level $\bar{A}$, respectively:

$$\bar{I} = \frac{1}{T}\int_0^T I(t)dt = \frac{1}{T}\left(\frac{M(T_2 + T_3)}{2}\right);$$

$$\bar{A} = \frac{1}{T}\int_0^T A(t)dt = \frac{1}{T}\left(\frac{m(T_1 + T_4)}{2}\right).$$

Moreover, since

$$m = (r - d)T_1;$$

$$M = (r - d)T_2;$$

$$M = dT_3;$$

$$m = dT_4,$$

the time lapses T_1, T_2, T_3 and T_4 are given by

$$T_1 = \frac{m}{r - d};$$

$$T_2 = \frac{M}{r - d};$$

$$T_3 = \frac{M}{d};$$

$$T_4 = \frac{m}{d}.$$

Consequently,

$$\bar{I} = \frac{M^2}{2q(1-d/r)} = \frac{\left[q(1-d/r)-m\right]^2}{2q(1-d/r)}; \tag{5.24}$$

$$\bar{A} = \frac{m^2}{2q(1-d/r)}. \tag{5.25}$$

Finally, using Equations (5.20), (5.24) and (5.25), Equation (5.23) can be rewritten as

$$\mu(q,m) = kd/q + cd + \frac{h\left[q(1-d/r)-m\right]^2}{2q(1-d/r)} + umd/q + \frac{vm^2}{2q(1-d/r)}. \tag{5.26}$$

If shortages are allowed, the minimum point (q^*, m^*) of the convex function $\mu(q,m)$ can be obtained by solving the equations

$$\left.\frac{\partial}{\partial q}\mu(q,m)\right|_{q=q^*,m=m^*} = 0;$$

$$\left.\frac{\partial}{\partial m}\mu(q,m)\right|_{q=q^*,m=m^*} = 0.$$

As a result,

$$q^* = \sqrt{\frac{h+v}{v}}\sqrt{\frac{2kd}{h(1-d/r)} - \frac{(ud)^2}{h(h+v)}}, \tag{5.27}$$

and

$$m^* = \frac{(hq^* - ud)(1-d/r)}{(h+v)}. \tag{5.28}$$

If shortages are not allowed (see Figure 5.27), Equation (5.26) can be simplified since $m = 0$:

$$\mu(q) = kd/q + cd + \frac{hq(1-d/r)}{2}. \tag{5.29}$$

Hence, a single equation has to be solved:

$$\left.\frac{d}{d\,q}\mu(q)\right|_{q=q^*} = 0,$$

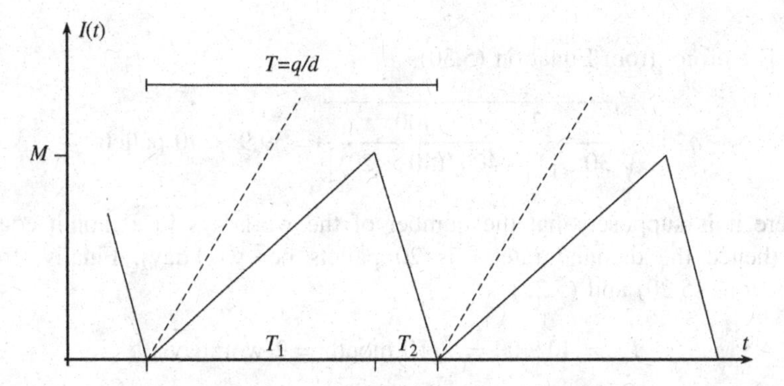

Figure 5.27 Inventory level as a function of time when shortage is not allowed.

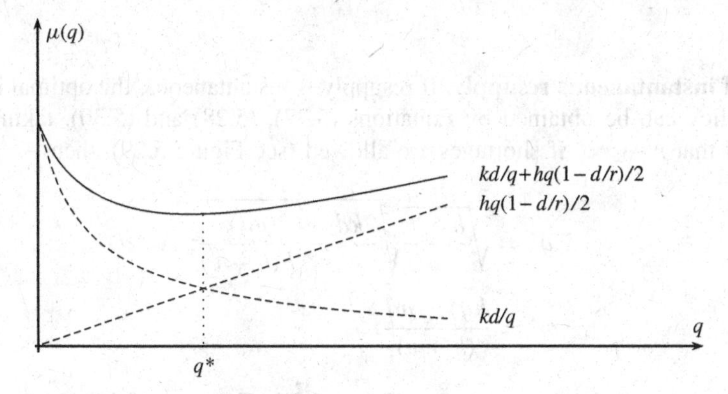

Figure 5.28 Average costs as a function of q.

Finally, the optimal order size q^* is (see Figure 5.28):

$$q^* = \sqrt{\frac{2kd}{h(1 - d/r)}}. \tag{5.30}$$

Golden Food distributes tinned foodstuff in Great Britain. In a warehouse located in Birmingham, the demand rate d for tomato puree is 400 pallets a month. The value of a pallet is $c =$ £ 2 500 and the annual interest rate p is 14.5% (including warehousing costs). Issuing an order costs £ 30. The replenishment rate r is 40 pallets per day. Shortages are not allowed. The holding cost is given by

$$h = 0.145 \times 2\,500 = \text{£ } 362.5 \text{ (year per pallet)} =$$
$$= \text{£ } 30.2 \text{ (month per pallet)}.$$

Therefore, from Equation (5.30):

$$q^* = \sqrt{\frac{2 \times 30 \times 400}{30.2\left[1 - 400/(40 \times 20)\right]}} = 39.9 \approx 40 \text{ pallets},$$

where it is supposed that the number of the workdays in a month equals 20 (hence the demand rate d is 20 pallets per workday). Finally, from Equations (5.20) and (5.22),

$$T^* = 40/400 = 1/10 \text{ month} = 2 \text{ workdays};$$

$$T_r^* = 40/40 = 1 \text{ workday}.$$

Case of instantaneous resupply. If resupply is instantaneous, the optimal inventory policy can be obtained by Equations (5.27), (5.28) and (5.30), taking into account that $r \to \infty$. If shortages are allowed (see Figure 5.29), then

$$q^* = \sqrt{\frac{h+v}{v}}\sqrt{\frac{2kd}{h} - \frac{(ud)^2}{h(h+v)}};$$

$$m^* = \frac{hq^* - ud}{(h+v)}.$$

If shortages are not allowed (see Figure 5.30), Equation (5.29) becomes

$$\mu(q) = kd/q + cd + \frac{hq}{2} \tag{5.31}$$

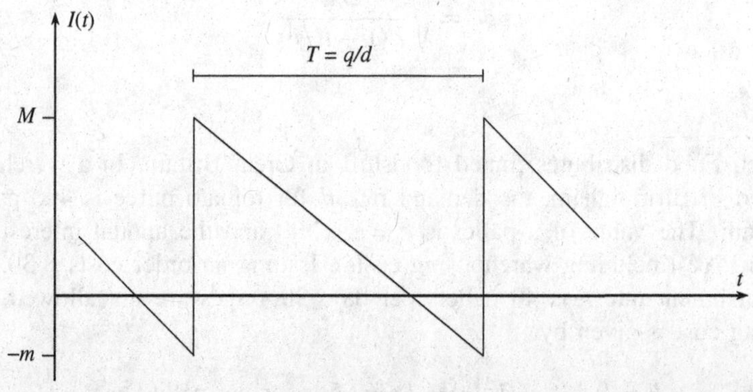

Figure 5.29 Inventory level as a function of time in the instantaneous resupply case.

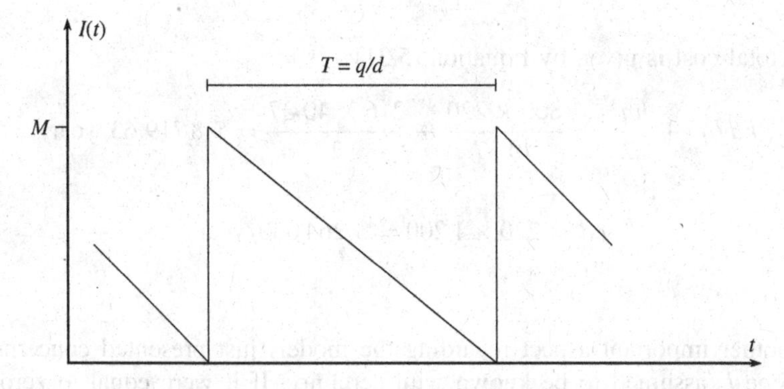

Figure 5.30 Inventory level as a function of time in the EOQ model.

and the optimal order size is given by

$$q^* = \sqrt{\frac{2kd}{h}}. \tag{5.32}$$

This is the classical (EOQ) *Economic Order Quantity* model introduced by F.W. Harris in 1913. The total cost per time unit of an EOQ policy is

$$\mu(q^*) = \sqrt{2kdh} + cd.$$

Optimal policies with no backlog (in particular, the EOQ policy) satisfy the ZIO (*Zero Inventory Ordering*) property which states that an order is received exactly when the inventory level falls to zero.

Al-Bufeira Motors manufactures spare parts for aircraft engines in Saudi Arabia. Its component Y02PN, produced in a plant located in Jiddah, has a demand of 220 units per year and a production cost of \$ 1 200. Manufacturing this product requires a time-consuming set-up that costs \$ 800. The current annual interest rate p is 18%, including warehousing costs. Shortages are not allowed. The holding cost is

$$h = 0.18 \times 1\,200 = \$\ 216 \text{ (year per unit)}.$$

Therefore, from Equation (5.32),

$$q^* = 40.4 \text{ units},$$

and, from Equation (5.20),

$$T^* = 40.37/220 = 0.18 \text{ years} = 66.8 \text{ days}.$$

The total cost is given by Equation (5.31):

$$kd/q^* + \frac{hq^*}{2} = \frac{800 \times 220}{40.37} + \frac{216 \times 40.37}{2} = \$\ 8\,719.63/\text{year},$$

plus

$$cd = 220 \times 1\,200 = \$\ 264\,000/\text{year}.$$

Another important aspect regarding the models just presented concerns the lead time l, assumed to be known with certainty. If it were equal to zero, the reorder could be issued at the moment when the inventory level becomes equal to $-m$. If, however, $l > 0$, the replenishment must be anticipated by l time units so that the shortage level does not become less than $-m$. Starting from l, the reorder point s can also be calculated (s is defined as the value of $I(t)$ in correspondence of which a new replenishment is needed):

$$s = (l - \lfloor l/T \rfloor T)d - m,$$

where $\lfloor l/T \rfloor$ is the number of delay periods included in the lead time (see Figure 5.31).

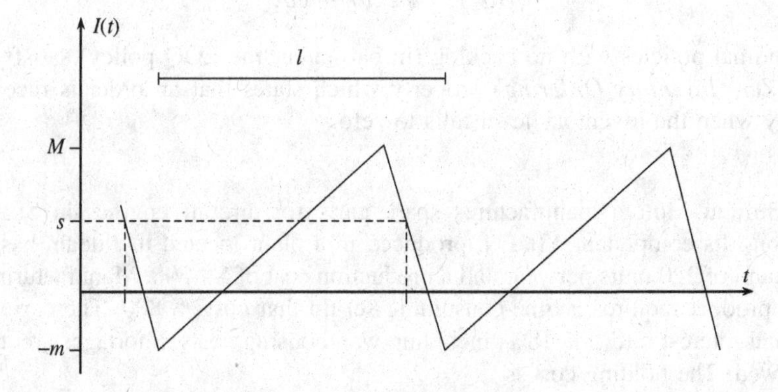

Figure 5.31 Reorder point.

In the Al-Bufeira Motors problem, a set-up has to be planned seven days in advance. Assuming $T = 67$ days, the reorder point s is:

$$s = (7 - \lfloor 7/67 \rfloor \times 67)\ \frac{220}{365} - 0 \approx 4 \text{ units}.$$

Deterministic models with a time-varying demand rate In the case where demand rate is deterministic but time-varying, the following procedure can be adopted. Let $1, \ldots, T_H$ be a finite and discrete time horizon. In addition, let d_t, $t = 1, \ldots, T_H$, be the demand at time period t, k the fixed reorder cost, and h the holding cost. The problem is to decide how much to order in each time period in such a way that the sum of reorder costs plus holding costs is minimized. No backlog is allowed. In 1958, H.M. Wagner and T.M. Whithin formulated this problem as follows. The decision variables are the amounts q_t, $t = 1, \ldots, T_H$, ordered at the beginning of time period t, the inventory level I_t, $t = 1, \ldots, T_H$, at the end of time period t; in addition let y_t, $t = 1, \ldots, T_H$, be a binary decision variable equal to 1 if an order is placed in time period t, 0 otherwise. The problem is then

$$\text{Minimize} \sum_{t=1,\ldots,T_H} (ky_t + hI_t) \tag{5.33}$$

subject to

$$I_t = I_{t-1} + q_t - d_t, t = 1, \ldots, T_H \tag{5.34}$$

$$q_t \leq y_t \sum_{r=t,\ldots,T_H} d_r, t = 1, \ldots, T_H \tag{5.35}$$

$$I_0 = 0 \tag{5.36}$$

$$I_t \geq 0, t = 1, \ldots, T_H$$

$$q_t \geq 0, t = 1, \ldots, T_H$$

$$y_t \in \{0, 1\}, t = 1, \ldots, T_H,$$

where the objective function (5.33) is the total cost. Equations (5.34) are the inventory-balance constraints, inequalities (5.35) state that for each time period $t = 1, \ldots, T_H$, q_t is zero if y_t is zero, and equation (5.36) specifies the initial inventory.

An optimal solution of the Wagner-Within model can be obtained in $O(T_H^2)$ time through a dynamic programming algorithm. This algorithm is based on the theoretical result which states that any optimal policy satisfies the ZIO property, that is,

$$q_t I_{t-1} = 0, t = 1, \ldots, T_H.$$

The proof is left to the reader as an exercise (see Problem 5.16). A corollary of the previous proposition is that in an optimal policy, the amount ordered at each time period is the total demand of a set of consecutive subsequent time periods.

The algorithm is as follows. Let $G = (V, A)$ be a directed acyclic graph, where $V = \{1, \ldots, T_H, T_H + 1\}$ is a vertex set and $A = \{(t, t') : t = 1, \ldots, T_H, t' = t + 1, \ldots, T_H + 1\}$ is an arc set. With each arc (t, t') is associated the cost of ordering in time period t to satisfy the demands in time periods $t, t + 1, \ldots, t' - 1$:

$$g_{tt'} = k + h \sum_{r=t,\ldots,t'-1} (r - t)d_r.$$

Then the shortest path between vertices 1 and $T_H + 1$ corresponds to a least cost inventory policy.

Sao Vincente Chemical is a Portuguese company producing lubricants. In the next year its product Serrado Oil is expected to have a demand of 720, 1 410, 830 and 960 pallets in winter, spring, summer and autumn, respectively. Manufacturing this product requires a time-consuming set-up that costs € 8 900. The current annual interest rate p is 7.5%, including warehousing costs. The variable production cost amounts to € 350 per pallet while the initial inventory is zero. The holding cost is

$$h = 0.075 \times 350/4 = \text{€ } 6.56/(\text{season per unit}).$$

Let $t = 1, 2, 3, 4$, represent the winter, spring, summer and autumn periods, respectively. By solving the Wagner-Within model, it follows that the optimal policy is to produce at the beginning of winter and summer of the next year. In particular, $y_1 = y_3 = 1$, $y_2 = y_4 = 0$, $q_1 = 2\,130$, $q_2 = 0$, $q_3 = 1\,790$, $q_4 = 0$, $I_1 = 1\,410$, $I_2 = 0$, $I_3 = 960$, $I_4 = 0$. Total holding and set-up costs amount to € 33 353.

Deterministic models with quantity discounts In the previous models, it has been assumed that the value of an item is always constant (equal to c). In practice, quantity discounts offered by suppliers, or economies of scale in the manufacturing processes, make the value of an item dependent on the order size q. In the following, the most practical applications of quantity discounts are examined under the EOQ hypothesis: (a) *quantity-discounts-on-all-units* and (b) *incremental quantity discounts*. In case (a) (quantity-discounts-on-all-units), function $f(q)$ is assumed to be piecewise linear (see Figure 5.32):

$$f(q) = c_i q, \; q_{i-1} \leq q < q_i, \; i = 1, 2, \ldots,$$

where $q_0 = 0, q_1, \ldots$ are known *discount breaks* ($q_i < q_{i+1}, i = 1, 2, \ldots$), $f(q_0) = 0$, and $c_i > c_{i+1}$, $i = 1, 2 \ldots$. Hence, if the order size q is included between discount breaks q_{i-1} and q_i, the value of *every* item is c_i, $i = 1, 2, \ldots$. It is worth noting that, depending on c_i coefficients, $f(q)$ can be greater than $f(q')$ for $q < q'$ (see Figure 5.32). In practice, the *effective* cost function is

$$f(q) = \min\{c_i q, c_{i+1} q_i\}, q_{i-1} \leq q < q_i, \; i = 1, 2, \ldots$$

The total average cost function $\mu(q)$ can be written as

$$\mu(q) = \mu_i(q), \; q_{i-1} \leq q < q_i, \; i = 1, 2, \ldots,$$

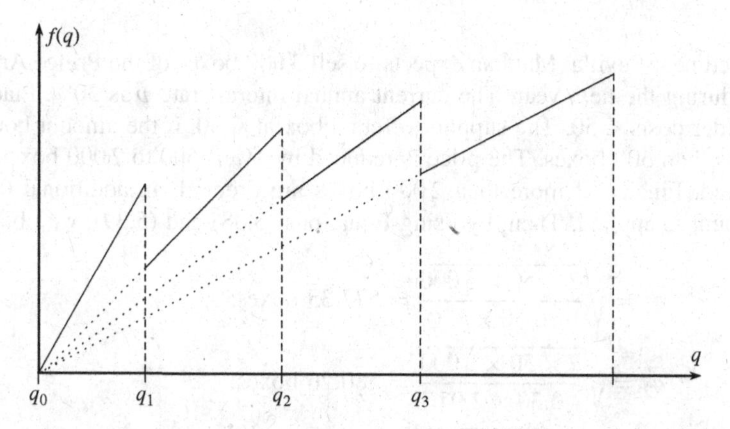

Figure 5.32 The value of q items in the case of quantity-discounts-on-all-units.

where, as $h_i = pc_i$, $i = 1, 2, \ldots$,

$$\mu_i(q) = kd/q + c_i d + \frac{h_i q}{2}, \quad i = 1, 2, \ldots \tag{5.37}$$

Then, the optimal order size q^* can be obtained through the following procedure.

Step 1. By imposing

$$\left.\frac{d\mu_i(q_i)}{d\,q_i}\right|_{q_i = q_i'} = 0, \quad i = 1, 2, \ldots,$$

determine the order size q_i', $i = 1, 2, \ldots$, that minimizes $\mu_i(q)$,

$$q_i' = \sqrt{\frac{2kd}{h_i}}, \quad i = 1, 2, \ldots \tag{5.38}$$

Let

$$q_i^* = \begin{cases} q_{i-1}, & \text{if } q_i' < q_{i-1} \\ q_i', & \text{if } q_{i-1} \le q_i' \le q_i \\ q_i, & \text{if } q_i' > q_i. \end{cases}, \quad i = 1, 2, \ldots \tag{5.39}$$

Step 2. Compute the optimal solution $q^* = q_{i^*}^*$, where

$$i^* = \arg \min_{i=1,2,\ldots} \{\mu(q_i^*)\}.$$

Maliban runs more than 200 stationery outlets in Spain. The firm buys its products from a restricted number of suppliers and stores them in a warehouse located near Sevilla. Maliban expects to sell 3 000 boxes of the Prince Arthur pen during the next year. The current annual interest rate p is 30%. Placing an order costs € 50. The supplier offers a box at € 50, if the amount bought is less than 500 boxes. The price is reduced by 1% if 500 to 2 000 boxes are ordered. Finally, if more than 2 000 boxes are ordered, an additional 0.5% discount is applied. Then, by using Equations (5.38) and (5.39), we obtain

$$q_1' = \sqrt{\frac{2 \times 50 \times 3000}{0.30 \times 3}} = 577.35 \text{ boxes;}$$

$$q_2' = \sqrt{\frac{2 \times 50 \times 3000}{0.30 \times 2.97}} = 580.26 \text{ boxes;}$$

$$q_3' = \sqrt{\frac{2 \times 50 \times 3000}{0.30 \times 2.955}} = 581.73 \text{ boxes;}$$

$$q_1^* = 500 \text{ boxes}, q_2^* = 580.26 \text{ boxes}, q_3^* = 2000 \text{ boxes}.$$

By comparing the corresponding annual average costs given by Equations (5.37), the optimal order size is $q^* = 580$ boxes ($\approx q_2^*$), corresponding to an annual cost of € 9 427.

In case (b) (incremental quantity discounts), function $f(q)$ is assumed to be depend on q as follows (see Figure 5.33):

$$f(q) = f(q_{i-1}) + c_i(q - q_{i-1}),\ q_{i-1} \le q < q_i,\ i = 1, 2, \ldots, \tag{5.40}$$

where $q_0 = 0, q_1, \ldots$ are known discount breaks ($q_i < q_{i+1}, i = 1, 2, \ldots$), $f(q_0) = 0$, and $c_i > c_{i+1},\ i = 1, 2 \ldots$. Consequently, if the order size q is included between discount breaks q_{i-1} and q_i, the value of $(q - q_{i-1})$ items is

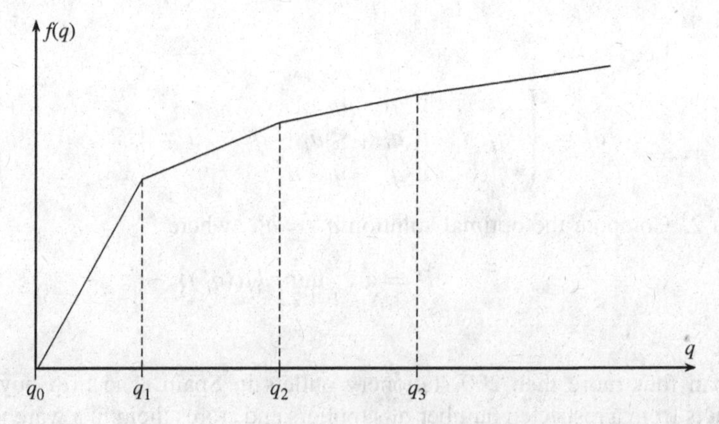

Figure 5.33 The value of q items in the case of incremental quantity discounts.

c_i, the value of $(q_{i-1} - q_{i-2})$ items is c_{i-1} etc. The average total cost function $\mu(q)$ is

$$\mu(q) = \mu_i(q), \ q_{i-1} \leq q < q_i, \ i = 1, 2, \ldots,$$

where, on the basis of Equation (5.31),

$$\mu_i(q) = kd/q + f(q)d/q + p\frac{f(q)}{q}q/2, \ i = 1, 2, \ldots$$

Using Equation (5.40), $\mu_i(q), \ i = 1, 2, \ldots$, can be rewritten as

$$\mu_i(q) = kd/q + \left[f(q_{i-1}) + c_i(q - q_{i-1})\right]d/q$$
$$+\frac{p}{2}\left[f(q_{i-1}) + c_i(q - q_{i-1})\right], \ i = 1, 2, \ldots \quad (5.41)$$

The optimal order size q^* can be computed through a procedure very similar to that used in the previous case.

Step 1. Determine the value $q_i^{'}, \ i = 1, 2, \ldots$, that minimizes $\mu_i(q)$ by imposing that

$$\left.\frac{d\mu_i(q_i)}{d \ q_i}\right|_{q_i = q_i^{'}} = 0, \ i = 1, 2, \ldots$$

Hence,

$$q_i^{'} = \sqrt{\frac{2d\left[k + f(q_{i-1}) - c_i q_{i-1}\right]}{pc_i}}, \ i = 1, 2, \ldots \quad (5.42)$$

If $q_i^{'} \notin [q_{i-1}, q_i]$, then let $\mu_i(q_i^{'}) = \infty, i = 1, 2, \ldots$

Step 2. Compute the optimal solution $q^* = q_{i^*}^*$, where

$$i^* = \arg \min_{i=1,2,\ldots} \{\mu(q_i^*)\}.$$

If Maliban (see the previous problem) applies an incremental quantity discount policy, then, by using Equation (5.42),

$$q_1^{'} = \sqrt{\frac{2 \times 3\,000 \times 50}{0.30 \times 3}} = 577.4 \text{ boxes;}$$

$$q_2^{'} = \sqrt{\frac{2 \times 3\,000\left[50 + (3 \times 500) - (2.97 \times 500)\right]}{0.30 \times 2.97}} = 661.6 \text{ boxes;}$$

$$q_3' = \sqrt{\frac{2 \times 3\,000\{50 + [(3 \times 500) + (2.97 \times 1\,500)] - (2.955 \times 2\,000)\}}{0.30 \times 2.955}} =$$
$$= 801.9 \text{ boxes.}$$

Consequently, as $q_1' > 500$ and $q_3' < 2\,000$, the optimal order size is $q^* = 662$ boxes ($\approx q_2'$), corresponding to an annual average cost, given by Equation (5.41), equal to € 9 501.67.

5.3.2.2 Stochastic models

Inventory problems with uncertain demand or lead times have quite a complex mathematical structure. In this section, a restricted number of stochastic models are illustrated. We first examine the classical *newsboy problem*, where a one-shot reorder decision has to be made. Then, (s, S) *policies* are introduced for a variant of the newsboy problem. Finally, the most common inventory policies used by practitioners (namely, the *reorder point*, the *reorder cycle*, the (s, S) and the *two-bin* policies) are reviewed and compared. The first three policies make use of data forecasts, whereas the fourth policy does not require any data estimate.

The newsboy problem In the newsboy problem, a resupply decision has to be made at the beginning of a period for a single commodity whose demand is not known in advance. The demand d is modelled as a random variable with a continuous cumulative distribution function $F_d(\delta)$. Let c be the purchasing cost or the variable manufacturing cost, depending on whether the goods are bought from an external supplier or produced by the company. Moreover, let r and u be the selling price and the salvage value per unit of commodity, respectively. Of course,

$$r > c > u.$$

There is no fixed reorder cost or an initial inventory. In addition, shortage costs are assumed to be negligible. If the company orders q units of commodity, the *expected revenue* $\rho(q)$ is

$$\rho(q) = r\int_0^\infty \min(\delta, q) dF_d(\delta) + u\int_0^\infty \max(0, q-\delta) dF_d(\delta) - cq =$$
$$= r\left(\int_0^q \delta dF_d(\delta) + q\int_q^\infty dF_d(\delta)\right) + u\int_0^q (q-\delta) dF_d(\delta) - cq.$$

By adding and subtracting $r\int_q^\infty \delta dF_d(\delta)$ to the right-hand side, $\rho(q)$ becomes

$$\rho(q) = rE[d] + r\int_q^\infty (q-\delta) dF_d(\delta) + u\int_0^q (q-\delta) dF_d(\delta) - cq, \qquad (5.43)$$

where $E[d]$ is the *expected demand*. It is easy to show that $\rho(q)$ is concave for $q \geq 0$, and $\rho(q) \to -\infty$ for $q \to \infty$. As a result, the maximum expected revenue is achieved when the derivative of $\rho(q)$ with respect to q is zero. Hence, by applying the Leibnitz rule, the optimality condition becomes

$$r(1 - F_d(q)) + uF_d(q) - c = 0,$$

where, by definition, $F_d(q)$ is the probability $Pr(d \leq q)$ that the demand does not exceed q. As a result, the *optimal order quantity* S satisfies the following condition:

$$Pr(d \leq S) = \frac{r - c}{r - u}. \tag{5.44}$$

Emilio Tadini & Sons is a hand-made shirt retailer located in Rome, Italy, close to Piazza di Spagna. This year Mr Tadini faces the problem of ordering a new, brightly colour shirt made by a Florentine firm. He assumes that the demand is uniformly distributed between 200 and 350 units. The purchasing cost is $c =$ € 18 while the selling price is $r =$ € 52 and the salvage value is $u =$ € 7. According to Equation (5.44), $Pr(d \leq S) = (S - 200)/(350 - 200)$ for $200 \leq S \leq 350$. Hence, Mr Tadini should order $S = 313$ units. According to Equation (5.43), the expected revenue is

$$\rho(q) = 52 \times 275 + 52 \int_{200}^{350} (q - \delta)\frac{1}{350 - 200} d\delta - 18q = 34q,$$

for $0 \leq q \leq 200$;

$$\rho(q) = 52 \times 275 + 52 \int_{q}^{350} (q - \delta)\frac{1}{350 - 200} d\delta +$$

$$+7 \int_{200}^{q} (q - \delta)\frac{1}{350 - 200} d\delta - 18q = -0.15q^2 + 94q - 6\,000,$$

for $200 < q \leq 350$, and

$$\rho(q) = 52 \times 275 + 7 \int_{200}^{350} (q - \delta)\frac{1}{350 - 200} d\delta - 18q = -11q + 12\,375,$$

for $q > 350$. Hence, the maximum expected revenue is equal to $\rho(313) =$ € 8 726.65.

The (s, S) policy for single-period problems If there is an initial inventory q_0 and a fixed reorder cost k, the optimal replenishment policy can be obtained as follows. If $q_0 \geq S$, no reorder is needed. Otherwise, the best policy is to order $S - q_0$, provided that the expected revenue associated with this choice is greater than the expected revenue associated with not producing anything. Hence, two cases can occur:

1. if the expected revenue $\rho(S) - k - cq_0$ associated with reordering is greater than the expected revenue $\rho(q_0) - cq_0$ associated with not reordering, then $S - q_0$ units have to be reordered;

2. otherwise, no order has to be placed.

As a consequence, if $q_0 < S$, the optimal policy consists of ordering $S - q_0$ units if $\rho(q_0) \leq \rho(S) - k$. In other words, if s is the number such that

$$\rho(s) = \rho(S) - k,$$

the optimal policy is to order $S - q_0$ units if the initial inventory level q_0 is less than or equal to s, otherwise not to order. Policies like this are known as (s, S) policies. Parameter s acts as a reorder point, while S is sometimes called the *order-up-to-level*.

If $q_0 = 50$ and $k =$ € 400 in the Emilio Tadini & Sons problem, $\rho(s) = \rho(S) - k =$ € 8 526.65 so that $s = 277$. As $q_0 < s$, the optimal policy is to order $S - q_0 = 253$ units.

The reorder point policy In the reorder point policy (or *fixed order quantity policy*), the inventory level is kept under observation in an almost continuous way. As soon as its net value $I(t)$ (the amount in stock minus the demand unsatisfied plus the orders placed but not received yet) reaches an order point s, a constant quantity q is ordered (see Figure 5.34).

The reorder size q is computed through the procedures illustrated in the previous sections, by replacing d with $\bar{d}$. In particular, under the EOQ hypothesis:

$$q = \sqrt{\frac{2k\bar{d}}{h}}. \tag{5.45}$$

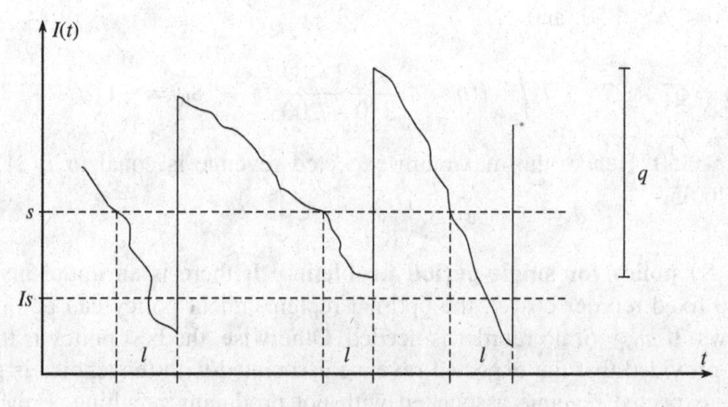

Figure 5.34 Reorder point policy.

The reorder point s is obtained by requiring that the inventory level be non-negative during the lead time l, with probability α. This is equivalent to assuming that demand should not exceed m during the time interval l. In the following, it is assumed that

- the demand rate d is distributed according to a normal distribution with mean $\bar{d}$ and standard deviation σ_d;
- $\bar{d}$ and σ_d are constant in time;
- the lead time l is deterministic or is distributed according to a normal distribution with mean $\bar{l}$ and standard deviation σ_l;
- the demand rate and the lead time are statistically independent.

The average demand rate $\bar{d}$ can be forecasted with one of the methods illustrated in Chapter 2, while the standard deviation σ_d can be estimated as the square root of *MSE*. Analogous procedures can be adopted for the estimation of $\bar{l}$ and σ_l.

Let z_α be the value under which a standard normal random variable falls with probability α (for example, $z_\alpha = 2$ for $\alpha = 0.9772$ and $z_\alpha = 3$ for $\alpha = 0.9987$). If l is deterministic, then

$$s = \bar{d}l + z_\alpha \sigma_d \sqrt{l}, \tag{5.46}$$

where $\bar{d}l$ and $\sigma_d\sqrt{l}$ are the mean and the standard deviation of the demand in a time interval of duration l, respectively. If l is random, then

$$s = \bar{d}\,\bar{l} + z_\alpha \sqrt{\sigma_d^2 \bar{l}^2 + \sigma_l^2 \bar{d}^2},$$

where $\bar{d}\,\bar{l}$ and $\sqrt{\sigma_d^2 \bar{l}^2 + \sigma_l^2 \bar{d}^2}$ are the mean and the standard deviation of the demand in a time interval of random duration l, respectively.

The reorder point s, minus the average demand in the reorder period, constitutes a *safety stock* I_S. For example, in case l is constant, the safety stock is

$$I_S = s - \bar{d}\,l = z_\alpha \sigma_d \sqrt{l}. \tag{5.47}$$

Papier is a French retail chain. At the outlet located in downtown Lyon, the expected demand for mouse pads is 45 units per month. The value of an item in stock is € 4, and the fixed reorder cost is equal to € 30. The annual interest rate is 20%. The demand forecasting *MSE* is 25. Lead time is one month and a service level equal to 97.7% is required. On the basis of Equation (5.21), the holding cost is

$$h = 0.2 \times 4 = € \, 0.8 \text{ (year per item)} = € \, 0.067 \text{ (month per item)}.$$

Therefore, since $\bar{d} = 45$, from Equation 5.45,

$$q^* = \sqrt{\frac{2 \times 30 \times 45}{0.067}} = 200.74 \approx 201 \text{ items.}$$

Moreover, σ_d can be estimated as

$$\sigma_d = \sqrt{25} = 5.$$

Since $l = 1$, from Equation (5.46), the reorder point s is

$$s = 45 + 2 \times 5 = 55 \text{ units.}$$

Consequently, the safety stock I_S is equal to

$$I_S = 55 - 45 = 10 \text{ units.}$$

The reorder cycle policy In the reorder cycle policy (or *periodic review* policy) the stock level is kept under observation periodically at time instants t_i ($t_{i+1} = t_i + T$, $T \geq 0$). At time t_i, $q_i = S - I(t_i)$ units are ordered (see Figure 5.35). Parameter S (referred to as the order-up-to-level) represents the maximum inventory level in case lead time t_l is negligible.

The review period T can be chosen using procedures analogous to those used for determining q^* in the deterministic models. For instance, under the EOQ hypothesis,

$$T = \sqrt{\frac{2k}{h\bar{d}}}. \tag{5.48}$$

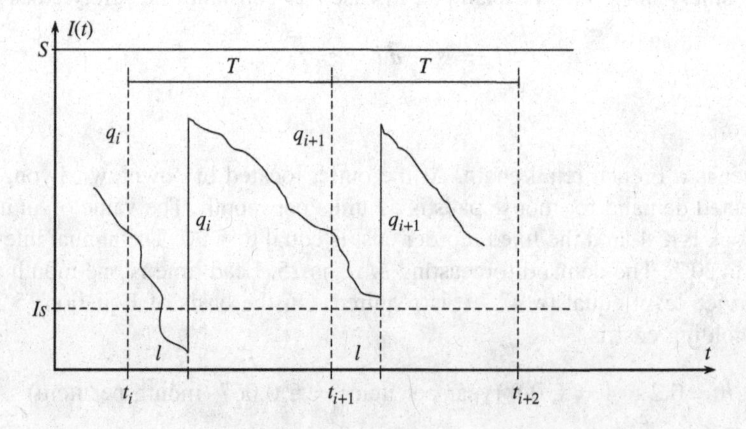

Figure 5.35 Reorder cycle policy.

Parameter S is determined in such a way that the probability that the inventory level becomes negative does not exceed a given value $(1-\alpha)$. Since the *risk interval* is equal to T plus l, S is required to be greater than or equal to the demand in $T+l$, with probability equal to α. If lead time l is deterministic, then

$$S = \bar{d}(T+l) + z_\alpha \sigma_d \sqrt{T+l}, \tag{5.49}$$

where $\bar{d}(T+l)$ and $\sigma_d\sqrt{T+l}$, are the mean and the standard deviation of the demand in $T+l$, respectively. If lead time is a random variable, then

$$S = \bar{d}(T+\bar{l}) + z_\alpha \sqrt{\sigma_d^2 (T+\bar{l}) + \sigma_l^2 \bar{d}^2},$$

where $\bar{d}(T+\bar{l})$ and $\sqrt{\sigma_d^2 (T+\bar{l}) + \sigma_l^2 \bar{d}^2}$ are the mean and the standard deviation of the demand in $T+\bar{l}$, respectively.

The difference between S and the average demand in $T+\bar{l}$ makes up a safety stock I_S. For example, if the lead time is deterministic,

$$I_S = z_\alpha \sigma_d \sqrt{T+l}. \tag{5.50}$$

Comparing Equation (5.50) with Equation (5.47), it can be seen that the reorder cycle inventory policy involves a higher level of safety stock. However, such a policy does not require a continuous monitoring of the inventory level.

In the Papier inventory problem, the parameters of the reorder cycle policy, computed through Equations (5.48) and (5.49), are

$$T = \sqrt{\frac{2 \times 30}{0.067 \times 45}} = 4.47 \text{ months},$$

$$S = 45 \times (4.47 + 1) + 2 \times 5 \times \sqrt{4.47+1} = 269.54 \text{ units}.$$

The associated safety stock, given by Equation (5.50), is equal to

$$I_S = 2 \times 5 \times \sqrt{4.47+1} = 23.39 \text{ units}.$$

The (s, S) policy The (s, S) inventory policy is a natural extension of the (s, S) policy illustrated for the single-period case. At time t_i, $S - I(t_i)$ items are ordered if $I(t_i) < s$ (see Figure 5.36). If s is large enough ($s \to S$), the (s, S) policy is similar to the reorder cycle policy. On the other hand, if s is small ($s \to 0$), the (s, S) policy is similar to a reorder point policy with a reorder point equal to s and a reorder quantity $q \cong S$. On the basis of these observations, the (s, S) policy can be seen as a good compromise between the reorder point and reorder cycle policies. Unfortunately, parameters T, S and s are difficult to determine analytically. Therefore simulation is often used in practice.

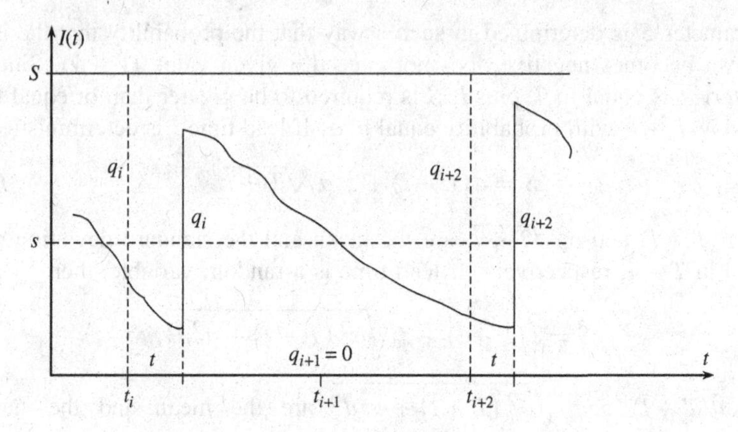

Figure 5.36 *(s, S) policy.*

Pansko, a Bulgarian chemical firm located in Plovdiv, supplies chemical agents to state clinical laboratories. Its product Merofosphine has a demand of 400 packages per week, a variable cost of 100 levs per unit and a profit of 20 levs per unit. Every time the manufacturing process is set up, a fixed cost of 900 levs is incurred. The annual interest rate p is 20%. If the commodity is not available in stock, a sale is lost. In this case, a cost equal to the profit of the lost sale is incurred. The *MSE* forecast equals 2 500. The lead time can be assumed to be constant and equal to a week. The inventory

Table 5.16 Average cost per week (in levs) in the Pansko problem. The average fixed cost, the average variable cost and the average shortage costs are reported in brackets.

s	S = 1 500	S = 2 000	S = 2 500
800	1 120.8 (337.8 + 168.5 + 614.5)	625.0 (224.8 + 236.8 + 163.3)	994.9 (152.3 + 330.2 + 512.3)
900	644.7 (447.6 + 184.6 + 12.4)	622.9 (225.0 + 236.8 + 161.0)	908.9 (162.9 + 339.3 + 406.6)
1 000	625.0 (450.0 + 184.9 + 0.0)	623.0 (225.0 + 236.8 + 161.0)	724.3 (197.9 + 375.7 + 150.5)
1 100	635.0 (450.0 + 184.0 + 0.0)	612.7 (229.7 + 239.1 + 143.9)	634.6 (222.2 + 403.3 + 9.0)
1 200	635.0 (450.0 + 185.0 + 0.0)	622.7 (291.2 + 276.3 + 55.1)	631.8 (224.9 + 406.8 + 0.0)

is managed by means of an (s, S) policy with a period T of two weeks. The values s and S are selected by simulating the system for all combinations of s (equal to 800, 900, 1 000, 1 100 and 1 200, respectively) and S (equal to 1 500, 2 000 and 2 500, respectively). According to the results reported in Table 5.16, $s = 1\,100$ and $S = 2\,000$ are the best choice. This would result in an average cost per week equal to 612.7 levs.

The two-bin policy The two-bin policy can be seen as a variant of the reorder point policy where no demand forecast is needed, and the inventory level does not have to be monitored continuously. The items in stock are supposed to be stored in two identical bins. As soon as one of the two becomes empty, an order is issued for an amount equal to the bin capacity.

Browns supermarkets make use of the two-bin policy for managing the inventory of tomato juice bottles. The capacity of each bin is 400 boxes, containing 12 bottles each. In a supermarket close to Los Alamos, New Mexico, United States, the inventory level on last December 1 was 780 boxes of 12 bottles each. Last December 6, the inventory level was less than 400 boxes and an order of 400 boxes was issued (see Table 5.17). The order was fulfilled the subsequent day.

Table 5.17 Daily sales of tomato juice (in bottles) during the first week of last December in a Browns supermarket.

Day	Sales	Inventory level
1 Dec	850	8 510
2 Dec	576	7 934
3 Dec	932	7 002
4 Dec	967	6 035
5 Dec	945	5 090
6 Dec	989	4 101
7 Dec	848	3 253

5.3.2.3 Selecting an inventory policy

It is quite common for a warehouse to contain several hundreds (or even thousands) of items. In such a context, goods having a strong impact on the total cost have to be inventory-managed carefully while for less important goods it is wise to resort to simple and low-cost inventory policies.

The problem is generally tackled by clustering the goods into three categories using the *ABC* classification introduced in Section 1.2.3, on the basis of the average value of the goods in stock. In particular, category *A* is made up of products corresponding to a high percentage (e.g. 80%) of the total warehouse value. Category *B* is constituted by a set of items associated to an additional 15% of the warehouse value, while category *C* is formed by the remaining items. The goods of categories *A* and *B* should be managed with inventory policies based on forecasts and a frequent monitoring (e.g. category *A* by means of the reorder point policy and category *B* through the reorder cycle policy). Products in category *C* can be managed using the two-bin policy that does not require any forecast.

The Walloon Transport Consortium (WTC) operates a Belgian public transport service in the Walloon region. Buses are maintained in a facility located in Ans, close to a vehicle depot. The average inventory levels, the unit values and the total average value of the spare parts kept in stock are reported in Table 5.18. It was decided to allocate the products corresponding approximately to the first 80% of the total inventory value to category *A*, the items associated to the following 15% to category *B*, and the remaining commodities to category *C* (see Table 5.19). It is worth noting that category *A* contains about 30% of the goods, while each of the categories *B* and *C* accounts for about 35% of the inventory. It was decided to manage the inventory of products of category *A* by using the reorder point policy, whereas the inventory of *B* and *C* products were managed by using the the reorder cycle policy and the (s, S) policy, respectively.

Table 5.18 Spare parts stocked by WTC.

Product code	Average inventory	Average unit value (€)	Total average value (€)
AX24	137	50	6 850
BR24	70	2 000	140 000
BW02	195	250	48 750
CQ23	6	6 000	36 000
CR01	16	500	8 000
FE94	31	100	3 100
LQ01	70	2 500	175 000
MQ12	18	200	3 600
MW20	75	500	37 500
NL01	15	1 000	15 000
PE39	16	3 000	48 000
RP10	20	2 200	44 000
SP00	13	250	3 250
TA12	100	2 500	250 000
TQ23	10	5 000	50 000
WQ12	30	12 000	360 000
WZ34	30	15	450
ZA98	70	250	17 500

Table 5.19 *ABC* classification of the spare parts in the WTC problem.

Product code	Average stock	Total average value (€)	Cumulated average inventory (%)	Total cumulated value (%)	Class
WQ12	30	360 000	3.25	28.87	*A*
TA12	100	250 000	14.10	48.92	*A*
LQ01	70	175 000	21.69	62.95	*A*
BR24	70	140 000	29.28	74.18	*A*
TQ23	10	50 000	30.37	78.19	*A*
BW02	195	48 750	51.52	82.10	*B*
PE39	16	48 000	53.25	85.95	*B*
RP10	20	44 000	55.42	89.47	*B*
MW20	75	37 500	63.56	92.48	*B*
CQ23	6	36 000	64.21	95.37	*B*
ZA98	70	17 500	71.80	96.77	*C*
NL01	15	15 000	73.43	97.98	*C*
CR01	16	8 000	75.16	98.62	*C*
AX24	137	6 850	90.02	99.17	*C*
MQ12	18	3 600	91.97	99.45	*C*
SP00	13	3 250	93.38	99.72	*C*
FE94	31	3 100	96.75	99.96	*C*
WZ34	30	450	100.00	100.00	*C*
Total	922	1 247 000			

5.3.2.4 Multiproduct inventory models

When several products are kept in stock, their inventory policies are intertwined because of common constraints and joint costs, as we now discuss in two separate cases. In the first case, a limit is placed on the total investment in inventory, or on the warehouse space. In the second case, products share joint ordering costs. For the sake of simplicity, both analysis will be performed under the EOQ model hypothesis.

Models with capacity constraints Let n be the number of products in stock and q_j, $j = 1, \ldots, n$, the amount of product j ordered at each replenishment. The inventory management problem can be formulated as

$$\text{Minimize } \mu(q_1, \ldots, q_n) \tag{5.51}$$

subject to

$$g(q_1, \ldots, q_n) \leq b \tag{5.52}$$

$$q_1, \ldots, q_n \geq 0, \tag{5.53}$$

where the objective function (5.51) is the total average cost per time unit. Under the EOQ hypothesis, the objective function $\mu(q_1, \ldots, q_n)$ can be written as

$$\mu(q_1, \ldots, q_n) = \sum_{j=1}^{n} \mu_j(q_j),$$

where, on the basis of Equation (5.31),

$$\mu_j(q_j) = k_j d_j / q_j + c_j d_j + \frac{h_j q_j}{2}, \quad j = 1, \ldots, n,$$

and quantities d_j, c_j, h_j, $j = 1, \ldots, n$, are the demand rate, the value and the holding cost of product j, respectively. As is customary, $h_j = p_j c_j$, $j = 1, \ldots, n$, where p_j is the interest rate of product j.

Equation (5.52) is a side constraint (referred to as a 'capacity constraint') representing both a budget constraint and a warehouse constraint. It can usually be considered as linear:

$$\sum_{j=1}^{n} a_j q_j \leq b, \tag{5.54}$$

where a_j, $j = 1, \cdots, n$, and b are constants. As a result, problem (5.51)–(5.53) has to be solved through iterative algorithms for non-linear programming problems, such as the conjugate gradient algorithm. Alternatively, the following simple heuristic can be used if the capacity constraint is linear and the interest rates are identical for all the commodities ($p_j = p$, $j = 1, \ldots, n$).

Step 1. Using Equation (5.32), compute the EOQ order sizes q_j', $j = 1, \ldots, n$:

$$q_j' = \sqrt{\frac{2k_j d_j}{p c_j}}, \quad j = 1, \ldots, n. \tag{5.55}$$

If capacity constraint (5.54) is satisfied, *STOP*, the optimal order size for each product j, $j = 1, \ldots, n$, has been determined.

Step 2. Increase the interest rate p of a δ quantity to be determined. Then, the order sizes become:

$$q_j(\delta) = \sqrt{\frac{2k_j d_j}{(p + \delta) c_j}}, \quad j = 1, \ldots, n. \tag{5.56}$$

Determine the value δ^* satisfying the equation.

$$\sum_{j=1}^{n} a_j q_j(\delta^*) = b.$$

Hence,

$$\delta^* = \left(\frac{1}{b}\sum_{j=1}^{n}\left(a_j\sqrt{\frac{2k_j d_j}{c_j}}\right)\right)^2 - p. \tag{5.57}$$

Insert δ^* in Equation (5.56) in order to determine the order sizes $\bar{q}_j$, $j = 1, \ldots, n$.

New Frontier distributes knapsacks and suitcases in most US states. Its most successful models are the Preppie knapsack and the Yuppie suitcase. The Preppie knapsack has a yearly demand of 150 000 units, a value of \$ 30 and a yearly holding cost equal to 20% of its value. The Yuppie suitcase has a yearly demand of 100 000 units, a value of \$ 45 and a yearly holding cost equal to 20% of its value. In both cases, placing an order costs \$ 250. The company management requires the average capital invested in inventories does not exceed \$ 75 000. This condition can be expressed by the following constraint:

$$30q_1/2 + 45q_2/2 \leq 75\,000,$$

where it is assumed, as a precaution, that the average inventory level is the sum of the average inventory levels of the two products. The EOQ order sizes, given by Equation (5.55),

$$q_1' = \sqrt{\frac{2 \times 250 \times 150\,000}{0.2 \times 30}} = 3\,535.53 \text{ units};$$

$$q_2' = \sqrt{\frac{2 \times 250 \times 100\,000}{0.2 \times 45}} = 2\,357.02 \text{ units},$$

do not satisfy the budget constraint. Applying the conjugated gradient algorithm starting from the initial values $(q_1, q_2) = (1, 1)$, the following solution is obtained after 300 iterations:

$$\bar{q}_1 = 2\,500 \text{ units};$$

$$\bar{q}_2 = 1\,666.66 \text{ units},$$

whose total cost is \$ 9 045 000. Applying the heuristic procedure, by using Equation (5.57), the same solution is obtained. In effect:

$$\delta^* = \left[\frac{1}{75\,000}\left(\frac{30}{2}\sqrt{\frac{2 \times 250 \times 150\,000}{30}} + \frac{45}{2}\sqrt{\frac{2 \times 250 \times 100\,000}{45}}\right)\right]^2 +$$

$$-0.2 = 0.2,$$

hence:

$$\bar{q}_1 = \sqrt{\frac{2 \times 250 \times 150\,000}{(0.2 + 0.2) \times 30}} = 2\,500 \text{ units;}$$

$$\bar{q}_2 = \sqrt{\frac{2 \times 250 \times 150\,000}{(0.2 + 0.2) \times 45}} = 1\,666.66 \text{ units.}$$

Models with joint costs For the sake of simplicity, we assume in this section that only two products are kept in inventory. Let k_1 and k_2 be the fixed costs for reordering the two products at different moments in time, and let k_{1-2} be the fixed cost for ordering both products at the same time ($k_{1-2} < k_1 + k_2$). In addition, let T_1 and T_2 be the time lapses between consecutive replenishments of products 1 and 2, respectively (see Figure 5.37). Then,

$$q_1 = d_1 T_1; \tag{5.58}$$

$$q_2 = d_2 T_2. \tag{5.59}$$

The periodicity of a joint replenishment policy is

$$T = \max\{T_1, T_2\}.$$

In each period T, the orders issued for the two products are

$$N_1 = T/T_1;$$

$$N_2 = T/T_2.$$

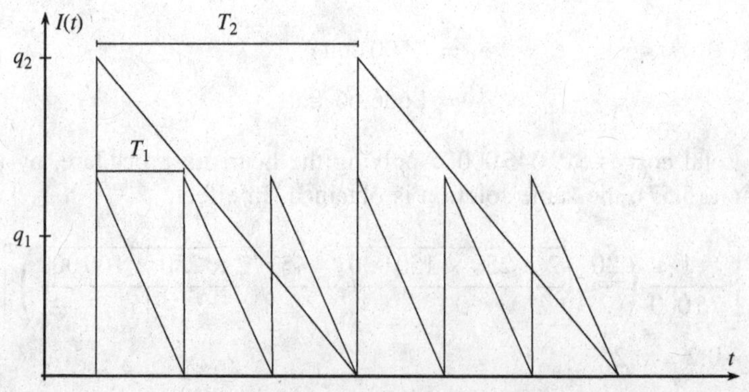

Figure 5.37 Inventory level as a function of time in the case of synchronized orders.

Here, N_1 and N_2 are positive integer numbers, one of them being equal to 1 (in the situation depicted in Figure 5.37, $N_1 = 3$ and $N_2 = 1$). During each time period T, two products are ordered simultaneously exactly once. Moreover, $N_j - 1$ single orders are placed for each product j, $j = 1, 2$. Hence, the total average cost per time unit is

$$\mu(T, N_1, N_2) = \frac{k_{1-2} + (N_1 - 1)k_1 + (N_2 - 1)k_2}{T} + c_1 d_1 + c_2 d_2 + \frac{h_1 d_1 T}{2N_1} + \frac{h_2 d_2 T}{2N_2}. \quad (5.60)$$

By solving the equation

$$\left.\frac{\partial}{\partial T}\mu(T, N_1, N_2)\right|_{T=T*} = 0,$$

the value $T^*(N_1, N_2)$ that minimizes $\mu(T, N_1, N_2)$ is obtained:

$$T^*(N_1, N_2) = \sqrt{\frac{2N_1 N_2[k_{1-2} + (N_1 - 1)k_1 + (N_2 - 1)k_2]}{h_1 d_1 N_2 + h_2 d_2 N_1}}, \quad (5.61)$$

as a function of N_1 and N_2.

Shamrock Microelectronics is an Irish company which assembles printed circuit boards (PCBs) for a number of major companies in the appliance sector. The Y23 PCB has an annual demand of 3 000 units, a value of € 30 and a holding cost equal to 20% of its value. The Y24 PCB has an annual request of 5 000 units, a value of € 40 and a holding cost equal to 25% of its value. The cost of issuing a joint order is € 300 while ordering a single item costs € 250. If no joint orders are placed, the order sizes are, according to Equation (5.32),

$$q_1^* = \sqrt{\frac{2 \times 250 \times 3\,000}{0.2 \times 30}} = 500 \text{ units},$$

$$q_2^* = \sqrt{\frac{2 \times 250 \times 5\,000}{0.25 \times 40}} = 500 \text{ units}.$$

From Equations (5.58) and (5.58),

$$T_1^* = 500/3\,000 = 1/6,$$

$$T_2^* = 500/5\,000 = 1/10.$$

This means that Shamrock would issue $1/T_1^* = 6$ orders per year of Y23 PCB and $1/T_2^* = 10$ orders per year of the Y24 PCB. Since

$$\mu_1(q_1^*) = \frac{250 \times 3\,000}{500} + 30 \times 3\,000 + \frac{0.2 \times 30 \times 500}{2} = € \, 93\,000/\text{year},$$

$$\mu_2\left(q_2^*\right) = \frac{250 \times 5\,000}{500} + 40 \times 5\,000 + \frac{0.25 \times 40 \times 500}{2} = € \ 205\,000/\text{year},$$

the average annual cost is € 298 000/year. If a joint order is placed and $N_1 = 1$, $N_2 = 2$, the periodicity of joint orders is, according to Equation (5.61),

$$T^* = \sqrt{\frac{2 \times 1 \times 2 \times (300 + 250)}{0.2 \times 30 \times 3\,000 \times 2 + 0.25 \times 40 \times 5\,000 \times 1}} = 0.16.$$

Shamrock would issue $1/T^* = 6.25$ joint orders per year. The annual average cost, computed through Equation (5.60), is equal to

$$\mu(T^*, 1, 2) = \frac{300+250}{0.16} + 30 \times 3\,000 + 40 \times 5\,000 + \frac{0.2 \times 30 \times 3\,000 \times 0.16}{2} + \frac{0.25 \times 40 \times 5\,000 \times 0.16}{2 \times 2} = € \ 296\,877.5/\text{year}$$

5.4 Operational decisions for warehouse logistics management

Operational decisions mainly concern the organization of material-handling activities in the warehouse. In particular, because the shipping activities are related to received orders which are intrinsically random, it is clear that they must be managed in real time. These include, among others, the picking of the packages from the storage zone to satisfy customer orders and the consolidation of the packages into load units.

5.4.1 Package picking from the storage zone

The picking of packages from the storage zone to satisfy customer orders is a particularly burdensome activity, both time-consuming and costly, with respect to other warehouse logistics activities.

This activity can be carried out in two different ways: 'order-to-picker', and 'picker-to-product'. The first mode is achieved by means of automatic material-handling systems (such as AS/RS) and is convenient when the number of items in the warehouse is large (more than 10 000). The second mode, which is more widespread, entails the use of pickers for retrieval. In this case, it is necessary to determine the picking list for each operator, composed of the packages to be retrieved from the storage zone, and a least-cost picker route.

In many solutions adopted in practice, the picking list is composed of packages belonging to just one order. This ensures greater reliability and robustness of the picking activities in the warehouse, but reduces the level of productivity, since the number of trips to the storage zone increases.

Otherwise, the picking list is composed of packages belonging to several orders, yielding significant time savings and thus increasing productivity. The order grouping can be obtained by identifying a seed order and adding other orders to it in an iterative fashion, without ever exceeding the capacity of the retrieval material-handling system. The seed order can be chosen randomly, or can be the one with the highest number of packages. The orders to be aggregated are chosen so that the corresponding packages are 'close' in the storage zone. For this reason, it is necessary to define a suitable distance metric between two orders. A possible choice is the Euclidean distance between the centres of gravity (see Section 3.3.1) of the two orders, where the centre of gravity $(\bar{x}, \bar{y})$ of an order composed of n packages having Cartesian coordinates (x_i, y_i), $i = 1, \ldots, n$, is given by

$$\bar{x} = \sum_{i=1}^{n} x_i/n \tag{5.62}$$

and

$$\bar{y} = \sum_{i=1}^{n} y_i/n. \tag{5.63}$$

In the storage zone of the SISA DC at Gioia Tauro, two picking orders were received whose Cartesian coordinates (see Figure 5.38) are reported in Tables 5.20 and 5.21, respectively. The DC has a layout with six 3 m wide cross aisles and a 5 m wide central aisle. Each package is assigned to a (1.5×1.5) m^2 area. The centre of gravity $g^{(1)}$ of the first order corresponds to the point in the warehouse of coordinates $(\bar{x}^{(1)}, \bar{y}^{(1)})$, calculated by means of (5.62) and (5.63):

$$\bar{x}^{(1)} = \sum_{i=1}^{6} x_i^{(1)}/6 = 8.25 \text{ m};$$

Table 5.20 List of packages and corresponding Cartesian coordinates (in m) of the first picking order in the storage zone of the SISA DC.

Package	Abscissa	Ordinate
1	0.75	3.75
2	0.75	18.75
3	6.75	6.75
4	6.75	12.75
5	11.25	18.75
6	23.25	14.25

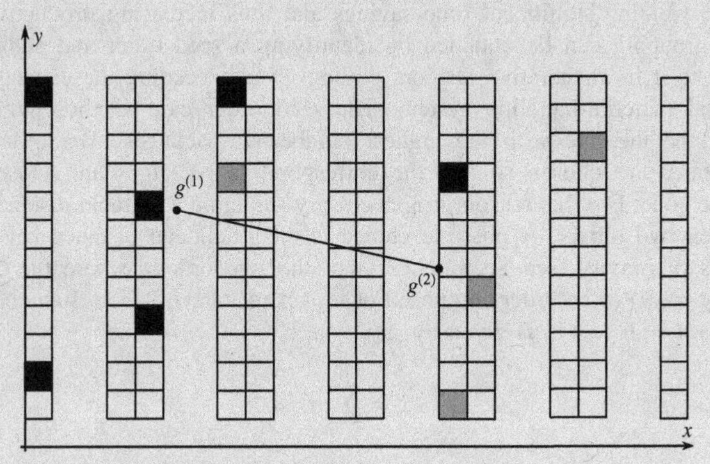

Figure 5.38 Layout of the SISA storage zone and storage locations (in black) of the packages of the first picking order and of the second order (in grey). The centres of gravity of the two picking orders are identified by the points $g^{(1)}$ and $g^{(2)}$.

Table 5.21 List of packages and corresponding Cartesian coordinates (in m) of the second picking order in the storage zone of the SISA DC.

Package	Abscissa	Ordinate
1	11.25	14.25
2	23.25	2.25
3	24.75	8.25
4	30.75	15.75

$$\bar{y}^{(1)} = \sum_{i=1}^{6} y_i^{(1)}/6 = 12.50\,\text{m}.$$

The centre of gravity $g^{(2)}$ of the second order has the following coordinates $(\bar{x}^{(2)}, \bar{y}^{(2)})$:

$$\bar{x}^{(2)} = \sum_{i=1}^{4} x_i^{(2)}/4 = 22.50\,\text{m};$$

$$\bar{y}^{(2)} = \sum_{i=1}^{4} y_i^{(2)}/4 = 10.125\,\text{m}.$$

The Euclidean distance between the two centres of gravity is therefore equal to 14.45 m.

Another method used to group orders into a single picking list is based on the logic of savings, which also characterizes a heuristic used for vehicle-routing problems (see Section 6.8.2.2). For more details, the reader should refer to the specialized scientific literature on this topic.

The zone-picking algorithm constitutes an alternative way of forming the picking list, typically adopted when the storage zone is large. In this case, the storage zone is ideally divided into different subzones and a picking list is formed of packages belonging to different orders but stocked in the same subzone. The picking operations are carried out more quickly, due to the traffic reduction in the picking aisles. The disadvantage of this algorithm, compared to the previous ones, is that relaxing the constraint that the packages belong to the same picking order increases the risk of errors in the subsequent package consolidation phase.

The second decision-making problem related to the picking phase is the determination of the minimum-time picker route in the storage zone (it should be noted that the time associated with picker routes within the storage zone represents about 70% of the time spent completing the picking activities).

Picker routing is part of a large class of combinatorial optimization problems, the vehicle-routing problems, which will be examined extensively in Chapter 6 in the context of distribution logistics. In this section a single picker problem, known as the road travelling salesman problem (RTSP), is illustrated. The RTSP is a slight variant of the classical travelling salesman problem (TSP) (see Section 6.8.1) and consists of determining a minimum-time tour including a subset of vertices of a graph. The RTSP is NP-hard, but, in the case of warehouses, it is often solvable in polynomial time due to the particular characteristics of the underlying network.

For example, with reference to the storage zone represented in Figure 5.39, the RTSP could be modelled on a graph $G = (A \cup B \cup V, E)$, where $A = \{a_1, \ldots, a_r\}$ and $B = \{b_1, \ldots, b_r\}$ are the sets of vertices representing,

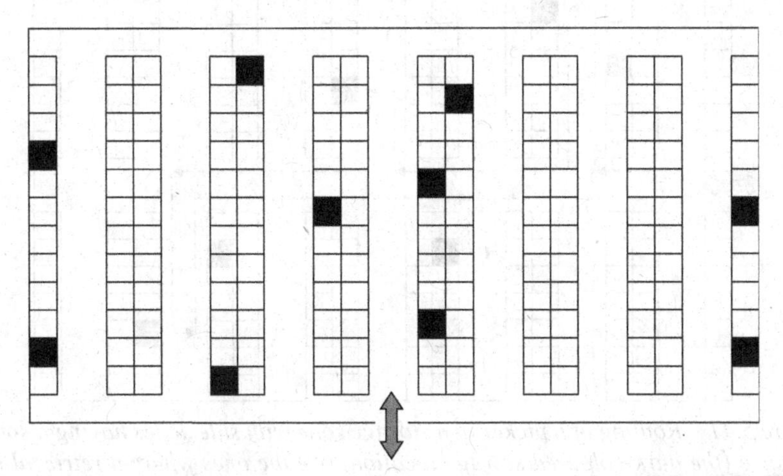

Figure 5.39 Layout of a storage zone. The storage locations are dark-coloured.

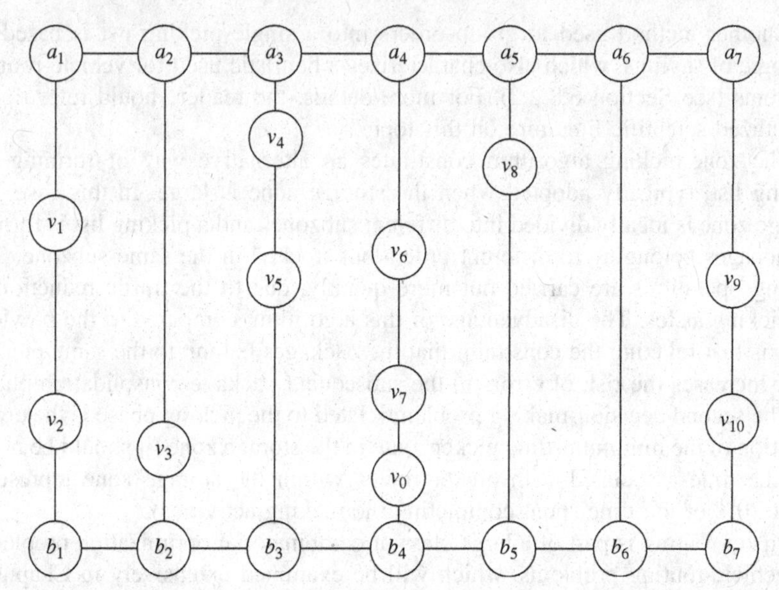

Figure 5.40 Graph $G = (A \cup B \cup V, E)$ associated to the RTSP in the storage zone of Figure 5.39.

respectively, the upper and lower ends of the r side aisles of the storage zone; $V = \{v_0, v_1, \ldots, v_n\}$ is the set of the n storage locations and of v_0, corresponding to the I/O point of the storage zone; E is the set of edges joining storage locations, the ends of the side aisles and the I/O point. The graph corresponding to the storage zone of Figure 5.39 is reported in Figure 5.40.

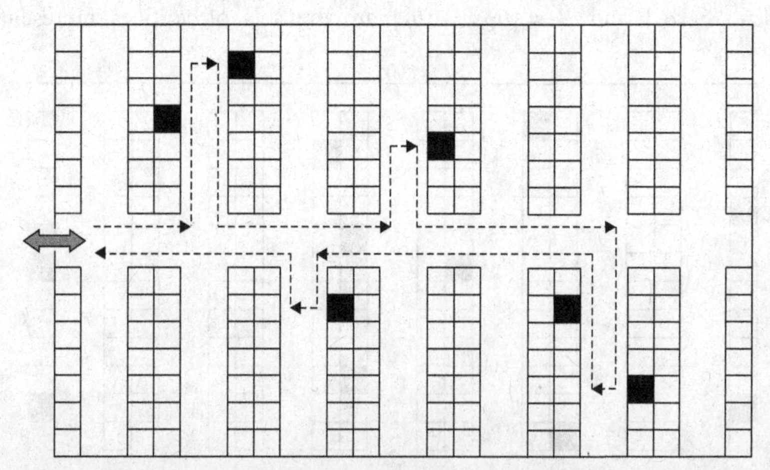

Figure 5.41 Routing of a picker in a storage zone with side aisles having a single entrance (the dark-coloured storage locations are the ones where a retrieval has to be performed).

If each aisle has a single entrance, the minimum-time tour is obtained by first visiting all required storage locations placed in the upper side aisles and then all required storage locations situated in the lower side aisles (see Figure 5.41). On the other hand, if the side aisles have some interruptions (i.e. if there are more than one cross aisles), the problem can be solved to optimality by using a dynamic programming algorithm, whose worst-case computational complexity is a linear function of the number of side aisles. However, if there are several cross aisles, the dynamic programming procedure becomes impracticable.

Therefore, in what follows, four heuristics are illustrated:

- S-shaped heuristic;
- largest gap heuristic;
- combined heuristic;
- aisle-by-aisle heuristic.

To describe these heuristics, the storage zone is assumed to have a general layout like that in Figure 5.42, consisting of a single I/O point, of a number of

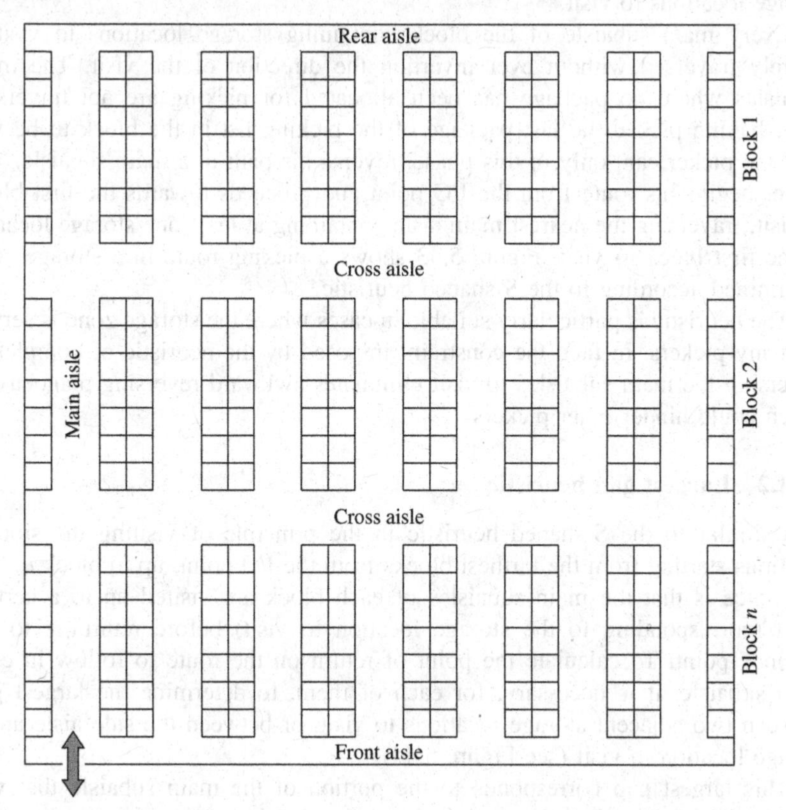

Figure 5.42 Layout of the storage zone divided into blocks.

blocks numbered from 1 to n (1 is the farthest block from the I/O point, n is the nearest), linked together by main aisles. The storage locations are generally at both sides of a generic main aisle. Between each pair of blocks there is a cross aisle which can be used to go from one main aisle to another, or from one block to the next. Transversally, there is the rear aisle in correspondence to block 1, just as the front aisle is transversal connecting all the main aisles to block n. It is assumed, moreover, that the pickers can cross the aisles in both directions and can also change directions within them. In fact, the aisles are sufficiently wide to allow movements from both sides without having to change positions.

5.4.1.1 S-shaped heuristic

The general idea of the S-shaped heuristic is that the visit to the storage locations for retrievals take place block by block, starting from block 1 (or from the first block containing storage locations to visit), gradually approaching block n, the closest to the position of the I/O point. The visit to the storage locations of each block takes place following an S-shaped route (hence the name of the heuristic), beginning from the main aisle most to the left or most to the right containing storage locations to visit.

Every main subaisle of the block containing storage locations to visit is entirely traversed without ever inverting the direction of the visit. The main subaisles where no package has been allocated for picking are not traversed. After having picked the last package of the picking list in the block to be visited, the picker can, only in this phase, reverse his path in a main subaisle. The picker begins his route from the I/O point and proceeds towards the first block to visit, traversing the nearest main aisle containing at least one storage location of the first block to visit. Figure 5.43 shows a picking route in a storage zone determined according to the S-shaped heuristic.

The heuristic is particularly suitable in cases where the storage zone is served by many pickers. In fact, the constraint imposed by the heuristic of completely traversing the main subaisles to visit eliminates awkward reversing manoeuvres which could hinder other pickers.

5.4.1.2 Largest gap heuristic

It is similar to the S-shaped heuristic in the principle of visiting the storage locations starting from the farthest blocks from the I/O point, up to block n. The difference is that the main subaisles of each block are visited up to a certain point (corresponding to the storage location to visit) before returning to the entrance point. To calculate the point of return on the route to follow in each main subaisle, it is necessary, for each of them, to determine the largest gap between two adjacent storage locations to visit, or between the side aisle and a storage location to visit (see Figure 5.44).

This largest gap corresponds to the portion of the main subaisle that will not be visited. The fact that a return can be made to the entrance point of a

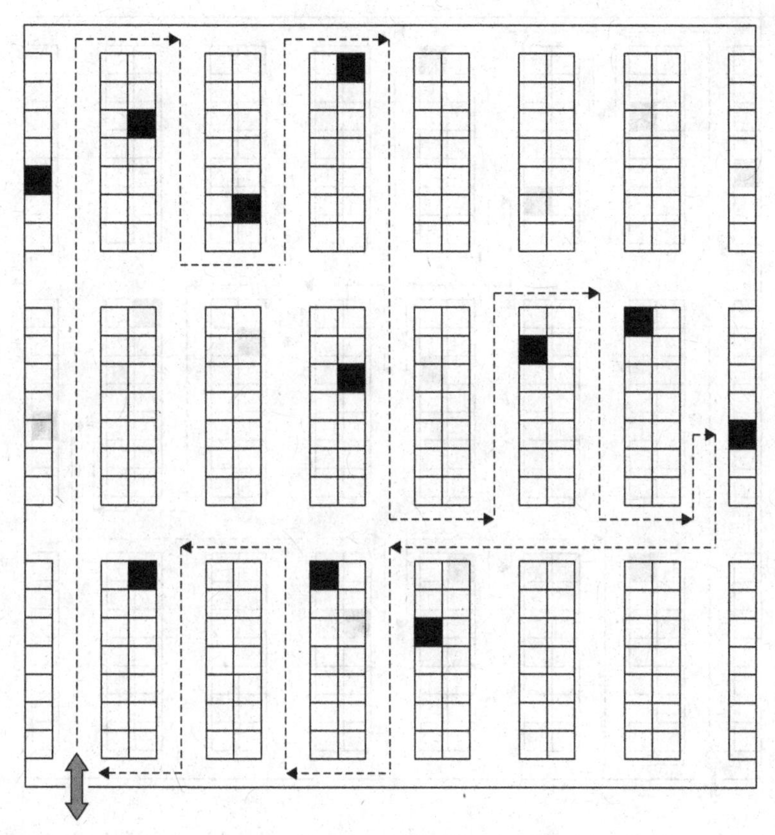

Figure 5.43 Picking route in a storage zone determined by using the S-shaped heuristic.

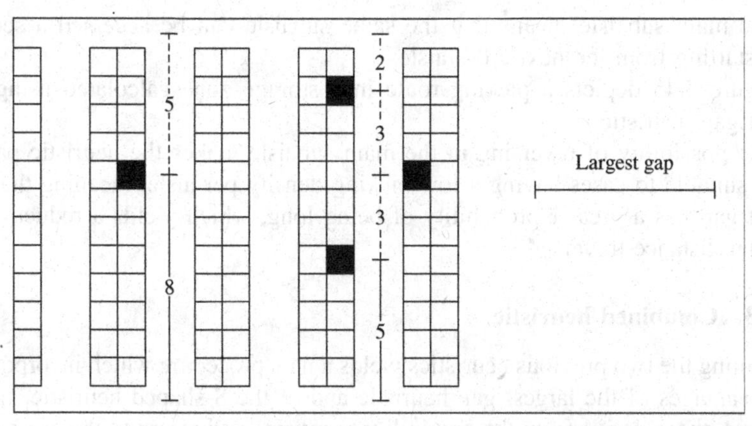

Figure 5.44 Computation of the largest gap in two main subaisles of the storage zone.

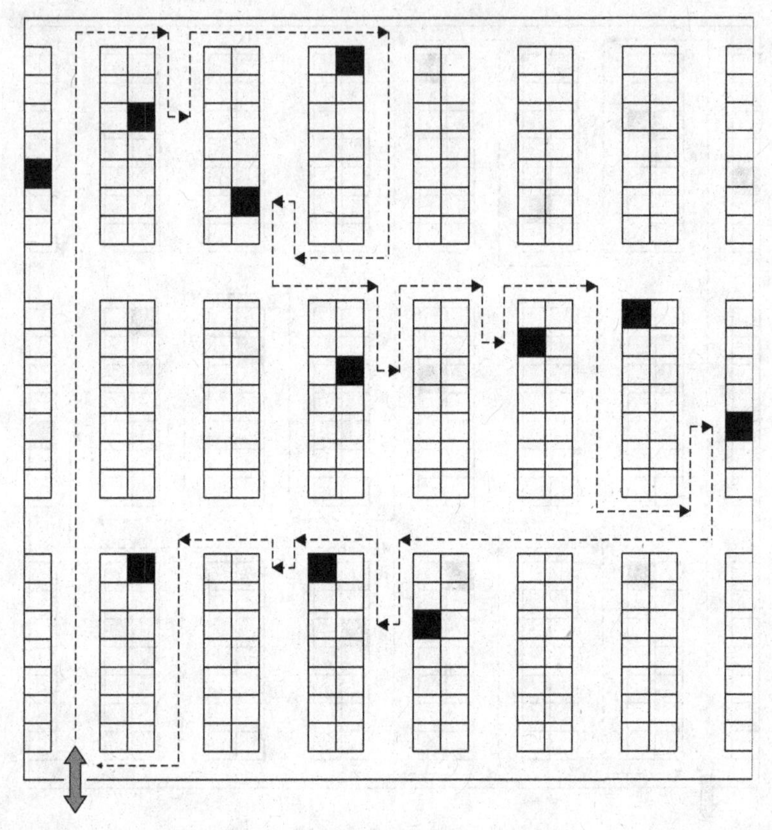

Figure 5.45 Picking route in a storage zone determined by using the largest gap heuristic.

generic main subaisle means that the same subaisle can be accessed a second time, starting from the next cross aisle.

Figure 5.45 depicts a picking route in a storage zone calculated using the largest gap heuristic.

The possibility of reversing in the main subaisle makes the heuristic particularly suitable to cases having a low picking density per aisle, meaning that the largest gap has a greater probability of being long, which yields a reduction of the total distance traversed.

5.4.1.3 Combined heuristic

Combining the two previous heuristics yields a new procedure which incorporates the advantages of the largest gap heuristic and of the S-shaped heuristic. In the first case, these derive from the possibility to reverse and return to the cross aisle used for entrance. In the second case, they derive from the complete traversal of the main subaisles containing storage locations to visit. In the combined heuristic, the main subaisle containing storage locations for the picking of packages is

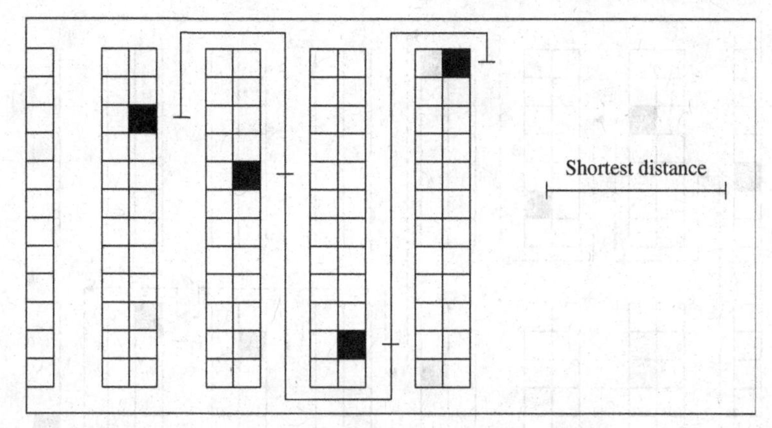

Figure 5.46 Determination of routes between two consecutive storage locations using the combined heuristic.

visited just once. The objective is to create a minimum gap to traverse between the last storage postion of the main subaisle currently visited and the first storage position of the next aisle (see Figure 5.46).

Figure 5.47 shows a picking route in a storage zone calculated by means of the combined heuristic.

The risk of aisle congestion caused by reversing traffic is less than with the largest gap heuristic.

5.4.1.4 Aisle-by-aisle heuristic

This heuristic is based on the general principle that every main aisle is traversed just once. The route begins from the I/O point and proceeds towards the nearest main aisle containing at least one storage location to visit. When all storage locations in this main aisle have been visited, the picker identifies the cross aisle that allows her to reach the next main aisle. In each case, the cross aisles to use (to go from one main aisle to the next) are chosen so as to minimize the distance travelled.

Figure 5.48 shows a picking route in a storage zone determined by using the aisle-by-aisle heuristic.

5.4.2 Package consolidation in load units

This is an operational problem, typically faced in the shipping zone, which consists of consolidating the dispatches, that is, loading different items retrieved from the storage zone to make up load units of the same kind (pallets, containers, generally indicated in the sequel as bins). These are then considered as indivisible entities for the next movement. From a mathematical point of view, packing problems are mostly NP-hard, so that in most cases heuristics are used. On the other hand, it is also difficult to design heuristics that will yield feasible

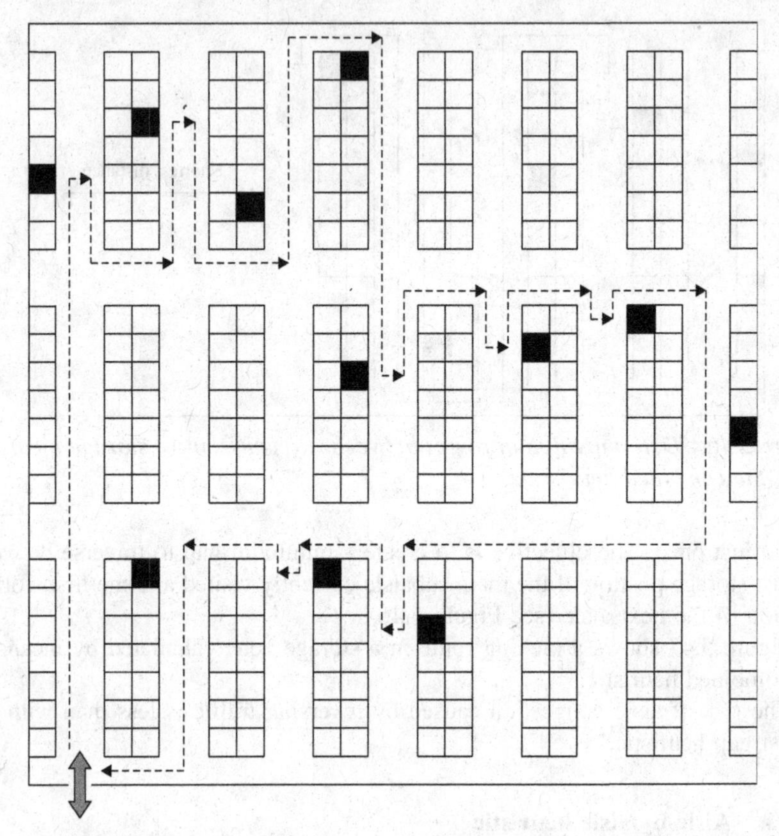

Figure 5.47 Picking route in a storage zone determined by using the combined heuristic.

solutions since numerous constraints for packing problems (such as those connected to load stability) are complicated to formalize. In these cases, general-type algorithms (such as those that will be illustrated below) can serve as a basis for the design of a decision support system applicable to practical situations.

Classification In some packing problems, not all physical characteristics of the items have to be considered when packing. For instance, when loading high-density goods onto a truck, items can be characterized by their weight only, without any concern for their length, width and height. As a result, packing problems can be classified according to the number of parameters needed to characterize an item.

- *One-dimensional packing problems*. One-dimensional problems often arise when dealing with high-density items, in which case weight is binding.
- *Two-dimensional packing problems*. Two-dimensional problems usually arise when loading a pallet with items having the same height.

- *Three-dimensional packing problems.* Three-dimensional problems occur when dealing with low-density items, in which case volume is binding.

In the following, it is assumed, for the sake of simplicity, that items are rectangles in two-dimensional problems, and that their sides must be parallel or perpendicular to the sides of the bins in which they are loaded. Similarly, in three-dimensional problems, the items are assumed to be parallelepipeds and their surfaces are parallel or perpendicular to the surfaces of the bins in which they are loaded. These assumptions are satisfied in most settings.

Packing problems are usually classified as *off-line* and *on-line* problems, depending on whether the items to be loaded are all available or not when packing starts. In the first case, item characteristics can be preprocessed (e.g. items can be sorted by nondecreasing weights) in order to improve heuristic performance. A heuristic using such preprocessing is said to be *off-line*, otherwise it is called *on-line*. Clearly, an on-line heuristic can be used for solving an off-line problem, but an off-line heuristic cannot be used for solving an on-line problem.

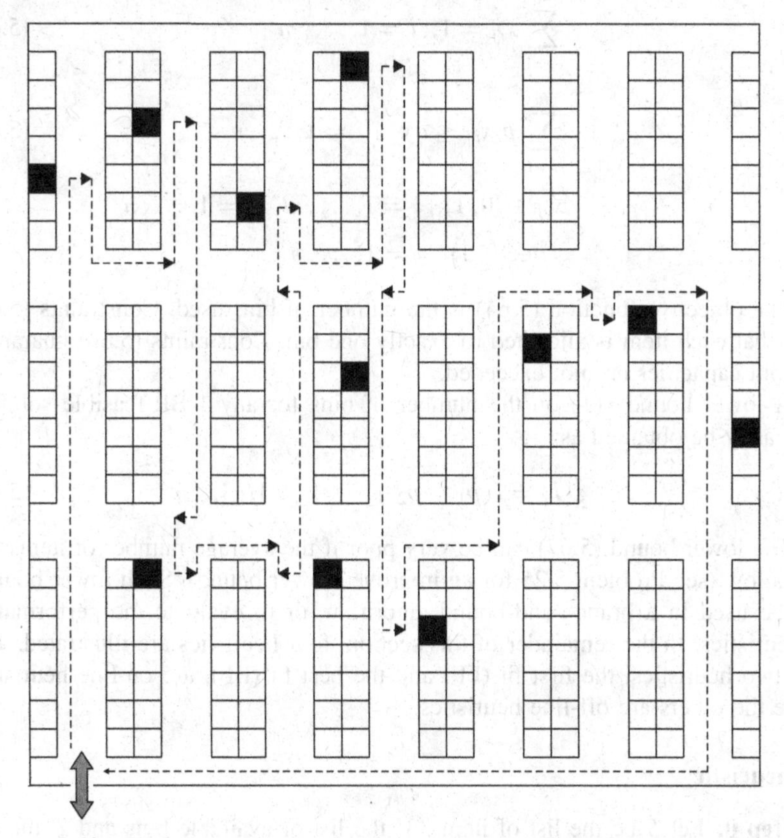

Figure 5.48 Picking route in a storage zone determined by using the aisle-by-aisle heuristic.

5.4.2.1 One-dimensional packing problems

The simplest one-dimensional packing problem is known as the *bin packing* (1-BP) problem. It amounts to determining the least number of identical capacitated bins in which a given set of weighted items can be accommodated. Let m be the number of items to be loaded; n the number of available bins (or an upper bound on the number of bins in an optimal solution); p_i, $i = 1, \ldots, m$, the weight of item i, and q ($\geq p_i$, $i = 1, \ldots, m$) the capacity of each bin j, $j = 1, \ldots, n$. The problem can be modelled by means of binary decision variables x_{ij}, $i = 1, \ldots, m$, $j = 1, \ldots, n$, each of them equal to 1 if item i is assigned to bin j or 0 otherwise, and binary decision variables y_j, $j = 1, \ldots, n$, equal to 1 if bin j is used or 0 otherwise. The 1-BP problem can then be modelled as follows:

$$\text{Minimize} \sum_{j=1}^{n} y_j \tag{5.64}$$

subject to

$$\sum_{j=1}^{n} x_{ij} = 1, \; i = 1, \ldots, m \tag{5.65}$$

$$\sum_{i=1}^{m} p_i x_{ij} \leq q y_j, \; j = 1, \ldots, n \tag{5.66}$$

$$x_{ij} \in \{0, 1\}, \; i = 1, \ldots, m, \; j = 1, \ldots, n$$

$$y_j \in \{0, 1\}, \; j = 1, \ldots, n.$$

The objective function (5.64) is the number of bins used. Constraints (5.65) state that each item is allocated to exactly one bin. Constraints (5.66) guarantee that bin capacities are not exceeded.

A lower bound $\underline{z}(I)$ on the number of bins for any 1-BP feasible solution can easily be obtained as

$$\underline{z}(I) = \lceil (p_1 + p_2 + \cdots + p_m)/q \rceil. \tag{5.67}$$

The lower bound (5.67) can be very poor if the average number of items per bin is low (see Problem 5.25 for an improved lower bound). Such lower bounds can be used in a branch-and-bound algorithm, or to evaluate the performance of heuristics. In the remainder of this section, four heuristics are illustrated. The first two heuristics, the first fit (FF) and the best fit (BF), are on-line heuristics while the others are off-line heuristics.

FF heuristic

Step 0. Let S be the list of items, V the list of available bins and T the list of bins already used. Initially T is empty.

Step 1. Extract an item i from the top of list S and insert it into the first bin $j \in T$ having a residual capacity greater than or equal to p_i. If no such bin exists, extract a new bin k from the top of list V and put it at the bottom of T; insert item i into bin k.

Step 2. If $S = \emptyset$, *STOP*, all items have been loaded. Then T is the list of bins used, while V provides the list of bins unused. If $S \neq \emptyset$, go back to Step 1.

BF heuristic

Step 0. Let S be the list of items V the list of available bins and T the list of bins already used. Initially T is empty.

Step 1. Extract an item i from the top of list S and insert it into the bin $j \in T$ whose residual capacity is greater than or equal to p_i, and closer to p_i. If no such bin exists, extract a new bin k from the top of V and put it at the bottom of T; insert item i into bin k.

Step 2. If $S = \emptyset$, *STOP*, all items have been loaded. Then T is the list of bins used, while V provides the list of bins unused. If $S \neq \emptyset$, go back to Step 1.

Both of the two heuristics can be implemented so that the computational complexity is equal to $O(m \log m)$.

It is useful to characterize the performance ratios of such heuristics. Recall that the performance ratio R^H of a heuristic H is defined as follows:

$$R^H = \sup_I \left\{ \frac{z^H(I)}{z^*(I)} \right\},$$

where I is a generic instance of the problem, $z^H(I)$ is the objective function value of the solution provided by heuristic H for instance I and $z^*(I)$ represents the optimal solution value for the same instance.

This means that

(a) $z^H(I)/z^*(I) \leq R^H, \ \forall\ I$;

(b) there are some instances I such that $z^H/z^*(I)$ is arbitrarily close to R^H.

Unfortunately, the performance ratios of the FF and BF heuristics are not known, but it has been proved that

$$R^{\mathrm{FF}} \leq 7/4,$$

and

$$R^{\mathrm{BF}} \leq 7/4.$$

The FF and BF heuristics can be easily transformed into off-line heuristics, by preliminary sorting of the items by non-increasing weights, yielding the first fit decreasing (FFD) and the best fit decreasing (BFD) heuristics. Their complexity is still equal to $O(m \log m)$, while their performance ratios are

$$R^{\mathrm{FFD}} = R^{\mathrm{FBFD}} = 3/2.$$

Indeed, it can be proved that this is the minimum performance ratio that a polynomial 1-BP heuristic can have (see Problem 5.23).

Al Bahar is an Egyptian trucking company located in Alexandria which must plan the shipment of 17 parcels, whose characteristics are reported in Table 5.22. For these shipments the company can use a single van whose capacity is 600 kilograms. Applying the BFD heuristic, the parcels are sorted by non-increasing weights (see Table 5.23) and the solution reported in Table 5.24 is obtained. The number of trips is six. The lower bound on the number of trips given by Equation (5.67) is $\lceil 3\,132/600 \rceil = 6$. Hence, the BFD heuristic solution is optimal.

Table 5.22 Weight of the parcels (in kilograms) in the Al Bahar problem.

Number of parcels	Weight
4	252
3	228
3	180
3	140
4	120

Table 5.23 Sorted list of the parcels in the Al Bahar problem (parcel weights are computed in kilograms).

Parcel	Weight	Parcel	Weight
1	252	10	180
2	252	11	140
3	252	12	140
4	252	13	140
5	228	14	120
6	228	15	120
7	228	16	120
8	180	17	120
9	180		

Table 5.24 Parcel-to-trip allocation in the optimal solution of the Al Bahar problem (parcel weights are computed in kilograms).

Parcel	Weight	Trip	Parcel	Weight	Trip
1	252	1	10	180	5
2	252	1	11	140	3
3	252	2	12	140	5
4	252	2	13	140	5
5	228	3	14	120	5
6	228	3	15	120	6
7	228	4	16	120	6
8	180	4	17	120	6
9	180	4			

5.4.2.2 Two-dimensional packing problems

The simplest two-dimensional packing (2-BP) problem consists of determining the least number of identical rectangular bins in which a given set of rectangular items can be accommodated. It is also assumed that no item rotation is allowed. Let L and W be the length and the width of a bin, respectively, and let l_i and w_i, $i = 1, \dots, m$, be the length and the width of item i. The 2-BP problem can be formulated as a set-partitioning problem. In this case, the set (indicated by V) is formed by all feasible loading plans of a subset of items in a single bin. Let a_{ij}, $i = 1, \dots, m$, $j \in V$, be a binary constant, equal to 1 if the item i belongs to the loading plan j, 0 otherwise. Indicating with x_j, $j \in V$, the binary decision variable, with value 1 if the jth loading plan is chosen (which means that a bin is loaded according to modality j), 0 otherwise, the 2-BP problem can be formulated in the following way:

$$\text{Minimize} \sum_{j \in V} x_j \tag{5.68}$$

subject to

$$\sum_{j \in V} a_{ij} x_j = 1, \; i = 1, \dots, m \tag{5.69}$$

$$x_j \in \{0, 1\}, \; j \in V. \tag{5.70}$$

The objective function (5.68) computes the number of bins used, whereas constraints (5.69) ensure that each item i, $i = 1, \dots, m$, is loaded in exactly one bin. The model (5.68)–(5.70) is valid also for the 1-BP case, for which the feasibility check of a loading plan j simply corresponds to verify that $\sum_{i=1}^{m} p_i a_{ij} \leq q$.

Despite its simplicity of formulation, the problem defined by (5.68)–(5.70) is NP-hard. The number of feasible loading plans to be generated depends on the particular instance and increases exponentially with the increase of the number of items. Therefore, the only way to solve the problem (5.68)–(5.70) is to dynamically generate the feasible loading plans when necessary. In this case, it is useful to have good lower bounds on the number of bins, one of which is easily obtained by letting

$$\underline{z}(I) = \lceil (l_1 w_1 + l_2 w_2 + \cdots + l_m w_m)/LW \rceil.$$

Most heuristics for the 2-BP problem are based on the idea of forming layers of items inside the bins. Each layer has a width W, and a length equal to that of its longest item. All the items of a layer are located on its bottom, which corresponds to the level of the longest item of the previous layer (see Figure 5.49).

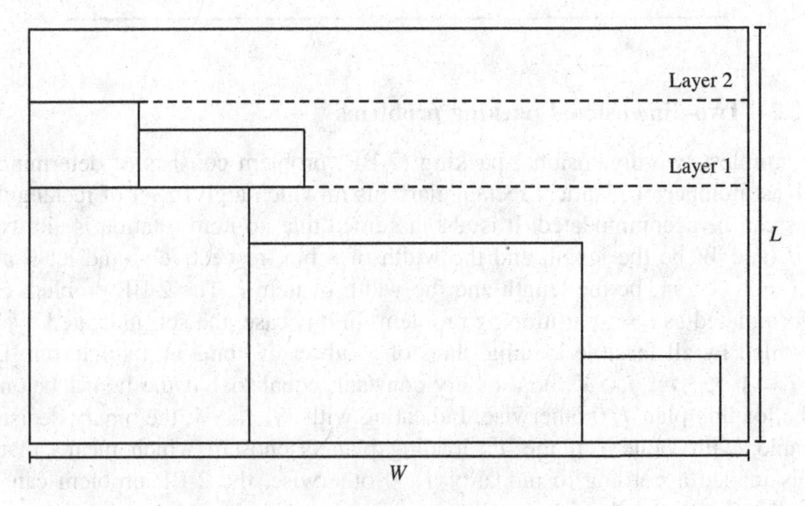

Figure 5.49 Layers of items inside a bin.

Here we illustrate two off-line heuristics, named finite first fit (FFF) and finite best fit (FBF) heuristics.

FFF heuristic

Step 0. Let S be the list of items, sorted by non-increasing lengths, V the list of bins and T the list of bins used. Initially T is empty.

Step 1. Extract an item i from the top of S and insert it into the leftmost position of the first layer (which can accommodate it) of the first bin $j \in T$. If no such layer exists, create a new one in the first bin of T (which can accommodate it) and introduce item i in the leftmost position of the layer. If there is no bin of T which can accommodate the layer, extract from the top of V a new bin k and put it at the bottom of T, and load item i into the leftmost position at the bottom of bin k.

Step 2. If $S = \emptyset$, *STOP*, all items have been loaded. Then, T represents the list of bins used, while V provides the list of the unused bins. If $S \neq \emptyset$, go back to Step 1.

FBF heuristic

Step 0. Let S be the list of items, sorted by non-increasing lengths, V the list of bins and T the list of bins used. Initially T is empty.

Step 1. Extract an item i from the top of S and insert it into the leftmost position of the layer of a bin $j \in T$ whose residual width is greater than or equal to, and closer to, item width. If no such layer exists, create a new one in the bin of T whose residual length is greater than or equal to, and closer to, the length of item i. Then, introduce item i in the leftmost position of the layer. If there is no bin of T which can accommodate the layer, extract from the top of V a new bin k and put it at the bottom of T, and load item i into the leftmost position at the bottom of bin k.

Step 2. If $S = \emptyset$, *STOP*, all items have been loaded. Then, T represents the list of bins used, while V provides the list of the unused bins. If $S \neq \emptyset$, go back to Step 1.

Such layer heuristics have a low computational complexity, since the effort for selecting the layer where an item has to be inserted is quite small. However, they can turn out to be inefficient if the average number of items per bin is relatively small. In such a case, the following bottom-left (BL) heuristic usually provides better solutions.

BL heuristic

Step 0. Let S be the list of items, sorted by non-increasing lengths, V the list of bins and T the list of bins used. Initially T is empty.

Step 1. Extract an item i from the top of list S and insert it into the leftmost position at the bottom of the first bin $j \in T$ able to accommodate it. If no such bin exists, extract a new bin k from the top of V, and put it at the bottom of T; load item i into the leftmost position at the bottom of bin k.

Step 2. If $S = \emptyset$, *STOP*, all items have been loaded, T represents the list of bins used, while V provides the list of bins unused. If $S \neq \emptyset$, go back to Step 1.

Kumi is a South Korean company manufacturing customized office furniture in Pusan. Outgoing products for overseas customers are usually loaded into containers ISO 40, whose characteristics are reported in Table 5.25. Once packaged, parcels are 2 m or 1 m high. They are loaded on wooden pallets so that they cannot be rotated at loading time. The list of parcels shipped on May 14 last is reported in Table 5.26. Parcels that are 1 m high are coupled in order to form six pairs of $(1 \times 1)\,\text{m}^2$ parcels and five pairs of $(0.8 \times 0.5)\,\text{m}^2$ parcels. Then, each such pair is considered as a single item. Applying the FBF algorithm, the solution reported in Figure 5.50 is obtained.

Table 5.25 Characteristics of the ISO 40 container.

Length (m)	Width (m)	Height (m)	Capacity (m^3)	Capacity (kg)
12.069	2.373	2.405	68.800	26 630

Table 5.26 Parcels shipped by Kumi company; dimensions are expressed in metres.

Quantity	Length	Width	Height
6	1.50	1.50	2.00
5	1.20	1.70	2.00
13	1.00	1.00	1.00
11	0.80	0.50	1.00

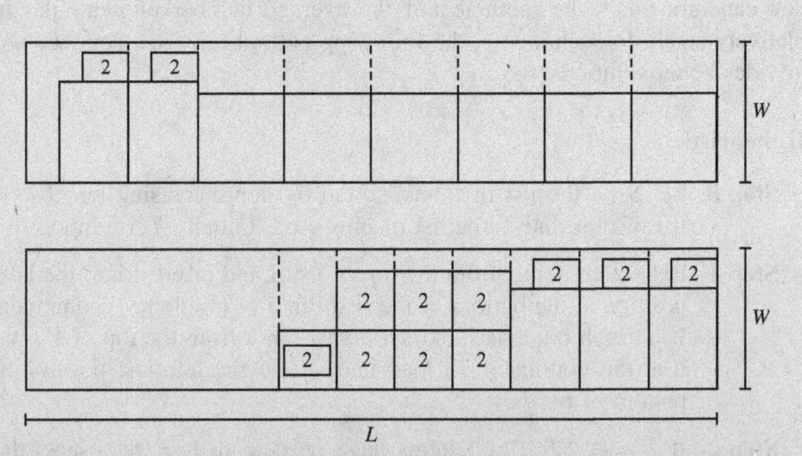

Figure 5.50 Parcels allocated to the two containers shipped by Kumi company (2 indicates two overlapping parcels).

5.4.2.3 Three-dimensional packing problems

The simplest three-dimensional packing (3-PB) problem consists of determining the least number of identical parallelepipedic bins in which a given set of parallelepipedic items can be accommodated. It is also assumed that no item rotation is allowed. Let L, W, and H be the length, the width and the height of a bin, respectively, and let l_i, w_i, and h_i, $i = 1, \ldots, m$, be the length, the width and the height of item i.

The problem can be formulated using the same set-partitioning model (5.68)–(5.70) illustrated in the two-dimensional case.

A lower bound $\underline{z}(I)$ on the number of bins is

$$\underline{z}(I) = \lceil l_1 w_1 h_1 + l_2 w_2 h_2 + \cdots + l_m w_m h_m)/LWH \rceil. \tag{5.71}$$

The simplest heuristics for 3-BP problems insert items sequentially into layers parallel to some bin surfaces (e.g. to $W\ H$ surfaces). In the sequel, a 3-BP-L heuristic, based on this principle, is illustrated.

3-BP-L heuristic

Step 0. Let S be the list of items.

Step 1. Solve the 2-BP problem associated to m items characterized by w_i, h_i, $i = 1, \ldots, m$, and bins characterized by W and H. Let k be the number of bidimensional bins used (referred to as *sections* in the following). The length of each section is equal to the length of the largest item loaded into it.

Step 2. Solve the 1-BP problem associated to the k sections, each of which has a weight equal to its length, while bins have a capacity equal to L.

If the items are all available when bin loading starts, it can be useful to sort list S by non-increasing values of the volume. However, unlike one-dimensional problems, more complex procedures are usually needed to improve solution quality.

McMillan Company is a motor carrier headquartered in Bristol, United Kingdom. The firm has recently made the procedure for allocating outgoing parcels to vehicles semi-automatic, using a decision support system. This software tool uses the 3-BP-L algorithm as a basic heuristic, and then applies a local search procedure. Last January 26, the parcels to be loaded were those reported in Table 5.27. The characteristics of the vehicles are indicated in Table 5.28. The parcels are loaded on pallets and cannot be rotated. First, the parcels are sorted by non-increasing volumes. Then, the 3-BP-L heuristic (in which 2-BP problems are solved through the BL heuristic) is used. The solution (see Tables 5.29 and 5.30) is made up of six (2.4×1.8) m^2 sections, loaded as reported in Figure 5.51. Finally, a 1-BP problem is solved by means of the BFD heuristic. In the solution (see Tables 5.29 and 5.30) three vehicles are used, the most loaded of which carries a weight of 1 220 kg, less than the weight capacity. It is worth noting that the lower bound provided by Equation (5.71) is $\lceil 35.963/28.08 \rceil = 2$.

Table 5.27 Parcels loaded at McMillan Company.

Type	Quantity	Length (m)	Width (m)	Height (m)	Volume (m^3)	Weight (kg)
1	2	2.50	0.75	1.30	2.4375	155
2	4	2.10	1.00	0.95	1.9950	140
3	7	2.00	0.65	1.40	1.8200	130
4	4	2.70	0.70	0.80	1.5120	115
5	3	1.20	1.50	0.80	1.4400	110

Table 5.28 Characteristics of the vehicles in the McMillan problem.

Length (m)	Width (m)	Height (m)	Capacity (m^3)	Capacity (kg)
6.50	2.40	1.80	28.08	1 230

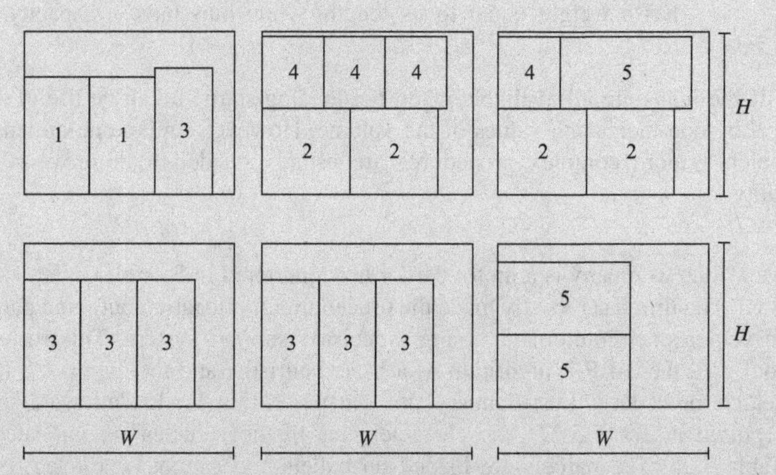

Figure 5.51 Sections generated at the end of Step 1 of the 3-BP-L heuristic in the McMillan problem.

Table 5.29 Width and weight of the sections generated at the end of Step 1 of the 3-BP-L heuristic in the McMillan problem.

Section	Width (m)	Weight (kg)
1	2.50	440
2	2.70	625
3	2.70	505
4	2.00	390
5	2.00	390
6	1.20	220

Table 5.30 Section allocation to vehicles at the end of Step 2 of the 3-BP-L heuristic in the McMillan problem.

Vehicle	Sections
1	2, 3
2	1, 4, 5
3	6

The 3-BP problem can be also faced using a more sophisticated heuristic based on the computation of the so-called extreme points. The idea is to load a new item into a bin exploiting to the maximum the empty space left by the items already there. When a new item i, $i = 1, \ldots, m$, sized (l_i, w_i, h_i) is loaded into a bin, in such a way that its rear vertex, the lowest to the left (south-west vertex), has coordinates (x_i, y_i, z_i), it will identify a series of extreme points upon which it is possible to base the positioning of the successive items (see Figure 5.52). These extreme points will be given by the intersection between the projections along the orthogonal axes of the bin of the vertices $(x_i + l_i, y_i, z_i)$, $(x_i, y_i + w_i, z_i)$ and $(x_i, y_i, z_i + h_i)$ and the items already present in the bin.

In more details, loading an item into the bin will generate new extreme points (see Figure 5.53) according to the following procedure.

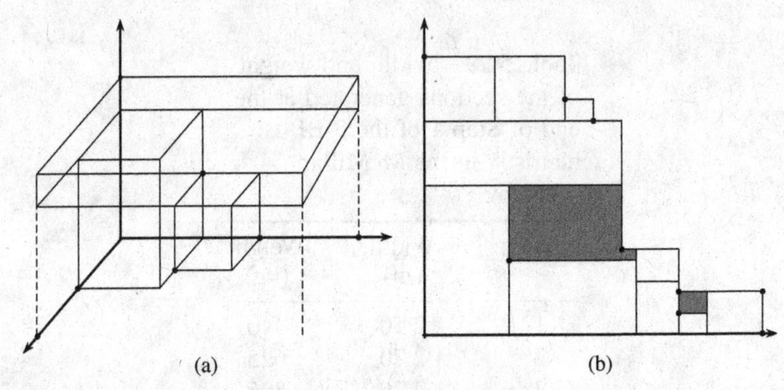

Figure 5.52 Examples of extreme points in three dimensions and in two dimensions.

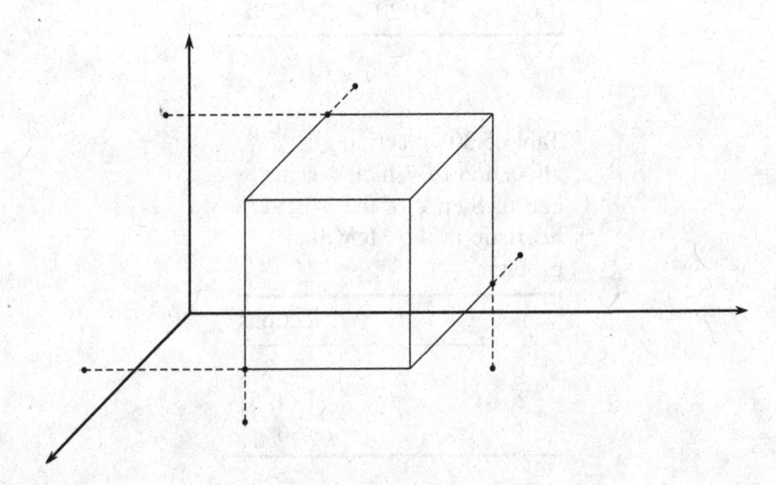

Figure 5.53 Projections of the vertices identifying the extreme points.

Updating procedure of the extreme points

Step 0. Let E_j be the list of the extreme points associated with the jth bin. The aim is to place an item i sized (l_i, w_i, h_i) into bin j.

Step 1. If bin j is empty, item i is placed with its south-west vertex in position (0, 0, 0) and three extreme points will be generated with coordinates (l_i, 0, 0), (0, w_i, 0) and (0, 0, h_i). These points are put into E_j.

Step 2. Otherwise, the item is placed with its south-west vertex in position (x_i, y_i, z_i) and the following points are projected:

- the point ($x_i + l_i$, y_i, z_i) towards plane xy and towards plane xz;
- the point (x_i, $y_i + w_i$, z_i) towards plane xy and towards plane yz;
- the point (x_i, y_i, $z_i + h_i$) towards plane xz and towards plane yz.

To the list of extreme points are added the points given by the intersection of these projections with the corner of the first item found in the bin or, in the absence of it, with the wall of the bin.

Heuristics based on extreme points differ among themselves on the basis of the criterion used to extract from the list of extreme points the point where to place the item to be loaded. Below, for the sake of brevity, a single heuristic is illustrated.

3-BP-FFD heuristic

Step 0. Let S ($\neq \emptyset$) be the list of items to be put into the bins. Let V be the list of bins available, and T the list of bins used; initially $T = \{1\}$. Let E_j, $j \in T$, be the list of extreme points associated with the jth bin used. Initially $E_j = \{(0, 0, 0)\}$, $j \in T$.

Step 1. Extract an item i from the top of list S and load it into the first compatible bin $j \in T$. A bin is compatible when in E_j there is an extreme point π which allows item i to be placed making the south-west vertex coincident with π without overlapping on any item already present in the bin (feasible extreme point). If bin j has several feasible extreme points, choose the one with the lowest value of the z coordinates for the placing of i, at parity of z choose the one with the lowest value of the y coordinates and at parity of z and y choose the one with the lowest x coordinate. If, instead, there is no compatible bin, from the top of V extract a new bin k and place it at the bottom of T. Initialize the list $E_k = \{(0, 0, 0)\}$. Load item i into bin k.

Step 2. Update the list of extreme points of the bin into which item i has been placed.

Step 3. If $S = \emptyset$, *STOP*, all the items have been inserted, T represents the list of bins used. If $S \neq \emptyset$, go back to Step 1.

LWK ships goods by sea to North America from its own warehouse situated in the port of Hamburg, using a standard-sized bin ship for transport. The shipping orders are a set of one or more products packed in parallelepiped-shaped parcels. Using suitable tables, the warehouse manager calculates the total weight and maximum volume that the bin to be shipped is able to transport and transmits the load list. It is not rare to find that the bin (whose size is shown in Table 5.31), once filled, is emptied and then filled again with a new loading plan, owing to parcels left out. Sometimes, despite having repeated the loading procedure, there are still unloaded parcels. The labour-intensiveness of this procedure has driven the company recently to adopt a more sophisticated decision support system. The basic procedure used to define the loading plan into the bins is the 3-BP-FFD heuristic. Note that the type of parcels forbids item rotation. The list S of parcels (see Table 5.32) to be loaded is sorted according to non-increasing values of the volume. So, we obtain:

$$S = \{16, 10, 15, 19, 20, 21, 6, 17, 14, 9, 2, 18, 3, 7, 8, 13, 11, 12, 4, 5, 1\}.$$

Table 5.31 Characteristics of the ISO 20 bin used for LWK loading operations.

Length (m)	Width (m)	Height (m)	Volume (m^3)
5.899	2.352	2.388	33.13

Table 5.32 Parcels to be loaded into the LWK bin.

Item	Length (m)	Width (m)	Height (m)	Volume (m^3)
1	0.30	0.35	0.90	0.094500
2	0.46	0.66	1.44	0.437184
3	0.60	0.66	0.95	0.376200
4	0.32	0.33	0.96	0.101376
5	0.32	0.33	0.96	0.101376
6	0.97	1.03	1.22	1.218902
7	0.68	0.68	0.68	0.314432
8	0.68	0.68	0.68	0.314432
9	0.64	0.64	1.15	0.471040
10	1.20	2.20	2.22	5.860800
11	0.44	0.55	0.65	0.157300
12	0.40	0.55	0.60	0.132000
13	0.54	0.64	0.70	0.241920
14	0.54	0.80	1.20	0.518400
15	1.20	2.20	2.21	5.834400
16	1.63	2.05	2.22	7.418130
17	0.40	0.69	1.90	0.524400
18	0.36	0.69	1.73	0.429732
19	1.00	1.40	1.50	2.100000
20	1.42	1.08	0.84	1.288224
21	1.16	1.37	0.78	1.239576

The first item (16) is placed into the first bin, with the rear vertex lowest and to the left, in the lowest left-hand corner of the bin (see Figure 5.54). The list of extreme points associated with the first bin after putting in item

16 therefore results in:

$$E_1 = \{(1.63, 0, 0), (0, 2.05, 0), (0, 0, 2.22)\}.$$

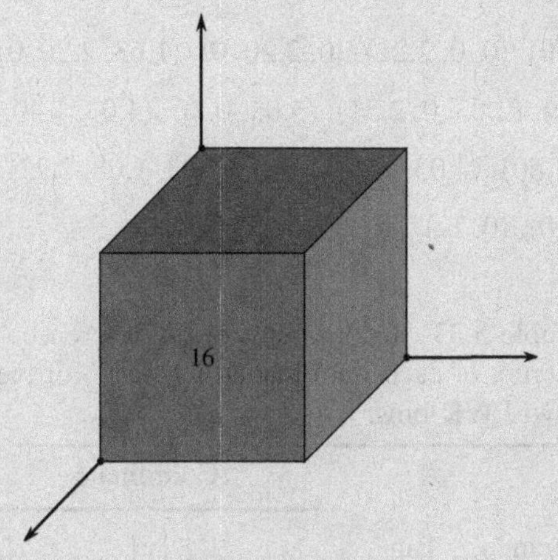

Figure 5.54 Loading of item 16 into the first LWK bin (extreme points are represented by black dots).

At the second iteration, also item 10 is placed in the first bin, as shown in Figure 5.55. The list of extreme points associated with the first bin, after putting in item 10, is modified in this way:

$$E_1 = \{(0, 2.05, 0), (0, 0, 2.22), (2.83, 0, 0), (0, 2.20, 0), (1.63, 2.20, 0), (1.63, 0, 2.22)\},$$

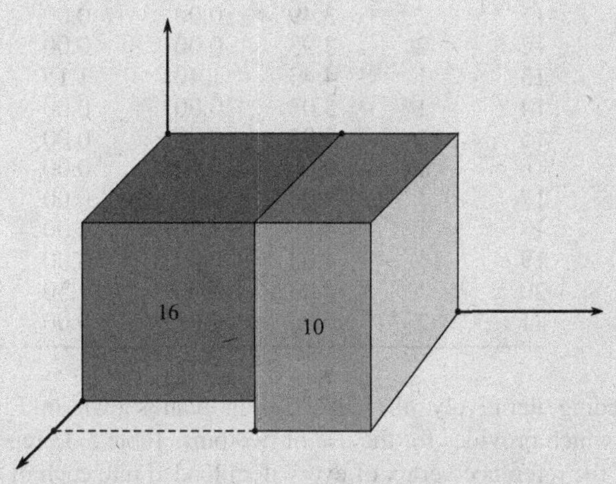

Figure 5.55 Loading of item 10 into the first LWK bin (extreme points are represented by black dots).

The third, fourth and fifth iterations lead to items 15, 19 and 20 being put into the first bin, whereas the successive item, (21) in list S, will be put into a new bin. Following these operations, the lists of extreme points associated with the two open bins are the following:

$$E_1 = \{(0, 2.05, 0), (0, 0, 2.22), (0, 2.20, 0), (1.63, 2.20, 0), (1.63, 0, 2.22), (2.83, 2.20, 0), (2.83, 0, 2.21), (5.03, 0, 0), (4.03, 1.40, 0), (5.45, 0, 2.22), (4.03, 1.08, 1.50), (4.03, 1.08, 1.50), (4.03, 1.08, 2.22), (4.03, 0, 2.34)\};$$

$$E_2 = \{(1.16, 0, 0), (0, 1.37, 0), (0, 0, 0.78)\}.$$

Table 5.33 Coordinates of the reference vertex of each item loaded into each of the two LWK bins.

		Coordinate		
Item	Bin	[x]	[y]	[z]
1	1	5.57	0.66	0.00
2	1	5.07	1.40	0.00
3	1	4.43	1.40	1.15
4	1	5.57	0.00	0.00
5	1	5.57	0.33	0.00
6	2	1.16	0.00	0.00
7	2	2.13	0.00	0.00
8	2	2.81	0.00	0.00
9	1	4.43	1.40	0.00
10	1	1.63	0.00	0.00
11	2	3.49	0.00	0.00
12	2	3.93	0.00	0.00
13	1	4.43	1.40	1.15
14	1	5.03	0.00	0.00
15	1	2.83	0.00	0.00
16	1	0.00	0.00	0.00
17	1	4.03	1.40	0.00
18	1	5.53	1.40	0.00
19	1	4.03	0.00	0.00
20	1	4.03	0.00	1.50
21	2	0.00	0.00	0.00

Proceeding iteratively, the full loading plan shown in Figure 5.56 is obtained, which provides for the use of two bins. Table 5.33 reports the coordinates of the reference vertex of every item loaded into each of the two bins.

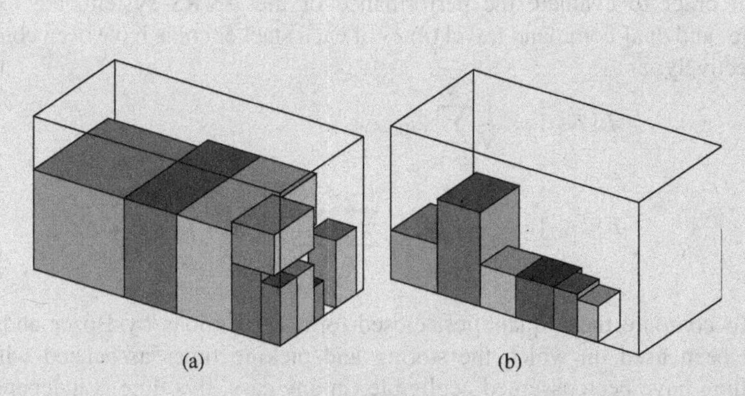

Figure 5.56 Full loading plan of the two LWK bins.

The 3-BP-FFD heuristic can be also adapted to the case in which rotation of the items is allowed.

5.5 Case study: Performance evaluation of an AS/RS system conducted by Wert Consulting

Wert Consulting is a Dutch company that specializes in the design, manufacturing and selling of shelvings, stacker cranes for AS/RS warehouses and other storage systems. In February 2012, Wert acted as consultant of a primary food company with the aim at evaluating the performance of a palletized AS/RS warehouse, in which each aisle will be served by a new stacker crane.

The stacker crane can operate according to one of two modes:

- *single cycle*: storage and retrieval operations are performed one at a time;
- *dual cycle*: pairs of storage and retrieval operations are made in sequence in an attempt to reduce the overall travel time.

We denote by M, the number of storage aisles; 0, the I/O point of the stacker crane; N, the total number of storage positions in the rack; L_x, the length of the rack; L_y, the height of the rack; v_x, the (constant) speed of the stacker crane along the x axis; v_y, the (constant) speed of the stacker crane along the y axis; t_{0i}, $i = 1, \ldots, N$, the one-way travel time between the I/O point and the storage position i ($t_{0i} = t_{i0}$); t_{ij}, $i, j = 1, \ldots, N$, the one-way travel time between the storage positions i and j ($t_{ij} = t_{ji}$); T_{SC}, the round-trip travel time of a single command (a random variable), and T_{DC} the round-trip travel time of a dual command (a random variable).

In order to evaluate the performance of the AS/RS system, the expected single- and dual-command travel times of each stacker crane have been computed, respectively, as

$$E[T_{SC}] = \frac{1}{N}\sum_{i=1}^{N} 2t_{0i};$$

$$E[T_{DC}] = \frac{1}{N(N-1)}\sum_{i=1}^{N-1}\sum_{j=i+1}^{N} 2(t_{0i} + t_{ij} + t_{j0}).$$

To compute these quantities, closed-form expressions by Bozer and White have been used, in which the storing and picking times associated with load handling have been assumed negligible (in any case, this time is independent of the rack shape and of the travel speed of the stacker crane) and the I/O point is located at the lower left-hand corner.

Let T_x be the horizontal travel time required to go to the farthest column from the I/O station. Likewise, let T_y be the vertical travel time required to go to the farthest row (level). Then, by definition, $T_x = L_x/v_x$ and $T_y = T_y/v_y$. Since the stacker crane travels simultaneously and independently in the horizontal and vertical directions, $T = \max(T_x, T_y)$ is the time required to reach the farthest storage location from the I/O point and $b = \min(T_x/T, T_y/T)$. In the sequel, without loss of generality, we assume that $T = T_x$ and that, consequently, $b = T_y/T$.

First of all, Wert computed the expected single-command travel time.

Let (x, y) be the Cartesian coordinate of a storage location. The stacker crane will take a time equal to $t_{xy} = \max(t_x, t_y)$ to reach (x, y), where $t_x = x/v_x$ and $t_y = y/v_y$. Under the hypothesis that each storage location is accessible with the same probability, the coordinates are independent random variables. Hence,

$$F_{t_{xy}}(t) = Pr(t_{xy} \leq t) = Pr(t_x \leq t, t_y \leq t) = Pr(t_x \leq t)Pr(t_y \leq t).$$

Moreover, the coordinates x and y are uniformly distributed:

$$Pr(t_x \leq t) = t/T,\ 0 \leq t \leq T;$$

$$Pr(t_y \leq t) = \begin{cases} t/bT, & \text{if } 0 \leq t \leq bT \\ 1, & \text{if } bT < t \leq T. \end{cases}$$

As a result, the cumulative probability distribution function of t_{xy} is

$$F_{t_{xy}}(t) = \begin{cases} t^2/bT^2, & \text{if } 0 \leq t \leq bT \\ t/T, & \text{if } bT < t \leq T. \end{cases}$$

The associated probability density function is

$$f_{t_{xy}}(t) = \begin{cases} 2t/bT^2, & \text{if } 0 \leq t \leq bT \\ 1/T, & \text{if } bT < t \leq T. \end{cases}$$

As a consequence, the expected single-command travel time is

$$E[T_{SC}] = 2\int_0^T t f_{t_{xy}}(t)\,dx = \frac{4}{bT^2}\int_0^{bT} t^2\,dx + \frac{2}{T}\int_{bT}^{T} t\,dx = (b^2/3+1)T.$$

Similarly, the expected dual-command travel time has been computed as

$$E[T_{DC}] = (4/3 + b^2/2 - b^3/30)T.$$

The stacker crane tested by Wert is characterized by the following data: $M = 3$, $L_x = 95$ m, $L_y = 20$ m, $v_x = 2.5$ m/sec and $v_y = 0.7$ m/sec.

Wert has computed the following parameters: $T_x = L/v_x = 38.00$ sec, $T_y = L_y/v_y = 28.57$ sec, $T = \max(T_x, T_y) = 38.00$ sec and $b = \min(T_x/T, T_y/T) = 0.752$. Since the stacker crane will operate only single commands, the expected travel time has been estimated as $(b^2/3+1)\times T = 45.16$ sec. Hence, the throughput of the stacker crane is approximately equal to 3 600/45.16, which corresponds to about $80 \times 3 = 240$ pallets per hour for the whole warehouse.

5.6 Case study: Inventory management at Wolferine

Wolferine is a division of the industrial group UOP Limited which manufactures copper and brass tubes. The company production processes take place in a factory located in London, Ontario, Canada, with highly automated systems operating with a very low work-in-process. The raw materials originate from mines located close to the factory. Consequently, the firm does not need to stock a large amount of raw materials (as a rule, no more than a two-week demand). As far as the finished-products inventories are concerned, Wolferine makes use of EOQ models (see Section 5.3.2.1). In the autumn of 1980, the firm operated with a production level close to the plant capacity. At that time the interest rate was around 10%. Using this value in the EOQ model, the company set the finished-goods inventory level equal to 833 tons. During the subsequent two years, an economic recession hit the industrialized countries. The interest rate underwent continuous and quick variations (up to 20% in August 1981), the demand of finished products went down by 20%, and the price level increased sharply. According to the EOQ model, the finished-goods inventory level should have been lower under those conditions. In order to illustrate this result, let n be the number of products; k an order fixed cost (assumed independent from the product); d_i, $i = 1, \ldots, n$, the annual demand of product i; p the interest rate (increased to take into account warehousing costs); c_i, $i = 1, \ldots, n$, the price of product i, and $\bar{I}$ the average stock level at time period t_0 (January 1981). On the basis of Equation (5.32),

$$\bar{I}(t_0) = \frac{1}{2}\sum_{i=1}^{n}\sqrt{\frac{2kd_i}{pc_i}}.$$

We can express parameter p as the sum of a bank interest rate p_1 and of a rate p_2 associated to warehousing costs:

$$p = p_1 + p_2.$$

Moreover, let δ_1, δ_2 and δ_3 be the variation rates (supposed equal for all products) of price, demand and interest rate at time period t. The average stock level is equal to

$$\bar{I}(t) = \frac{1}{2}\sum_{i=1}^{n}\sqrt{\frac{[k(1+\delta_1)][d_i(1+\delta_2)]}{[p_1(1+\delta_3)+p_2][c_i(1+\delta_1)]}} = \bar{I}(t_0)\sqrt{\frac{(1+\delta_2)p}{p_1(1+\delta_3)+p_2}}. \quad (5.72)$$

According to Equation (5.72), if the demand decreases ($\delta_2 < 0$) and interest rate increases ($\delta_3 > 0$), the stock level should be lower. However, the managers of Wolferine continued to operate as in 1981. As a result, the inventory turnover index (see Section 5.1.1) suddenly decreased. Moreover, to protect the manufacturing process against strikes at the mines, the firm decided to also hold a raw-material inventory. As a result, when the recession ended in 1983, the firm had an exceedingly large stock of both raw materials and finished products.

5.7 Case study: Airplane loading at FedEx

FedEx is one of the leading express carriers in the world, with a freight traffic estimated at about two million parcels per day. Its sales offices are located in more than 180 countries and the company uses a fleet of about 700 airplanes and more than 40 000 trucks and vans.

In the United States, parcels whose origin and destination exceed a given distance are consolidated in containers and sent by air. An airplane may fly between a pair of destinations or may follow a multistop route where containers are loaded or unloaded at intermediate stops.

In order to use airplane capacity efficiently, a key issue is to devise good loading plans, taking into account a number of aspects: the load must be balanced around the centre of gravity of the aircraft, the total weight in the various areas of the aircraft must not exceed given thresholds in order to limit the cutting forces on the plane etc. These aspects are especially critical for some planes, such as the Airbus A300, a low-fuel-consumption aircraft. In addition, when loading an airplane assigned to a multistop route, containers to be unloaded at intermediate stops must be positioned close to the exit.

In order to allow airplanes to take off on time, the allocation of containers on board must be performed in real time, that is, containers must be loaded on the aircraft as soon as they arrive at the airport. As a matter of fact, 30% to 50% of the containers are already on board when the ground staff has a complete knowledge of the features of the containers to be loaded. Of course, once some containers have been loaded, it may become impossible to load the subsequent containers. As a result, when new containers arrive at the airport, it is sometimes necessary to define a new loading plan in which some containers that were previously loaded are unloaded. This situation arises frequently when the total load is close to the capacity of the aircraft, as shown in Figure 5.57, where the percentage of success in loading an airplane is reported as a function of the percentage of the load capacity used.

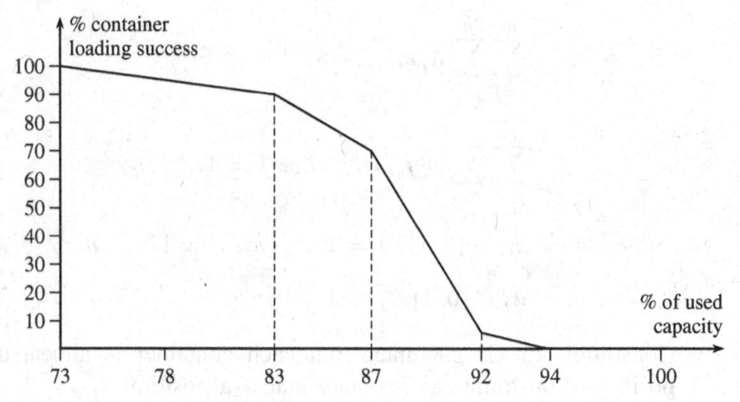

Figure 5.57 Percentage of success in loading an aircraft as a function of the percentage of load capacity (simulation made by FedEx.)

The objective pursued by FedEx consists of loading the largest number of containers as possible. If no container is yet loaded, the following solution procedure is implemented.

Step 1. Let m be the number of containers to be loaded; n the number of positions in the loading area; q the number of areas into which the plane is divided; p_i, $i = 1, \dots, m$, the weight of container i; P_j, $j = 1, \dots, n$, the maximum weight that can be loaded in position j; d_j, $j = 1, \dots, n$, the distance from position j to the centre of gravity O; M^{min} and M^{max} the minimum and maximum moments of the loads with respect to O; L_k, $k = 1, \dots, q$, the total maximum weight that can be placed in area k, and f_{jk}, $j = 1, \dots, n$, $k = 1, \dots, q$, the fraction of position j contained in area k. Also let x_{ij}, $i = 1, \dots, m$, $j = 1, \dots, n$, be a binary decision variable equal to 1 if container i is placed in position j, and 0 otherwise, and u_j, $j = 1, \dots, n$, a binary decision variable equal to 1 if position j is used, and 0 otherwise. A feasible solution is defined by the following set of constraints:

$$\sum_{j=1}^{n} x_{ij} = 1, \ i = 1, \dots, m; \tag{5.73}$$

$$\sum_{i=1}^{m} x_{ij} \leq m u_j, \; j = 1, \ldots, n; \tag{5.74}$$

$$\sum_{i=1}^{m} p_i x_{ij} \leq P_j u_j, \; j = 1, \ldots, n; \tag{5.75}$$

$$\sum_{i=1}^{m} \sum_{j=1}^{n} d_j p_i x_{ij} \leq M^{\max}; \tag{5.76}$$

$$\sum_{i=1}^{m} \sum_{j=1}^{n} d_j p_i x_{ij} \geq M^{\min}; \tag{5.77}$$

$$\sum_{i=1}^{m} \sum_{j=1}^{n} p_i f_{jk} x_{ij} \leq L_k, \; k = 1, \ldots, q; \tag{5.78}$$

$$x_{ij} \in \{0, 1\}, \; i = 1, \ldots, m, \; j = 1, \ldots, n; \tag{5.79}$$

$$u_j \in \{0, 1\}, \; j = 1, \ldots, n. \tag{5.80}$$

Constraints (5.73) guarantee that each container is allocated in a position. Constraints (5.74) state that if a position j, $j = 1, \ldots, n$, accommodates a container, then its corresponding u_j variable must be equal to 1. Constraints (5.75) ensure that the total weight loaded in any position does not exceed a pre-established upper bound. Constraints (5.76) and (5.77) impose that the total moment, with respect to point O, is within the pre-established interval. Constraints (5.78) ensure the respect of the weight bounds in each section.

Step 2. If problem (5.73)–(5.80) is infeasible, a container $\bar{i}$ is eliminated from the loading list and the problem is solved again (where, of course, $x_{\bar{i}j} = 0$, $j = 1, \ldots, n$). Step 2 is repeated until a feasible solution is found.

If some containers have already been loaded, the previous procedure is modified as follows. Let $\bar{I}$ be the set of the containers already loaded and let $j(i)$, $j = 1, \ldots, n$, $i \in \bar{I}$, be the position assigned to container i. Then, additional constraints

$$x_{ij_i} = 1, i \in \bar{I} \tag{5.81}$$

are added to (5.73)–(5.80). If this problem is feasible, the procedure stops since the loading plan partially executed can be completed. Otherwise, the constraint (5.81) associated with container $i \in \bar{I}$ allocated to the position closest to the entrance-exit is removed (this corresponds to unloading the container from the aircraft). This step is repeated until a feasible solution is obtained.

5.8 Questions and problems

5.1 In many industrialized countries, the average inventory turnover index is around 20 for dairy products and around five for household electrical appliances. Discuss these figures.

5.2 Consider a warehouse equipped with a stacker crane. The longitudinal and latitudinal speeds are 25 cm/sec and 40 cm/sec, while the corresponding aisle length and height are 20 m and 12 m. Supposing that the home position is the left most position at 1 m from floor, compute the maximum waiting time for the completion of a retrieval operation.

5.3 Which warehouse layout is the most suitable for a highly seasonable demand pattern?

5.4 Show that a warehouse can be modelled as a queueing system.

5.5 A warehouse stores nearly 20 000 pallets. Pallets turn around about five times a year. How much is the required labour force? Assume two eight-hour shifts per day and about 250 working days per year. (Hint: apply Little's Law stating that for a queueing system in steady state, the average length L_Q of the queue equals the average arrival rate λ times the average waiting time T_W, $L_Q = \lambda T_W$.)

5.6 Let d be the daily demand from all orders, l_C the average length of a rail car, q the capacity of a rail car and n_C the number of car changes per day. Estimate the length of rail dock l_D needed by a warehouse.

5.7 Foral distributes electrical products and plans to build a new distribution centre, where the storage zone will accommodate 60 000 boxes. Goods will be stocked on conventional racks and transported by means of traditional

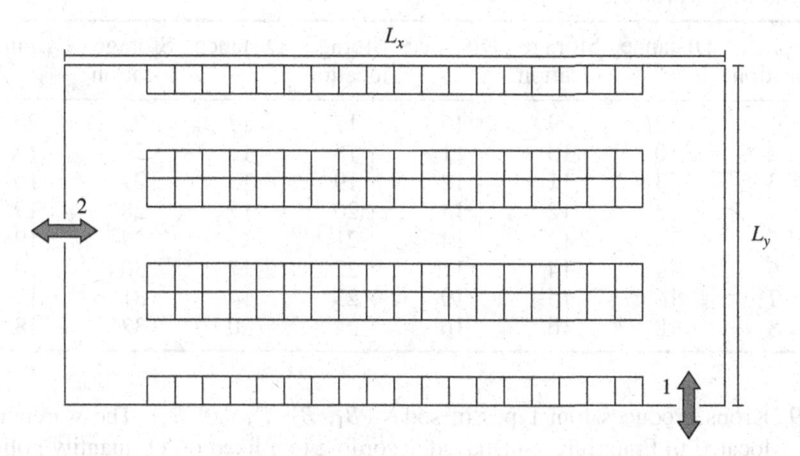

Figure 5.58 Storage zone layout in the Foral problem.

trolleys. Each rack has five shelves, and each of them has room for up to five stacked boxes. Racks will be arranged as in Figure 5.58, where vertical aisles are 1.5 m wide, while the horizontal aisles are 2.5 m wide. Each box requires a square area equal to $50 \times 50\,\text{cm}^2$. Size the storage zone, under the hypothesis that the probability that the picker enters the storage zone from the first I/O point is p ($0 \leq p \leq 1$), whereas the probability is $(1 - p)$ if he enters from the second I/O point.

5.8 Motif is a Turkish electrotechnical company which operates a warehouse in Ankara. The warehouse has a single I/O point and 32 storage locations. Tables 5.34 and 5.35 report the characteristics of the six products and the distances between the I/O point and the storage locations, respectively. Assuming that the travel time from the I/O point to storage location k ($k = 1, \ldots, 32$) is directly proportional to the corresponding distance, determine the optimal allocation of the products in the warehouse.

Table 5.34 Characteristics of the products in the Motif problem.

Product	Storage locations required	Daily number of storages and retrievals
1	6	133
2	5	128
3	8	198
4	5	126
5	4	128
6	3	98

Table 5.35 Distance (in m) between the storage locations and the I/O points in the Motif problem.

Storage location	Distance	Storage location	Distance	Storage location	Distance	Storage location	Distance
1	12	9	15	17	17	25	20
2	10	10	11	18	13	26	14
3	13	11	12	19	13	27	16
4	13	12	14	20	17	28	19
5	17	13	14	21	19	29	19
6	14	14	17	22	17	30	20
7	16	15	19	23	14	31	17
8	18	16	16	24	20	32	18

5.9 Krons produces four types of sodas (B_1, B_2, B_3 and B_4). The warehouse located in Frankfurt is managed according to a fixed order quantity policy. Table 5.36 reports the values of the order sizes (in pallets) and of the safety

stocks for each product. Figures 5.59–5.62 report the inventory level over time (in pallets) for the four products.
In order to rationalize the warehouse, its manager has chosen a two-class storage policy, where each class is made up of two products. Determine which is the most convenient combination of the products for each class, with the aim of minimizing the number of storage locations.

Table 5.36 Order size and safety stock (in pallets) for the four products in the Krons problem.

Product	Order size	Safety stock
B_1	400	100
B_2	700	150
B_3	600	120
B_4	300	80

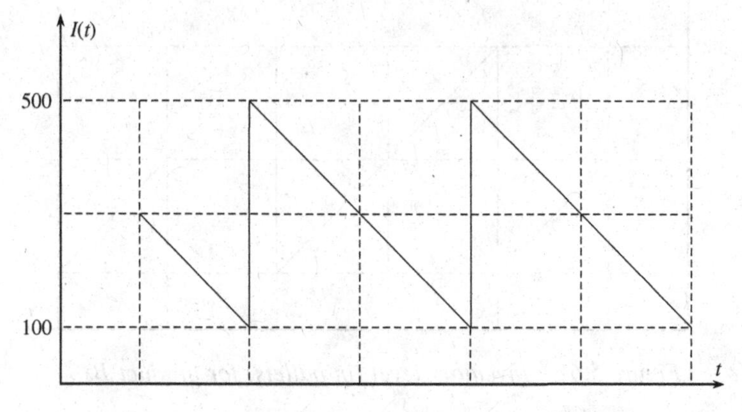

Figure 5.59 Inventory level (in pallets) for product B_1.

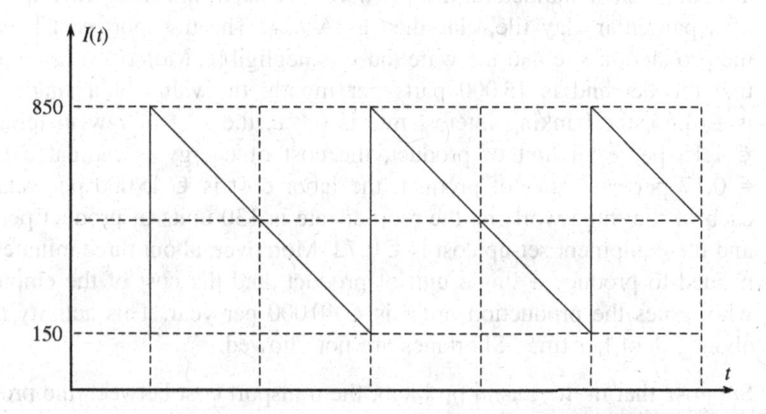

Figure 5.60 Inventory level (in pallets) for product B_2.

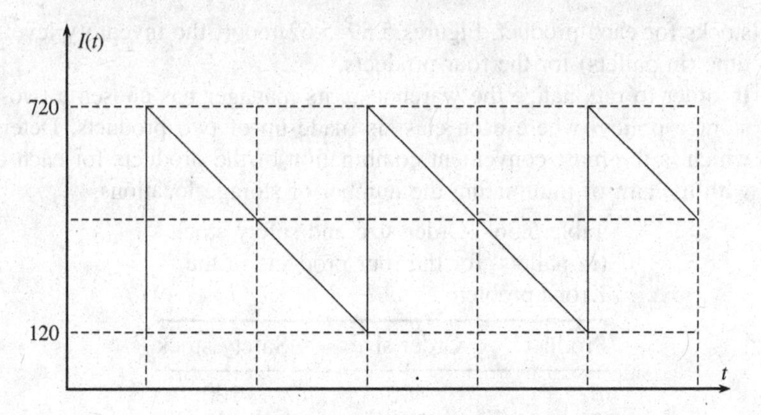

Figure 5.61 *Inventory level (in pallets) for product B_3.*

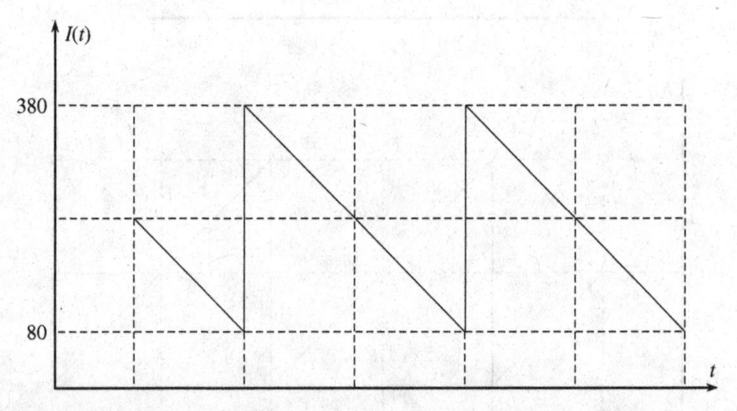

Figure 5.62 *Inventory level (in pallets) for product B_4.*

5.10 The Scottish manufacturer Lasim needs to determine the order quantity of a particular clay tile, classified as A-124. The transport cost between the production site and the warehouse is negligible. Moreover, it is known that the demand is 15 000 parts per month, the value of a single item is € 3.50, the banking interest rate is 6.8%, the cost of raw materials is € 1.13 per each unit of product, the cost of energy is estimated to be € 0.17 per each unit of product, the labor cost is € 45 000 per year for each of the three workers, the reorder rate is 130 units of product per day and the equipment set-up cost is € 0.72. Moreover, about three minutes are needed to produce a single unit of product, and the cost of the employee who issues the production order is € 40 000 per year. This activity takes about 5% of her time. Shortages are not allowed.

5.11 Suppose that in the Lasim problem, the transport cost between the production site and the warehouse is not negligible. In this case, the inventory replenishment of product A-124 has a transport cost of € 25. Determine the reorder quantity and level if the lead time is 10 days.

5.12 Modify the EOQ formula for the case where the warehouse has a finite capacity Q.

5.13 Modify the EOQ formula for the case where the holding cost is a concave function of the number of items kept in inventory.

5.14 Devise an optimal inventory policy for the EOQ model with a finite time horizon T_H.

5.15 Draw the auxiliary graph used for solving the Sao Vincente Chemical problem (see Section 5.3.2.1) as a shortest-path problem.

5.16 Modify the Wagner-Within model for the case where the warehouse is capacitated. Does the ZIO property still hold?

5.17 What is the optimal order quantity in the newsboy model if the warehouse is capacitated?

5.18 Extend the procedure for simultaneously managing the inventory of three products, characterized by a joint-order replenishment policy.

5.19 The material-handling systems for the Maltese distribution centre in La Valletta of Ekom makes use of hand transpallet trucks which can hold at most three roll containers, with a capacity of 15 packages each. Last January 9, at 10:15 AM, a picking list was created of eight orders to be retrieved from the storage zone, according to Table 5.37.

Table 5.37 Picking list in the Ekom problem.

Order	Number of packages
1	17
2	14
3	18
4	9
5	10
6	13
7	7
8	6

The Euclidean distances between the centres of gravity of such orders are reported in Table 5.38.
The Euclidean distance between the centres of gravity of order 7 and 8 can be obtained from Tables 5.39 and 5.40, reporting the list of packages and the corresponding Cartesian coordinates.

Table 5.38 Euclidean distances (in m) between the centres of gravity of orders in the Ekom problem.

	Order							
	1	2	3	4	5	6	7	8
1	0.00	31.56	19.62	31.48	28.60	23.17	25.90	28.77
2		0.00	28.30	20.50	29.49	22.10	25.39	25.34
3			0.00	29.02	21.40	19.57	31.68	21.31
4				0.00	25.06	31.65	28.50	26.33
5					0.00	31.29	22.51	23.98
6						0.00	25.42	19.85
7							0.00	?

Table 5.39 Package list and corresponding Cartesian coordinates for order 7 in the Ekom problem.

Package (i)	Abscissa ($x_i^{(7)}$)	Ordinate ($y_i^{(7)}$)
1	24.06	48.71
2	46.38	46.94
3	53.73	48.37
4	6.45	19.29
5	46.10	48.22
6	6.57	22.58
7	35.98	38.51

Table 5.40 Package list and corresponding Cartesian coordinates for order 8 in the Ekom problem.

Package (i)	Abscissa ($x_i^{(8)}$)	Ordinate ($y_i^{(8)}$)
1	1.61	9.47
2	41.71	46.90
3	4.25	34.96
4	32.02	17.49
5	19.24	23.35
6	17.37	26.01

Determine a lower bound on the number of grouped orders. Identify how to group orders, based on a seed order, choosing orders on the basis of their number of packages.

5.20 Determine the picker route in the storage zone of the French warehouse of the Guillen company (see Figure 5.63), using the S-shaped heuristic, the largest gap heuristic, the combined heuristic and the aisle-by-aisle heuristic.

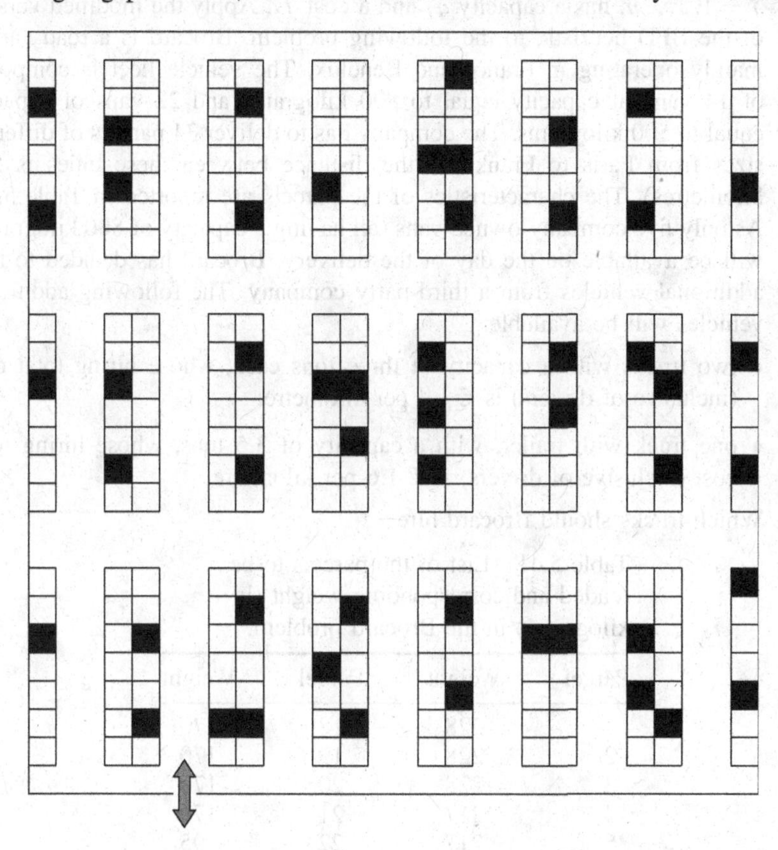

Figure 5.63 Storage zone layout in the Guillen problem.

5.21 Show that an optimal picker route cannot traverse an aisle (or a portion of an aisle) more than twice. Illustrate how this property can be used to devise a dynamic programming algorithm.

5.22 Demonstrate that the FF and BF heuristics for the 1-BP problem take $O(m \log m)$ steps.

5.23 Demonstrate that no polynomial heuristic H for the 1-BP problem can have $R^H < 3/2$ unless P = NP. (Hint: prove that solving the 2-partition problem, which is NP-hard, is equivalent to solving a particular instance of the 1-BP problem. The 2-partition problem consists of establishing whether there exists a partition (J_1, J_2) of a set J of non-negative numbers $t_1, t_2, \ldots, t_m$, such that $\sum_{j \in J_1} t_j = \sum_{j \in J_2} t_j = \frac{1}{2} \sum_{j \in J} t_j$.)

5.24 Devise a branch-and-bound algorithm based on formula (5.67).

5.25 Devise an improved 1-BP lower bound.

5.26 Modify the heuristics for the 1-BP problem for the case where each bin j, $j = 1, \ldots, n$, has a capacity q_j and a cost f_j. Apply the modified version of the BFD heuristic to the following problem. Brocard is a road carrier mainly operating in France and Benelux. The vehicle fleet is composed of 14 vans of capacity equal to 800 kilograms and 22 vans of capacity equal to 500 kilograms. The company has to deliver 34 parcels of different sizes from Paris to Frankfurt (the distance between these cities is 592 kilometres). The characteristics of the parcels are reported in Table 5.41. As only five company-owned vans (all having a capacity of 800 kilograms) will be available on the day of the delivery, Brocard has decided to hire additional vehicles from a third-party company. The following additional vehicles will be available:

- two trucks with a capacity of three tons each, whose hiring total cost (inclusive of drivers) is € 1.4 per kilometre;
- one truck with trailer, with a capacity of 3.5 tons, whose hiring total cost (inclusive of drivers) is € 1.6 per kilometre.

Which trucks should Brocard hire?

Table 5.41 List of the parcels to be loaded and corresponding weight (in kilograms) in the Brocard problem.

Parcel	Weight	Parcel	Weight
1	228	18	170
2	228	19	170
3	228	20	170
4	217	21	170
5	217	22	95
6	217	23	95
7	217	24	95
8	210	25	95
9	210	26	75
10	210	27	75
11	210	28	75
12	195	29	75
13	195	30	75
14	195	31	75
15	170	32	55
16	170	33	55
17	170	34	55

5.27 Determine a lower bound on the optimal solution cost in the Brocard problem by suitably modifying Equation (5.67). Also determine the optimal shipment decision by solving a suitable modification of the 1-BP problem.

Table 5.42 Characteristics of the packages in the Pertax problem (measures expressed in cm).

Package	Length	Width	Height
1	60	70	80
2	40	90	100
3	50	60	120
4	30	20	90
5	100	60	80
6	100	70	120
7	40	80	110
8	100	100	80
9	140	50	100
10	90	70	130
11	80	30	130
12	60	50	120
13	130	90	60
14	100	50	60
15	50	130	70
16	20	60	70
17	90	70	130
18	130	50	120
19	50	30	130
20	60	30	100
21	80	40	110

5.28 Modify the BL heuristic for solving a 2-BP problem in which item rotation is allowed.

5.29 Consider a 2-BP problem with $W = 11$, $L = 16$ and $n = 6$, where the characteristics (w_i, l_i) of each object i, $i = 1, \ldots, 6$, are

$$\{(6, 8), (6, 8), (7, 8), (4, 9), (5, 7), (4, 4), (5, 6)\}.$$

- Use the BL heuristic to determine the number of bins to use;
- determine a lower bound $\underline{z}(I)$ on the number of bins;
- according to the value of $\underline{z}(I)$ obtained before, determine whether the use of the BL heuristic allows one to obtain the minimum number of bins used.

5.30 Pertax has to consolidate some packages, whose characteristics are reported in Table 5.42. Items can be rotated. Moreover, the bins that can be used have the following dimensions: $L = 200\,\text{cm}$, $W = 150\,\text{cm}$ and $H = 200\,\text{cm}$. Adapt the 3-BP-FFD heuristics to solve this problem.

6

Managing freight transport

6.1 Introduction

Freight transport activities are crucial in logistics systems planning. The reason is twofold: firstly, they determine the most important part (often between one-third and two-thirds) of the logistics costs; and, secondly, they significantly affect the service level provided to customers. Providing efficient and inexpensive freight transport services yields an increase in the distance at which the facilities of the logistics system can be economically set up. In turn, this would allow the operation of production plants in locations where economies of scale can be exploited thanks to automated manufacturing processes, a low-cost skilled workforce and low energy prices. Other consequences are the possibility to feasibly supply geographically far markets with perishable products and an increased competition among companies at a global scale, with obvious advantages for the consumers.

Freight transport involves different players: manufacturing companies (*shippers*), at which the demand for transport originates and which, in some cases, carry out the transport services on their own, by using a private fleet of vehicles; the *carriers*, which provide transport services to customers, and *governments*, which construct and operate the transport infrastructures and set the transport policies at a regional, national or international level (an analysis of these policies is beyond the scope of this book).

6.1.1 Modes of transport

Freight transport is carried out by the following modes of transport: train, road vehicle (e.g. truck), aircraft, ship (for transport on oceans, seas, lakes, canals and rivers) and pipeline. Some modes of transport (e.g. air) do not allow a door-to-door connection between any origin and destination and, therefore, should be

Introduction to Logistics Systems Management, Second Edition. Gianpaolo Ghiani, Gilbert Laporte and Roberto Musmanno.

used jointly with other modes (*intermodal transport*). In intermodal transport, the commodities can be stored temporarily and then consolidated into different bins (railway wagons, containers and pallets).

Modes of transport differ with respect to two fundamental parameters: *cost* and *transit time* (the latter parameter being related to reliability). These parameters are used by companies when selecting transport services, either when they act as private carriers, or when they are purchased outside.

Cost Freight transport cost depends on whether transport is realized by the shipper, or entrusted to a carrier. In the former case, the cost is given by the sum of costs associated with the depreciation, maintenance and insurance of the vehicles (owned or hired), crews' wages, fuel consumption, loading, unloading and transshipment operations, administration and use of vehicle depots. Moreover, there can be additional costs which depend on the specific mode of transport: for example, maritime or fluvial shipping leads to port mooring charges, port costs (agencies and piloting), canal transit cost and so on. Some of the costs depend only on transport time (e.g. the cost of insurance of owned vehicles), and others depend only on the distance covered (e.g. fuel consumption); still others are dependent both on time and on distance (e.g. vehicle depreciation costs), whereas other costs (such as administration) are customarily allocated as a fixed annual charge.

When a shipper uses a carrier, the freight transport cost can be calculated on the basis of the rates published by the carrier. For customized transport, the cost of a full load depends on both the origin and destination of the movement, as well as on the size and equipment of the vehicle required. For consolidation-based transport, each shipment is given a rating (called a *class*), which depends on the physical characteristics (weight, density etc.) of the goods. For example, in North America the railway classification includes 31 classes, while the US National Motor Freight Classification (NMFC) comprises 23 classes. Rates are usually reported in tables, or can be calculated through *rating engines* available

Table 6.1 LTL rates ($ per 100 pounds) from New York to Los Angeles published by the NCC; classes 55 and 70 correspond to products having densities higher than 15 and 35 pounds per cubic foot, respectively.

Weight (W)	Class 55	Class 70
$0 \leq W < 500$	129.57	153.82
$500 \leq W < 1\,000$	104.90	124.60
$1\,000 \leq W < 2\,000$	89.43	106.10
$2\,000 \leq W < 5\,000$	75.17	89.24
$5\,000 \leq W < 10\,000$	64.82	76.95
$10\,000 \leq W < 20\,000$	53.13	63.05
$20\,000 \leq W < 30\,000$	46.65	55.37
$30\,000 \leq W < 40\,000$	40.15	47.67
$W \geq 40\,000$	37.58	44.64

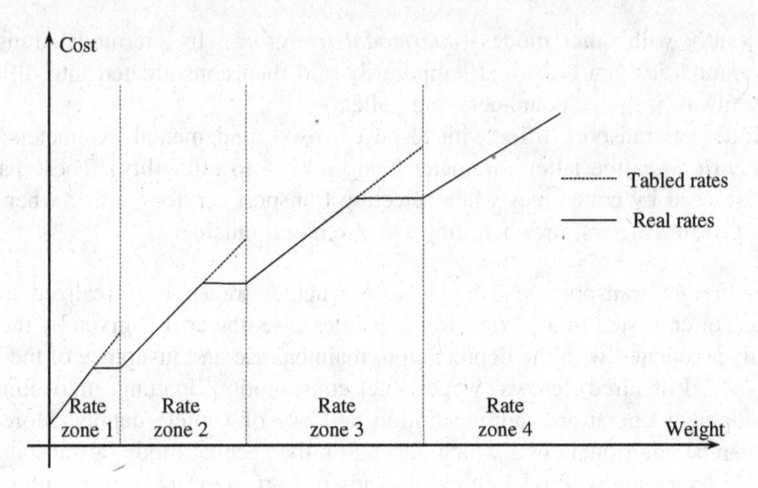

Figure 6.1 Transport rates for LTL trucking.

on the Internet. In Table 6.1, the LTL (see below in this section) rates for freight transport by road applied in the United States by the National Classification Committee (NCC) are reported; they are related to two classes. Rates are such that costs can be discontinuous, as illustrated in Figure 6.1 (cost may decrease by adding extra weight).

> Media Action, a US company, wants to transfer 9 400 pounds of class 55 freight from New York to Los Angeles. Using a carrier who applies the NCC rates shown in Table 6.1, it should pay $94 \times 64.82 = \$6\,093.08$. If the shipping load were 10 000 pounds, the cost would be $100 \times 53.13 = \$5\,313$. For this reason, the company declares to the carrier (which accepts) that it wishes to ship freight of 10 000 pounds, although the effective weight is less.

Air is the most expensive mode of transport, followed by road, rail, pipeline and water. According to recent surveys, transport by truck is approximately seven times more expensive than by train, which is usually four times more costly than by ship and by pipeline.

Transit time Transit time is the time a shipment takes to move from its origin to its destination. It also includes the time spent for any loading, unloading and transshipment operations. It is a random variable influenced by weather and traffic conditions, freight loading and unloading procedures and so on. The coefficient of variation (standard deviation over average transit time, see Section 1.5.1) of the transit time is a common measure of the reliability of a transport service.

On the basis of several statistical investigations carried out on different transport services, the most reliable mode of transport is, in general, the pipeline, followed by air, train, road vehicle and ship.

6.1.1.1 Rail transport

Rail transport is inexpensive (especially for long-distance movements), relatively slow and quite unreliable. This is due mainly to three reasons:

- convoys transporting freight have low priority compared to trains transporting passengers;
- direct train connections are quite rare;
- a convoy must include tens of cars in order to be worth operating.

As a result, the railroad is a slow mover of raw materials (coal, chemicals etc.) and of low-value finished products (steel, paper, sugar, tinned food etc.). With the aim of reducing the transfer cost within the rail terminals, it is preferred to ship multiple loads of the wagon capacity (*carload* transfers, or CL).

6.1.1.2 Road transport

Road transport of semi-finished and finished products is generally realized by using trucks. It can be *truckload* (TL) or *less-than-truckload* (LTL). A TL service moves a full load directly from its origin to its destination in a single trip (see Figure 6.2). If shipments add up to much less than the vehicle capacity (LTL loads), it is more convenient to resort to several trucking services in conjunction with cross-docking terminals rather than using direct shipments (see Figure 6.3). As a result, LTL trucking is slower than TL trucking. The main inconvenience of the road transport is the limited capacity of the trucks.

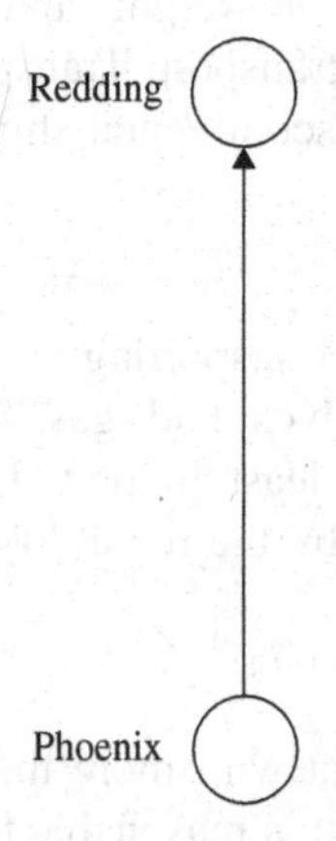

Figure 6.2 Example of TL transport.

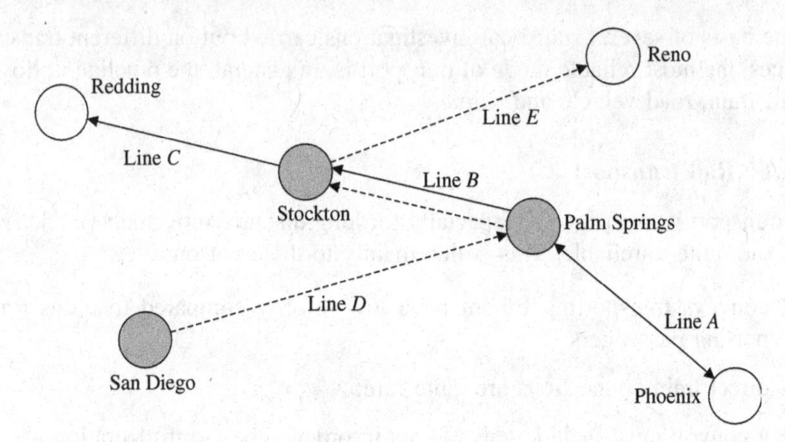

Figure 6.3 Example of LTL transport.

6.1.1.3 Air transport

Air transport is often used along with road transport in order to provide door-to-door services. While air transport is in principle very fast (the cruise speed of commercial flights is from 0.75 to 0.86 Mach), it is slowed down in practice by freight handling at airports. Consequently, air transport is not competitive for short- and medium-haul shipments. In contrast, it is quite popular for the transport of high-value products over long distances (about 20%, in value, of the world trade uses air as the mode of transport). Capacity (in terms of both weight and volume) of the aircrafts is relatively limited, compared to that of trains and ships.

6.1.1.4 Water transport

Transport by ship is used mainly in international trade to send bulk materials (cereals, petroleum, coal); 99% (in weight) and 50% (in value) of international trade occurs using this mode of transport. Transport by water is significantly less costly than air transport for transcontinental shipping.

6.1.1.5 Pipeline transport

Pipelines can be used only for transporting some specific categories of goods, such as petroleum, its derivatives and gas. The slowness of the transport (5–6 km/h) is compensated, at least in part, by the possibility of continuous provision (24 hours a day) and by the reliability of the pipelines and pumps.

6.1.1.6 Intermodal transport

The possibility of moving freight with more than one mode of transport allows hybrid services to be realized with a reasonable trade-off between cost and transit time. Although there are in principle several combinations of the five basic modes of transport, in practice only a few of them turn out to be convenient. The most frequent intermodal services are: aircraft-truck (*birdyback*) transport, train-truck (*piggyback*) transport and ship-truck (*fishyback*) transport. Containers are the most common load units in intermodal transport and can be moved in two ways:

- containers are loaded on a truck and the truck is then loaded onto a train, a ship or an aircraft (*trailer-on-flatcar*);
- containers are loaded directly on a train, a ship or an aircraft (*container-on-flatcar*). This solution proves particularly advantageous in the case of air transport, where the limited weight and volume capacity impedes in several cases the adoption of the trailer-on-flatcar solution.

6.1.2 Classification of transport problems

Transport problems can be classified on the basis of the distance between the origin and the destination of the shipment.

6.1.2.1 Long-haul transport problems

In *long-haul* freight transport, goods are moved over relatively long distances (which vary from some hundreds to some thousands of kilometres), between terminals or other facilities (plants, warehouses etc.). Long-haul transport services can provide for direct or indirect shipments. As mentioned in Section 6.1.1, with direct shipments, the freight is transferred from the origin to the destination without intermediate transshipments, whereas in the case of indirect shipments the movement is done by means of a sequence of trips which can be carried out with different vehicles if necessary. Direct long-haul transport services can be carried out by owned vehicles or by a shipper, whereas indirect services are generally assigned to shippers. Shippers' decision-making problems are generally more complex than those of companies executing their own transport. In fact, the latter, generally, have a need of direct transports from a limited number of origins to several destinations (*few-to-many* problems), whereas shippers, which serve many customers, usually have to transfer large quantities of freight from a high number of origins to many destinations (*many-to-many* problems).

Shippers can operate according to a schedule (*line services*) or on the basis of customers' requests. Under the former hypothesis, at a tactical level, the shipper should assign the freight traffic, on the basis of the transport demand (which is known), to the network of existing transport lines (*traffic assignment problem,* or TAP, see Section 6.2). Conversely, when the transport network has not already been designed or when the existing one proves to be inadequate, the shipper should solve, always at the tactical level, a *service network design problem* (SNDP, see Section 6.3), on the basis of the forecasted demand or on the basis of commercial agreements stipulated with the customers. Under the latter hypothesis, at the operational level, the shipper needs to assign his owned fleet dynamically (and the vehicles hired eventually by other shippers) to customer requests, so that these are satisfied and the cost of using the entire fleet is minimized. At the tactical level, another decision-making problem is the periodic repositioning of empty vehicles, so that the average response time to subsequent requests is kept as minimum as possible (*vehicle allocation problem*, or VAP, see Section 6.4). Other decision-making problems for shippers are, at the strategic level, the composition of the vehicle fleet to purchase, and, at the tactical level, the optimal crew scheduling (a problem in this category will be dealt with in Section 6.5). Finally, at the operational level, decision-making problems deal with the reassignment of vehicles and crews to take into account unexpected events, such as changes of the freight orders, vehicle breakdown, strikes or unfavourable weather conditions.

Regarding companies that act as private shippers, at the strategic level decision-making problems concern the purchase of their own fleet of vehicles, at the tactical level they concern the choice of vehicles to hire to integrate their own fleet (see Section 6.6) and at the operational level they address the optimal consolidation and dispatching of the shipping orders (see Section 6.7).

6.1.2.2 Short-haul transport problems

Short-haul transport includes movements having their origin and destination in a relatively small-sized geographic area (e.g. a city or a county). Problems of this type involve manufacturing companies, to supply their customers starting from DCs, using a fleet of their own vehicles (see Figure 6.4); local fast couriers, which transport loads between origin-destination pairs situated in the same area, and national or international carriers, which need to collect locally outbound parcels before sending them to a remote terminal as a consolidated load, and to locally distribute loads coming from remote terminals (see Figure 6.5).

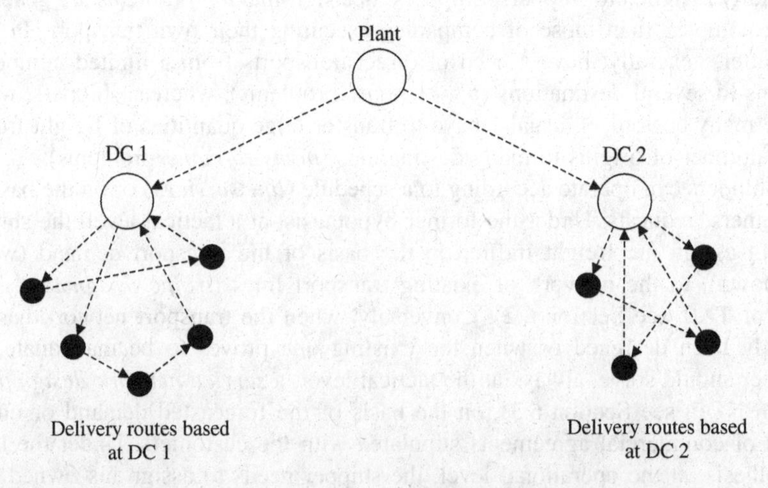

Figure 6.4 Freight delivery routes starting from two DCs.

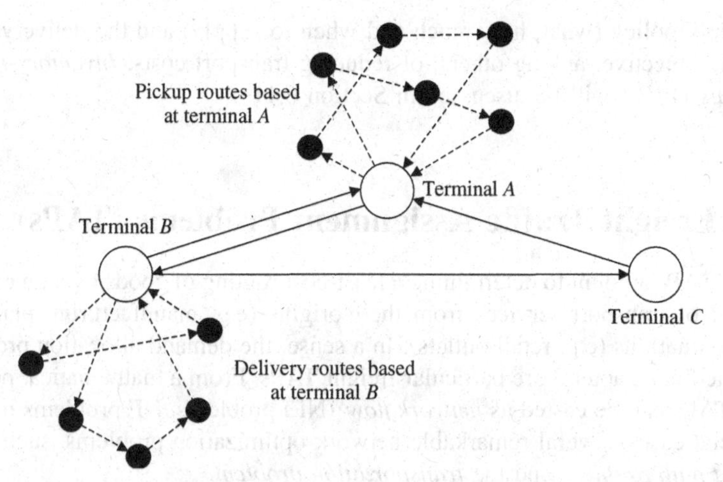

Figure 6.5 Pickup or delivery routes for parcels by a national or international carrier.

Short-haul transport problems also arise in garbage collection, mail delivery, appliance repair services and so on. The main decision-making problems concern, at the strategic level, the location of depots at which the routes originate (see Chapter 3 for this argument, although in some cases it is also necessary to take vehicle routing explicitly into account, in which case see Section 6.10); at the tactical level, the sizing of the vehicle fleet (see Section 6.6), and, at the operational level, the determination of vehicle routes to satisfy customer requests. In the simplest case of a single service request, the problem consists of determining a least-cost path from an origin to a destination; in the case of several simultaneous requests, a much more complicated *vehicle routing problem* (VRP) has to be solved, as shown in Section 6.8. Sometimes the vehicle routes can be planned in advance. This is the case, for example, of companies that plan product distribution on the basis of orders received from customers in previous days. Another case is solid urban waste collection, where the service routes are determined once or twice a year, assuming that the quantity of waste to collect daily remains almost constant for a certain number of months. Alternatively to the static case, there are situations where vehicle routing has to be determined dynamically, as soon as there is a new service request. This is the case, for example, of same-day urban courier companies, which have to pick up and deliver parcels, within a few hours, without previous notice. The availability of low-cost ICT tools (like GIS, GPS, cellular phones, traffic sensors etc.) allows data acquisition in real time. These data are then used to update vehicle routing dynamically. The main features of the *vehicle routing and dispatching problems* (VRDP) will be briefly illustrated in Section 6.9.

A different class of operational short-haul transport problem concerns companies that directly control the inventory level of products at the successor nodes of the logistics system (e.g. the retailers). In this case, the company could determine the supply policy (what, how much and when to supply) and the delivery route, with the objective, among others, of reducing transport costs. *Inventory-routing problems* (IRPs) will be discussed in Section 6.11.

6.2 Freight Traffic Assignment Problems (TAPs)

Freight TAPs amount to determining a least-cost routing of goods over an existing network of transport services from their origins (e.g. manufacturing plants) to their destinations (e.g. retail outlets). In a sense, the demand allocation problems illustrated in Chapter 3 are particular freight TAPs. From a mathematical point of view, TAPs can be casted as *network flow* (NF) problems. NF problems include, as special cases, several remarkable network optimization problems, such as the *shortest path problem* and the *transportation problem.*

TAPs can be classified as *static* or *dynamic.* Static models are suitable when the decisions to be made are not affected explicitly by time. They are formulated on a directed graph (or multigraph) $G = (V, A)$, where the vertex set V often corresponds to a set of facilities (terminals, plants and warehouses) and the arcs in the set A represent possible transport services linking the facilities. Some vertices represent *origins* of transport demand for one or several products, while others are *destinations*, or act as *transshipment points*. Let K be the set of traffic classes (or, simply, *commodities*). With each arc is associated a cost (possibly dependent on the amount of freight flow on the arc) and a capacity. Cost functions may represent both monetary costs and congestion effects arising at terminals.

In dynamic models, a time dimension is explicitly taken into account by modelling the transport services over a given planning horizon through a *time-expanded* directed graph. In a time-expanded directed graph, the planning horizon is divided into a number of time periods $t_1, t_2, \ldots$ and the physical network (containing terminals and other material resources) is replicated in each time period. Then, temporal links are added. A temporal link connecting two representations of the same terminal at two different time periods may represent freight waiting to be loaded onto an incoming vehicle, or the time required for freight classification at the terminal. On the other hand, a temporal link connecting two representations of different terminals may describe a transport service. Further vertices and arcs may be added to model the arrival of commodities at destinations and impose penalties in case of delays. With each link may be associated a capacity and a cost, similar to those used in static formulations. An example of a static transport service network is shown in Figure 6.6, while an associated time-expanded network is reported in Figure 6.7. In the static network there are three terminals (A, B, C) and four transport services operating from A to B, from B to A, from B to C and from C to A. The travel durations are equal to two, two, one and one days, respectively. If the planning horizon includes four days, a dynamic representation has four vertices for each terminal (A_i, $i = 1, \ldots, 4$, describes terminal A at the ith day). Some arcs (such as (A_1, B_3)) represent

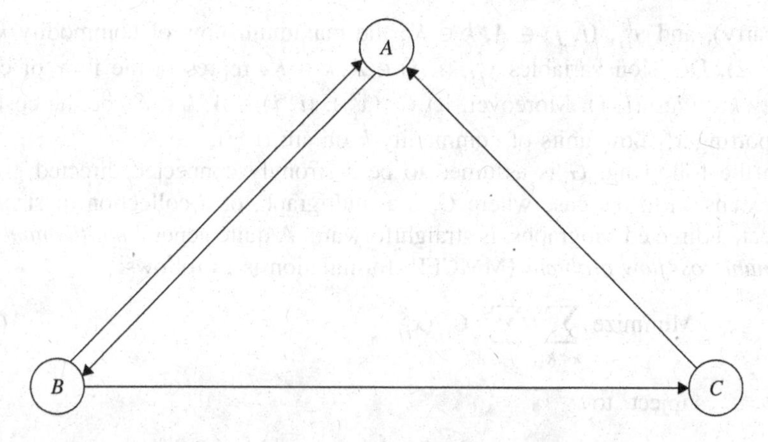

Figure 6.6 A static representation of a three-terminal transport system.

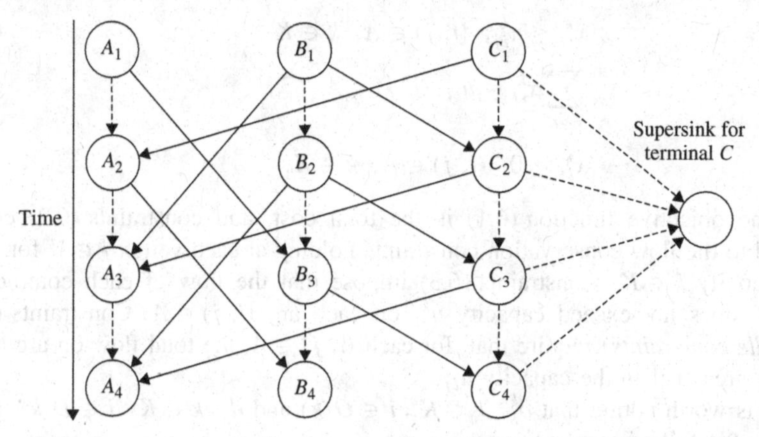

Figure 6.7 Dynamic network representation of the transport system illustrated in Figure 6.6.

transport services, while others (such as (B_2, B_3)) describe commodities standing idle at terminals. In addition, there may be supersinks (such as terminal C in Figure 6.7), for which the costs on the arcs entering the supersinks represent economic sanctions and penalties in case of transport service failure.

6.2.1 Minimum-cost flow formulation

Let $O(k)$, $k \in K$, be the set of origins of commodity k; $D(k)$, $k \in K$, the set of destinations of commodity k; $T(k)$, $k \in K$, the set of transshipment points with respect to commodity k; o_i^k, $i \in O(k)$, $k \in K$, the supply of commodity k of vertex i; d_i^k, $i \in D(k)$, $k \in K$, the demand of commodity k of vertex i; u_{ij}, $(i, j) \in A$, the capacity of arc (i, j) (i.e. the maximum flow that arc (i, j)

can carry), and u_{ij}^k, $(i, j) \in A$, $k \in K$, the maximum flow of commodity k on arc (i, j). Decision variables x_{ij}^k, $(i, j) \in A, k \in K$, represent the flow of commodity k on arc (i, j). Moreover, let $C_{ij}^k(x_{ij}^k)$, $(i, j) \in A$, $k \in K$, be the cost for transporting x_{ij}^k flow units of commodity k on arc (i, j).

In the following, G is assumed to be a strongly connected directed graph. The extension to the case where G is a multigraph, or a collection of strongly connected directed subgraphs, is straightforward. A quite general *multicommodity minimum-cost flow problem* (MMCFP) formulation is as follows:

$$\text{Minimize} \sum_{k \in K} \sum_{(i,j) \in A} C_{ij}^k(x_{ij}^k) \tag{6.1}$$

subject to

$$\sum_{\{j \in V:(i,j) \in A\}} x_{ij}^k - \sum_{\{j \in V:(j,i) \in A\}} x_{ji}^k = \begin{cases} o_i^k, & \text{if } i \in O(k) \\ -d_i^k, & \text{if } i \in D(k) \\ 0, & \text{if } i \in T(k) \end{cases} \quad i \in V,\ k \in K \tag{6.2}$$

$$x_{ij}^k \leq u_{ij}^k, \ (i, j) \in A, \ k \in K \tag{6.3}$$

$$\sum_{k \in K} x_{ij}^k \leq u_{ij}, (i, j) \in A \tag{6.4}$$

$$x_{ij}^k \geq 0, \ (i, j) \in A, \ k \in K.$$

The objective function (6.1) is the total cost, and constraints (6.2) correspond to the flow conservation constraints holding at each vertex $i \in V$ for each commodity $k \in K$. Constraints (6.3) impose that the flow of each commodity $k \in K$ does not exceed capacity u_{ij}^k on each arc $(i, j) \in A$. Constraints (6.4) (*bundle constraints*) require that, for each $(i, j) \in A$, the total flow on arc (i, j) is not greater than the capacity u_{ij}.

It is worth noting that o_i^k, $k \in K$, $i \in O(k)$ and d_i^k, $k \in K$, $i \in D(k)$, must satisfy the following conditions:

$$\sum_{i \in O(k)} o_i^k = \sum_{i \in D(k)} d_i^k, \ k \in K,$$

otherwise the problem is infeasible.

In the remainder of this section, some of the most relevant solution methods for some special cases of the MMCFP are illustrated.

6.2.2 Linear single-commodity minimum-cost flow problems

The *linear single-commodity minimum-cost flow problem* (LMCFP) can be formulated as follows:

$$\text{Minimize} \sum_{(i,j) \in A} c_{ij} x_{ij} \tag{6.5}$$

subject to

$$\sum_{\{j \in V:(i,j) \in A\}} x_{ij} - \sum_{\{j \in V:(j,i) \in A\}} x_{ji} = \begin{cases} o_i, & \text{if } i \in O \\ -d_i, & \text{if } i \in D, \quad i \in V \\ 0, & \text{if } i \in T \end{cases} \tag{6.6}$$

$$x_{ij} \leq u_{ij}, \; (i, j) \in A \tag{6.7}$$

$$x_{ij} \geq 0, \; (i, j) \in A. \tag{6.8}$$

The LMCFP is a structured LP problem and, as such, can be solved through the simplex algorithm or any other LP procedure. Instead of using a general-purpose algorithm, it is common to employ a tailored procedure, the (primal) *network simplex algorithm*, a specialized version of the classical simplex algorithm which takes advantage of the particular structure of the coefficient matrix associated with constraints (6.6) (corresponding to the vertex-arc incidence matrix of the directed graph G).

We first examine the case where there are no capacity constraints (6.7). In such a case, it is useful the following characterization of the basic solutions of the system of equations (6.6), which is stated without proof.

Property. The basic solutions of the system of equations (6.6) have $|V| - 1$ basic variables. Moreover, each basic solution corresponds to a tree spanning G and vice versa.

In order to find a basic solution of problem (6.5), (6.6), (6.8), it is therefore sufficient to select a tree spanning G, set to zero the variables associated to the arcs which are not part of the tree and then solve the system of linear equations (6.6). The latter step can be easily accomplished through a substitution method. Of course, the basic solution associated to a spanning tree is not always feasible, since the non-negativity constraints (6.8) may be violated (see Figure 6.8).

The network simplex algorithm has the same structure as the standard simplex algorithm. However, the optimality test and the pivot operations are performed in a simplified way. The sketch of the network simplex algorithm is reported below.

Step 1. Find an initial basic feasible solution $x^{(0)}$. Set $h = 0$.

Step 2. Determine the reduced costs $c'^{(h)}$ associated to $x^{(h)}$.

Step 3. If $c'^{(h)}_{ij} \geq 0$, $(i, j) \in A$, *STOP*, $x^{(h)}$ is an optimal solution; otherwise choose a variable x_{vw} such that $c'^{(h)}_{vw} < 0$.

Step 4. Select a variable x_{pq} coming out of the basis, make a pivot in order to substitute x_{pq} for x_{vw} in the basis, set $h = h + 1$ and go back to Step 2.

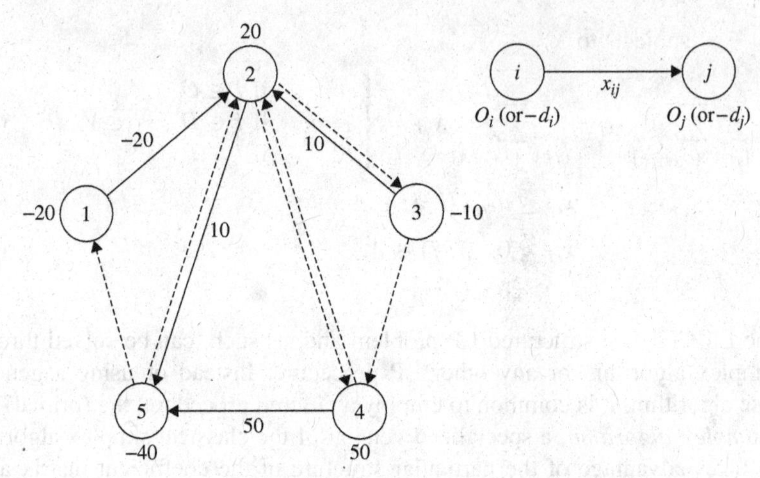

Figure 6.8 A spanning tree (full line arcs) of a directed graph and the associated (infeasible) basic solution (only the basic variables are reported).

The particular structure of problem (6.5), (6.6), (6.8) and of its dual,

$$\text{Maximize} \sum_{i \in O} o_i \pi_i - \sum_{i \in D} d_i \pi_i$$

subject to

$$\pi_i - \pi_j \leq c_{ij}, \ (i, j) \in A,$$

enables the execution of Steps 2 to 4 as follows. At Step 2, the reduced costs can be computed through the formula:

$$c'^{(h)}_{ij} = c_{ij} - \pi^{(h)}_i + \pi^{(h)}_j, \ (i, j) \in A, \tag{6.9}$$

where $\pi^{(h)} \in \mathfrak{R}^{|V|}$ can be determined by requiring that the reduced costs of the basic variables be zero:

$$c'^{(h)}_{ij} = c_{ij} - \pi^{(h)}_i + \pi^{(h)}_j = 0, \ (i, j) \in A: \ x^{(h)}_{ij} \text{ is a basic variable.}$$

At Step 3, if $c'^{(h)}_{ij} \geq 0$, $(i, j) \in A$, then $\pi^{(h)}_i - \pi^{(h)}_j \leq c_{ij}$, $(i, j) \in A$, that is, the solution $\pi^{(h)} \in \mathfrak{R}^{|V|}$ is feasible for the dual problem. Then, $x^{(h)}$ and $\pi^{(h)}$ are optimal for the primal and the dual problems, respectively.

On the other hand, if there is a variable x_{vw} whose reduced cost is negative at iteration h, then arc (v, w) does not belong to the spanning tree associated to iteration h. It follows that, by adding (v, w) to the tree, a single cycle Ψ is created. In order to decrease the objective function value as much as possible, the flow on arc (v, w) has to be increased as much as possible while satisfying constraints (6.6) and (6.8).

Let Ψ^+ be the set of arcs in Ψ oriented as (v, w), and let Ψ^- be the set of the arcs in Ψ oriented in the opposite direction (obviously, $\Psi = \Psi^+ \cup \Psi^-$). If the flow on arc (v, w) is increased by t units, then constraints (6.6) require that the flow on all arcs $(i, j) \in \Psi^+$ be increased by t units, and the flow on all arcs $(i, j) \in \Psi^-$ be decreased by the same amount.

The maximum increase of flow on (v, w) is therefore equal to the minimum flow on the arcs oriented in the opposite direction as (v, w), that is,

$$t = \min_{(i,j)\in\Psi^-} \{x_{ij}^{(h)}\}.$$

The arc $(p, q) \in \Psi^-$ for which such a condition holds determines which variable x_{pq} will come out from the basis.

The previous description shows that an iteration of the network simplex algorithm requires only a few additions and subtractions. As a result, this procedure is much faster than the standard simplex method and, in addition, does not make rounding errors.

In order to find a feasible solution (if any exists), the *big M method* can be used. A new vertex $i_0 \in T$ and $|V|$ dummy arcs between vertex i_0 and all the other vertices $i \in V$ are introduced. If $i \in O$, then a dummy arc (i, i_0) is inserted, otherwise an arc (i_0, i) is added. Let $A^{(a)}$ be the set of dummy arcs. With each dummy arc is associated an arbitrarily large cost M.

The dummy problem is as follows:

$$\text{Maximize} \sum_{(i,j)\in A} c_{ij}x_{ij} + M \sum_{(i,i_0)\in A^{(a)}} x_{ii_0} + M \sum_{(i_0,i)\in A^{(a)}} x_{i_0 i} \tag{6.10}$$

subject to

$$\sum_{\{j\in V:(i,j)\in A\cup A^{(a)}\}} x_{ij} - \sum_{\{j\in V:(j,i)\in A\cup A^{(a)}\}} x_{ji} = \begin{cases} o_i, & \text{if } i \in O \\ -d_i, & \text{if } i \in D, i \in V \cup \{i_0\} \\ 0, & \text{if } i \in T \end{cases} \tag{6.11}$$

$$x_{ij} \geq 0, \ (i, j) \in A \cup A^{(a)}. \tag{6.12}$$

Of course, the $|V|$ dummy arcs make up a spanning tree of the modified directed graph, corresponding to the following basic feasible solution of problem (6.10)–(6.12) (see Figure 6.9):

$$x_{ii_0}^{(0)} = o_i, \ i \in O;$$

$$x_{i_0 i}^{(0)} = d_i, \ i \in D;$$

$$x_{i_0 i}^{(0)} = 0, \ i \in T;$$

$$x_{ij}^{(0)} = 0, \ (i, j) \in A.$$

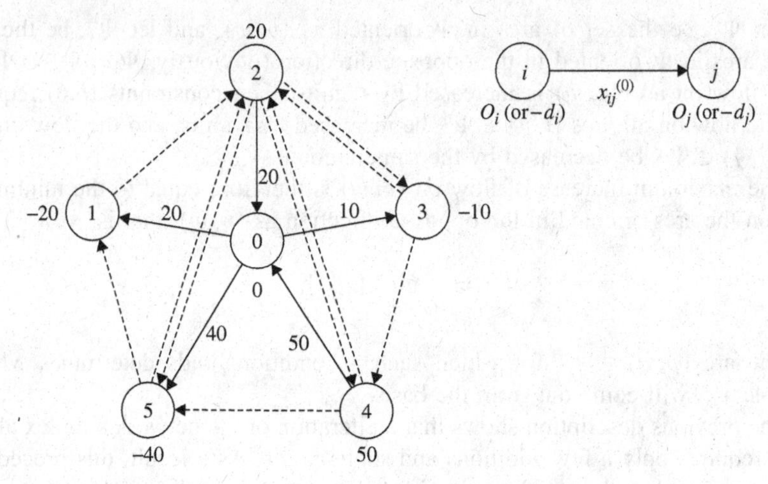

Figure 6.9 Dummy directed graph of the original directed graph in Figure 6.8 (0 is the dummy vertex. Full line arcs belong to the spanning tree. Only the flow associated to a feasible basic variable is reported).

By solving the dummy problem (6.10)–(6.12), a basic feasible solution to the original problem (6.5), (6.6), (6.8) is then obtained.

NTN is a Swiss freight intermodal carrier located in Lausanne. When a customer needs to transport goods between an origin and a destination, NTN supplies it with one or more empty containers in which the goods can be loaded. Once arrived at destination, the goods are unloaded and the empty containers have to be transported to the pickup points of new customers. As a result, NTN management needs to reallocate the empty containers periodically (in practice, on a weekly basis). Empty container transport is very expensive (its cost is nearly 35% of the total operating cost). Last May 13, several empty ISO 20 containers had to be reallocated among the terminals in Amsterdam, Berlin, Munich, Paris, Milan, Barcelona and Madrid. The number of empty containers available or demanded at the various terminals is reported, along with transport costs (in €/container), in Figure 6.10.

The problem can be formulated as follows:

$$\begin{aligned}\text{Minimize } & 30x_{12} + 30x_{21} + 40x_{13} + 40x_{31} + 20x_{14} + 20x_{41} + 30x_{23} + 30x_{32} \\ & +55x_{34} + 55x_{43} + 30x_{35} + 30x_{53} + 30x_{45} + 30x_{54} + 50x_{46} + 50x_{64} \\ & +70x_{47} + 70x_{74} + 30x_{56} + 30x_{65} + 25x_{67} + 25x_{76}\end{aligned}$$

subject to

$$x_{12} + x_{13} + x_{14} - x_{21} - x_{31} - x_{41} = -10$$

$$x_{21} + x_{23} - x_{12} - x_{32} = 20$$

$$x_{31} + x_{32} + x_{34} + x_{35} - x_{13} - x_{23} - x_{43} - x_{53} = 50$$

$$x_{41} + x_{43} + x_{45} + x_{46} + x_{47} - x_{14} - x_{34} - x_{54} - x_{64} - x_{74} = 20$$

$$x_{53} + x_{54} + x_{56} - x_{35} - x_{45} - x_{65} = -50$$

$$x_{64} + x_{65} + x_{67} - x_{46} - x_{56} - x_{76} = -20$$

$$x_{74} + x_{76} - x_{47} - x_{67} = -10$$

$$x_{12},\ x_{21},\ x_{13},\ x_{31},\ x_{14},\ x_{41},\ x_{23},\ x_{32},\ x_{34},\ x_{43},\ x_{35},$$
$$x_{53},\ x_{45},\ x_{54},\ x_{46},\ x_{64},\ x_{47},\ x_{74},\ x_{56},\ x_{65},\ x_{67},\ x_{76} \geq 0.$$

Using the network simplex method, the optimal solution illustrated in Figure 6.11 is obtained, whose cost is equal to €3 900.

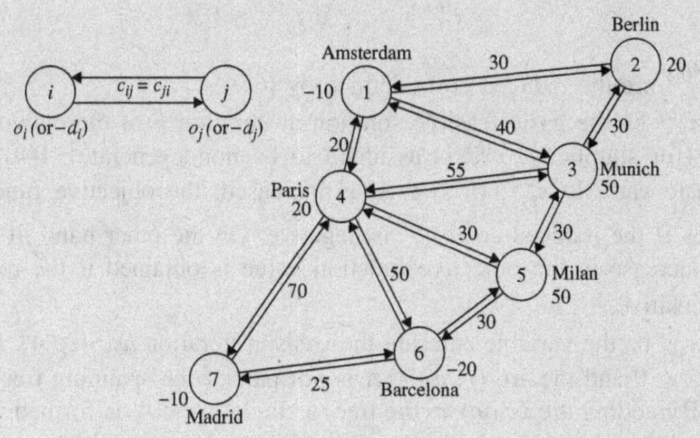

Figure 6.10 Graph representation of the NTN empty container allocation problem.

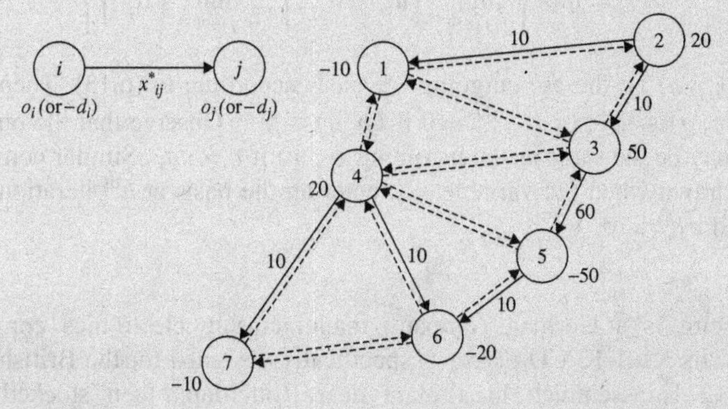

Figure 6.11 Optimal solution of the NTN empty container allocation problem (full line arcs belong to the spanning tree; for each basic variable, the associated flow is reported).

The algorithm, illustrated above, can be easily adapted to the case of capacitated arcs. To this purpose, constraints (6.7) are rewritten by introducing auxiliary variables $\gamma_{ij} \geq 0$:

$$x_{ij} + \gamma_{ij} = u_{ij}, \quad (i, j) \in A.$$

If variable x_{ij} is equal to u_{ij}, then the associated auxiliary variable γ_{ij} takes the value zero and is therefore out of the basis (if the solution is not degenerate). Based on this observation, the following optimality conditions can be derived (the proof is omitted for the sake of brevity).

Theorem. A basic feasible solution $x^{(h)}$ is optimal for LMCFP if, for each nonbasic variable $x_{ij}^{(h)}$, $(i, j) \in A$, the following conditions hold:

$$x_{ij}^{(h)} = 0, \text{ if } c_{ij}^{'(h)} \geq 0,$$

$$x_{ij}^{(h)} = u_{ij}, \text{ if } c_{ij}^{'(h)} \leq 0,$$

where $c_{ij}^{'(h)}$ are the reduced costs defined by (6.9).

Let $x^{(h)}$ be the basic feasible solution at iteration h of the network simplex method (for simplicity, $x^{(h)}$ is assumed to be nondegenerate). If the value of a nonbasic variable $x_{ij}^{(h)}$, $(i, j) \in A$, is increased, the objective function value decreases if the reduced cost $c_{ij}^{'(h)}$ is negative. On the other hand, if $x_{ij}^{(h)} = u_{ij}$, then a decrease in the objective function value is obtained if the reduced cost $c_{ij}^{'(h)}$ is positive.

Let x_{vw} be the variable entering the basis at iteration h (Step 4). If $x_{vw}^{(h)} = 0$, then $c_{vw}^{'(h)} < 0$ and the arc $(v, w) \in A$ is not part of the spanning tree associated to $x^{(h)}$. By adding arc (v, w) to the tree, a single cycle Ψ is formed. In the new basic feasible solution, variable x_{vw} will take a value t equal to:

$$t = \min\left\{ \min_{(i,j)\in\Psi^+} \left\{u_{ij} - x_{ij}^{(h)}\right\}, \min_{(i,j)\in\Psi^-} \left\{x_{ij}^{(h)}\right\}\right\}. \tag{6.13}$$

Let (p, q) be the arc outgoing the basis according to (6.13). Then, $x_{pq}^{(h+1)} = u_{pq}$ if $(p, q) \in \Psi^+$, or $x_{pq}^{(h+1)} = 0$ if $(p, q) \in \Psi^-$. Observe that the outgoing arc (p, q) may be the same as the incoming (v, w) if $t = u_{vw}$. Similar considerations can be drawn when the variable x_{vw} entering the basis at h^{th} iteration as $x_{vw}^{(h)} = u_{vw}$ with $c_{vw}^{'(h)} > 0$.

Boscheim is a German company manufacturing electronics convenience goods. Its KLR-12 CD player is specifically designed for the British market. KLR-12 is assembled in a plant near Rotterdam, then stocked in two warehouses located in Bristol and Middlesbrough and finally transported to the retailer outlets. The British market is divided into four sales districts whose centres of gravity are in London, Birmingham, Leeds and Edinburgh.

Yearly demands amount to 90 000, 80 000, 50 000 and 70 000 items, respectively. The transport costs per item from the assembly plant of Rotterdam to the warehouses of Bristol and Middlesbrough are € 24.5 and € 26.0, respectively, whereas the transport costs per item from the warehouses to the sales districts are reported in Table 6.2. Both warehouses have an estimated capacity of 15 000 items and are supplied 10 times a year. Consequently their maximum yearly throughput is 150 000 items.

Table 6.2 Transport costs (in €) per item from the warehouses to the sales districts in the Boscheim problem.

Warehouse	Sales districts			
	London	Birmingham	Leeds	Edinburgh
Bristol	9.6	7.0	15.2	28.5
Middlesbrough	19.5	13.3	5.0	11.3

The annual minimum-cost distribution plan can be obtained by solving the following LMCFP (see Figure 6.12):

$$\text{Minimize } 24.5x_{12} + 26.0x_{13} + 9.6x_{24} + 7.0x_{25} + 15.2x_{26} + 28.5x_{27} + 19.5x_{34} + 13.3x_{35} + 5.0x_{36} + 11.3x_{37}$$

subject to

$$x_{12} + x_{13} = 290\,000$$

$$x_{24} + x_{25} + x_{26} + x_{27} - x_{12} = 0$$

$$x_{34} + x_{35} + x_{36} + x_{37} - x_{13} = 0$$

$$-x_{24} - x_{34} = -90\,000$$

$$-x_{25} - x_{35} = -80\,000$$

$$-x_{26} - x_{36} = -50\,000$$

$$-x_{27} - x_{37} = -70\,000$$

$$x_{12} \leq 150\,000$$

$$x_{13} \leq 150\,000$$

$$x_{12}, x_{13}, x_{24}, x_{25}, x_{26}, x_{27}, x_{34}, x_{35}, x_{36}, x_{37} \geq 0.$$

By using the network simplex method, the optimal solution is determined: $x_{12}^* = 150\,000$, $x_{13}^* = 140\,000$, $x_{24}^* = 90\,000$, $x_{25}^* = 60\,000$, $x_{35}^* = 20\,000$,

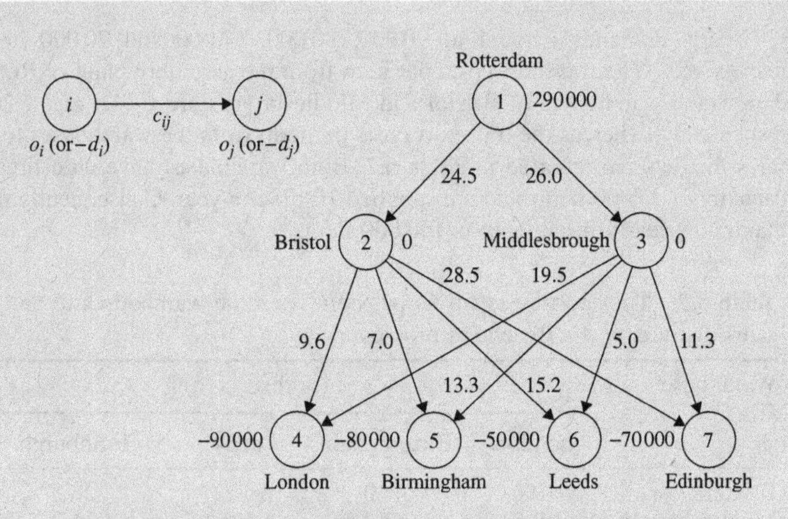

Figure 6.12 Graph representation of the Boscheim problem.

$x^*_{36} = 50\,000$ and $x^*_{37} = 70\,000$ (as usual, only nonzero variables are reported). It is worth noting that the district of London will be entirely served by the warehouse of Bristol, while the sales districts of Leeds and Edinburgh will be served by the Middlesbrough warehouse. The sales district of Birmingham is instead supplied by the warehouse of Bristol (75%), and by the warehouse of Middlesbrough (25%). The total transport cost is €9 906 000 per year.

6.2.3 Linear multicommodity minimum-cost flow problems

The *linear multicommodity minimum-cost flow problem* (LMMCFP) can be formulated as the following LP model:

$$\text{Minimize} \sum_{k\in K} \sum_{(i,j)\in A} c^k_{ij} x^k_{ij}$$

subject to

$$\sum_{\{j\in V:(i,j)\in A\}} x^k_{ij} - \sum_{\{j\in V:(j,i)\in A\}} x^k_{ji} = \begin{cases} o^k_i, & \text{if } i \in O(k) \\ -d^k_i, & \text{if } i \in D(k), \\ 0, & \text{if } i \in T(k) \end{cases} \quad i \in V, k \in K$$

$$x^k_{ij} \leq u^k_{ij},\ (i,j) \in A,\ k \in K$$

$$\sum_{k\in K} x^k_{ij} \leq u_{ij},\ (i,j) \in A \tag{6.14}$$

$$x^k_{ij} \geq 0,\ (i,j) \in A,\ k \in K.$$

The LMMCFP can be solved efficiently through a tailored Lagrangian procedure. Let λ_{ij}, $(i, j) \in A$ (≥ 0) be the Lagrangian multipliers attached to constraints (6.14). The Lagrangian relaxation of the LMMCFP is:

$$\text{Minimize} \sum_{k\in K} \sum_{(i,j)\in A} c_{ij}^k x_{ij}^k + \sum_{(i,j)\in A} \lambda_{ij} \left(\sum_{k\in K} x_{ij}^k - u_{ij} \right) \tag{6.15}$$

subject to

$$\sum_{\{j\in V:(i,j)\in A\}} x_{ij}^k - \sum_{\{j\in V:(j,i)\in A\}} x_{ji}^k = \begin{cases} o_i^k, & \text{if } i \in O(k) \\ -d_i^k, & \text{if } i \in D(k), \\ 0, & \text{if } i \in T(k) \end{cases} \quad i \in V, \; k \in K \tag{6.16}$$

$$x_{ij}^k \leq u_{ij}^k, \; (i, j) \in A, \; k \in K \tag{6.17}$$

$$x_{ij}^k \geq 0, \; (i, j) \in A, \; k \in K. \tag{6.18}$$

Relaxation (6.15)–(6.18), referred to in the sequel as R-LMMCFP, is made up of $|K|$ independent single-commodity minimum-cost flow problems, since $\sum_{(i,j)\in A} \lambda_{ij} u_{ij}$ in the objective function (6.15) is constant for a given set of Lagrangian multipliers λ_{ij}, $(i, j) \in A$. Therefore, the kth LMCF subproblem, $k \in K$,

$$\text{Minimize} \sum_{(i,j)\in A} \left(c_{ij}^k + \lambda_{ij} \right) x_{ij}^k \tag{6.19}$$

subject to

$$\sum_{\{j\in V:(i,j)\in A\}} x_{ij}^k - \sum_{\{j\in V:(j,i)\in A\}} x_{ji}^k = \begin{cases} o_i^k, & \text{if } i \in O(k) \\ -d_i^k, & \text{if } i \in D(k), \\ 0, & \text{if } i \in T(k) \end{cases} \quad i \in V \tag{6.20}$$

$$x_{ij}^k \leq u_{ij}^k, \; (i, j) \in A \tag{6.21}$$

$$x_{ij}^k \geq 0, \; (i, j) \in A, \tag{6.22}$$

can be solved through the network simplex algorithm.

Let $\text{LB}_{\text{LMCFP}}^k(\lambda)$, $k \in K$, be the optimal objective function value of the kth subproblem (6.19)–(6.22) and let $\text{LB}_{\text{R-LMMCFP}}(\lambda)$ be the lower bound provided by solving the R-LMMCFP.

For a given set of Lagrangian multipliers, $\text{LB}_{\text{R-LMMCFP}}(\lambda)$ is given by

$$\text{LB}_{\text{R-LMMCFP}}(\lambda) = \sum_{k\in K} \text{LB}_{\text{LMCFP}}^k(\lambda) - \sum_{(i,j)\in A} \lambda_{ij} u_{ij}.$$

Of course, $\text{LB}_{\text{R-LMMCFP}}(\lambda)$ varies as Lagrangian multipliers λ changes. The Lagrangian relaxation attaining the maximum lower bound value $\text{LB}_{\text{R-LMMCFP}}(\lambda)$ as λ varies is called *dual Lagrangian relaxation*. The following property follows from the LP theory.

Property. The lower bound provided by the dual Lagrangian relaxation is equal to the optimal objective function value of the LMMCFP model, that is,

$$\max_{\lambda \geq 0} \{LB_{R\text{-}LMMCFP}(\lambda)\} = z^*_{LMMCFP}.$$

Moreover, the dual Lagrangian multipliers λ^*_{ij}, $(i, j) \in A$, are equal to the optimal dual variables π^*_{ij}, $(i, j) \in A$, associated with the relaxed constraints (6.14).

In order to compute the dual Lagrangian multipliers, or at least a set of Lagrangian multipliers associated with a good lower bound, the classical subgradient procedure, already illustrated in Chapter 3, can be used.

Step 0. *Initialization.* Let H be a pre-established maximum number of subgradient iterations. Set $LB = -\infty$, $h = 1$ and $\lambda^{(h)}_{ij} = 0$, $(i, j) \in A$. Set UB equal to the cost of the best feasible solution if any is available, or set $UB = \infty$ otherwise.

Step 1. *Calculation of a new lower bound.* Solve the R-LMMCFP using $\lambda^{(h)}$ as a vector of Lagrangian multipliers. If $LB_{R\text{-}LMMCFP}(\lambda^{(h)}) > LB$, set $LB = LB_{R\text{-}LMMCFP}(\lambda^{(h)})$.

Step 2. *Checking the stopping criterion.* If solution $x^{k,(h)}_{ij}$, $(i, j) \in A$, $k \in K$, of the R-LMMCFP satisfies the relaxed constraints (6.14), and $\lambda^{(h)}_{ij}\left(\sum_{k \in K} x^{k,(h)}_{ij} - u_{ij}\right) = 0$, $(i, j) \in A$, *STOP*, the solution found is optimal ($z^*_{LMMCFP} = LB$). If $x^{k,(h)}_{ij}$, $(i, j) \in A$, $k \in K$, satisfies the relaxed constraints (6.14), update UB if necessary. If $UB = LB$, *STOP*, the feasible solution attaining UB is proved to be optimal. If $h = H$, *STOP*, LB represents the best lower bound available for z^*_{LMMCFP}.

Step 3. *Updating the Lagrangian multipliers.* Determine, for each $(i, j) \in A$, the subgradient of the relaxed constraint

$$s^{(h)}_{ij} = \sum_{k \in K} x^{k,(h)}_{ij} - u_{ij}, \quad (i, j) \in A.$$

Then set

$$\lambda^{(h+1)}_{ij} = \max\left(0, \lambda^{(h)}_{ij} + \beta^{(h)} s^{(h)}_{ij}\right), \quad (i, j) \in A,$$

where $\beta^{(h)}$ is a suitable coefficient:

$$\beta^{(h)} = \frac{\alpha}{h}, \quad (i, j) \in A, \tag{6.23}$$

with α arbitrarily chosen in the interval [0, 2]. Alternatively, if a feasible solution of the problem is available, set

$$\beta^{(h)} = \frac{\alpha(\text{UB} - \text{LB}_{\text{LMCFP}}(\lambda^{(h)}))}{\sum_{(i,j)\in A} \left(s_{ij}^{(h)}\right)^2}, \quad (i, j) \in A,$$

with UB equal to the objective function value of the best feasible solution available. Set $h = h + 1$ and go back to Step 1.

The subgradient algorithm converges to z^*_{LMMCFP} provided that the variations of the Lagrangian multipliers are 'small enough' (i.e. this assumption is satisfied if Equation (6.23) is used). However, this assumption makes the algorithm very slow.

If the solution $x_{ij}^{k,(h)}$, $(i, j) \in A$, $k \in K$, of the R-LMMCFP satisfies the relaxed constraints (6.14), it is not necessarily the optimal solution for the LMMCFP. A *sufficient* (but *not necessary*) condition for it to be optimal is given by the *complementary slackness* conditions

$$\lambda_{ij}^{(h)}\left(\sum_{k\in K} x_{ij}^{k,(h)} - u_{ij}\right) = 0, \quad (i, j) \in A. \tag{6.24}$$

If relations (6.24) are not satisfied, then the feasible solution $x_{ij}^{k,(h)}$, $(i, j) \in A$, $k \in K$, is simply a *candidate optimal* solution.

If $h = H$, the solution attaining LB could be infeasible for the LMMCFP, or, if feasible, may not satisfy the complementarity slackness conditions. In any case, if subproblems (6.19)–(6.22) are solved by means of the network simplex method, a basic (feasible or infeasible) solution for the LMMCFP is available. In fact, the basic variables of the $|K|$ subproblems (6.19)–(6.22) make up a basis of the LMMCFP. The basic solution obtained in this way can be used as the starting solution for the primal or dual simplex method depending on whether the solution is feasible for the LMMCFP or not. If initialized in this way, the simplex method is particularly efficient since the initial basic solution provided by the subgradient algorithm is a good approximation to the optimal solution.

Exofruit imports to the EU countries several varieties of tropical fruits, mainly coming from Northern Africa, Mozambique and Central America. The company purchases the products directly from farmers and transports them by sea to its warehouses in Marseille, France. The goods are then stored in refrigerated cells or at room temperature. Because purchase and selling prices vary during the year, Exofruit has to decide when and how much to buy in order to satisfy demand over the year. The problem can be modelled as an LMMCFP. In what follows, a simplified version of the problem is examined. It is assumed that a single source exists, products are grouped into two homogeneous groups (*macro-products*) and the planning horizon is divided into two semesters. Let d_{kt}, o_{kt}, and p_{kt}, $k = 1, 2$, $t = 1, 2$, be the demand, the maximum amount available(in tons) and the purchase prices (in €/tons) of macro-product k in semester t, respectively (see Table 6.3). The transport cost v of one ton of a macro-product is equal to € 100, while the stocking cost w of one ton of a macro-product is € 100 per semester. Finally, the maximum quantity q of goods that can be stored in a semester is 8 000 tons.

Table 6.3 Demand (in tons), maximum amounts available (in tons) and purchase prices (in €/ton), for each macro-product k, $k = 1, 2$, and for each semester t, $t = 1, 2$, in the Exofruit problem.

	d_{kt}		o_{kt}		p_{kt}	
	$t = 1$	$t = 2$	$t = 1$	$t = 2$	$t = 1$	$t = 2$
$k = 1$	18 000	18 000	26 000	20 000	500	700
$k = 2$	12 000	14 000	14 000	13 000	600	400

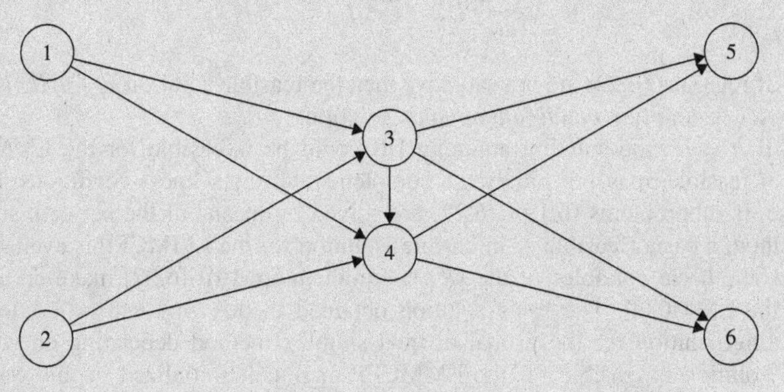

Figure 6.13 Graph representation of the Exofruit problem.

The problem can be formulated as an LMMCFP with two commodities (one for each macro-product) on the directed graph shown in Figure 6.13. In such a representation,

- vertices 1 and 5 represent the source and the destination of macro-product 1, respectively;
- vertices 2 and 6 represent the source and the destination of macro-product 2, respectively;
- vertices 3 and 4 represent the warehouse in the first and in the second semester, respectively;

- arc $(1, 3)$ has a cost per unit of flow equal to $c^1_{13} = p_{11} + v =$ €600/ton and a capacity equal to $u_{13} = o_{11} = 26\,000$ tons;
- arc $(1, 4)$ has a cost per unit of flow equal to $c^1_{14} = p_{12} + v =$ €800/ton and a capacity equal to $u_{14} = o_{12} = 20\,000$ tons;
- arc $(2, 3)$ has a cost per unit of flow equal to $c^2_{23} = p_{21} + v =$ €700/ton and a capacity equal to $u_{23} = o_{21} = 14\,000$ tons;
- arc $(2, 4)$ has a cost per unit of flow equal to $c^2_{24} = p_{22} + v =$ €500/ton and a capacity equal to $u_{24} = o_{22} = 13\,000$ tons;
- arc $(3, 4)$ represents the storage of goods for a semester and is therefore attached a cost per unit of flow equal to $c^1_{34} = c^2_{34} = w =$ €100/ton and a capacity of $u_{34} = q = 8\,000$ tons;
- arc $(3, 5)$ has a zero cost per unit of flow and a capacity equal to $u_{35} = d_{11} = 18\,000$ tons;
- arc $(4, 5)$ has a zero cost per unit of flow and a capacity equal to $u_{45} = d_{12} = 18\,000$ tons;
- arc $(3, 6)$ has a zero cost per unit of flow and a capacity equal to $u_{36} = d_{21} = 12\,000$ tons;
- arc $(4, 6)$ has a zero cost per unit of flow and a capacity equal to $u_{46} = d_{22} = 14\,000$ tons.

The problem is formulated as follows:

$$\text{Minimize } 600x^1_{13} + 800x^1_{14} + 100x^1_{34} + 700x^2_{23} + 500x^2_{24} + 100x^2_{34}$$

subject to

$$x^1_{35} + x^1_{45} = 36\,000$$
$$x^1_{13} - x^1_{34} - x^1_{35} = 0$$
$$x^1_{14} + x^1_{34} - x^1_{45} = 0$$
$$x^2_{36} + x^2_{46} = 26\,000$$
$$x^2_{23} - x^2_{36} - x^2_{34} = 0$$
$$x^2_{24} + x^2_{34} - x^2_{46} = 0$$
$$x^1_{13} \leq 26\,000$$
$$x^1_{14} \leq 20\,000$$
$$x^1_{35} \leq 18\,000$$
$$x^1_{45} \leq 18\,000$$

$$x^2_{23} \leq 14\,000$$
$$x^2_{24} \leq 13\,000$$
$$x^2_{36} \leq 12\,000$$
$$x^2_{46} \leq 14\,000$$
$$x^1_{34} + x^2_{34} \leq 8\,000$$
$$x^1_{13},\ x^1_{14},\ x^1_{34},\ x^1_{35},\ x^1_{45},\ x^2_{23},\ x^2_{24},\ x^2_{34},\ x^2_{36},\ x^2_{46} \geq 0.$$

By relaxing in a Lagrangian fashion constraint $x^1_{34} + x^2_{34} \leq 8\,000$ with a Lagrangian multiplier λ_{34}, the problem decomposes into two single-commodity linear minimum-cost flow problems. By initializing the subgradient algorithm with $\lambda^{(0)}_{34} = 0$ and using the updating formula (6.23) with $\alpha = 0.05$, the procedure provides, after 20 iterations, a lower bound LB = €40 197 387 ($\lambda^{(20)}_{34} = 99.887$). At the end, the procedure converges to $\lambda^*_{34} = 100$, which corresponds to an optimal objective value z^*_{LMMCFP} equal to € 40 200 000. Subproblem $k = 1$ has, for λ^*_{34}, two optimal basic solutions ($x^{1,*}_{13} = x^{1,*}_{14} = x^{1,*}_{35} = x^{1,*}_{45} = 18\,000$, $x^{1,*}_{34} = 0$) and ($x^{1,*}_{13} = 26\,000$, $x^{1,*}_{14} = 10\,000$, $x^{1,*}_{34} = 8\,000$, $x^{1,*}_{35} = 18\,000$, $x^{1,*}_{45} = 18\,000$), while subproblem $k = 2$ has a single optimal solution equal to: $x^{2,*}_{23} = x^{2,*}_{24} = 13\,000$, $x^{2,*}_{34} = 1\,000$, $x^{2,*}_{36} = 12\,000$, $x^{2,*}_{46} = 14\,000$. By combining the two partial solutions, the following two solutions are obtained: $x^{1,*}_{13} = x^{1,*}_{14} = x^{1,*}_{35} = x^{1,*}_{45} = 18\,000$, $x^{1,*}_{34} = 0$, $x^{2,*}_{23} = x^{2,*}_{24} = 13\,000$, $x^{2,*}_{34} = 1\,000$, $x^{2,*}_{36} = 12\,000$, $x^{2,*}_{46} = 14\,000$, and $x^{1,*}_{13} = 26\,000$, $x^{1,*}_{14} = 10\,000$, $x^{1,*}_{34} = 8\,000$, $x^{1,*}_{35} = 18\,000$, $x^{1,*}_{45} = 18\,000$, $x^{2,*}_{23} = x^{2,*}_{24} = 13\,000$, $x^{2,*}_{34} = 1\,000$, $x^{2,*}_{36} = 12\,000$, $x^{2,*}_{46} = 14\,000$). The former solution has an objective function value equal to €41 000 000 and is feasible, but does not satisfy the complementarity slackness conditions (6.24), while the latter is infeasible. It can be easy verified that the optimal solution is a convex combination of the two previous solutions and corresponds to: $x^{1,*}_{13} = 25\,000$, $x^{1,*}_{14} = 11\,000$, $x^{1,*}_{34} = 7\,000$, $x^{1,*}_{35} = 18\,000$, $x^{1,*}_{45} = 18\,000$, $x^{2,*}_{23} = 12\,000$, $x^{2,*}_{24} = 13\,000$, $x^{2,*}_{34} = 1\,000$, $x^{2,*}_{36} = 12\,000$, $x^{2,*}_{46} = 14\,000$.

6.3 Service network design problems

The design of a network of transport services is a tactical or operational decision particularly relevant to consolidation-based carriers. Given a set of terminals, the SNDP amounts to determining the features (frequency, number of intermediate stops etc.) of the routes to be operated, the traffic assignment along these routes, the operating rules at each terminal and possibly the relocation of empty vehicles and containers. The objective is the minimization of a generalized cost taking into account a combination of carrier's operating costs and customers' expectations. Figure 6.14 shows two alternative service networks for a three-terminal transport system in which it is assumed that each arc is associated to a line operated once a day. In the former network, each terminal is connected directly to every other terminal (so that each shipment takes one day) but this comes at the expense of a higher operating cost. In the latter network, operating costs are lower but the transport between certain origin-destination pairs may require two days (unless all lines are synchronized).

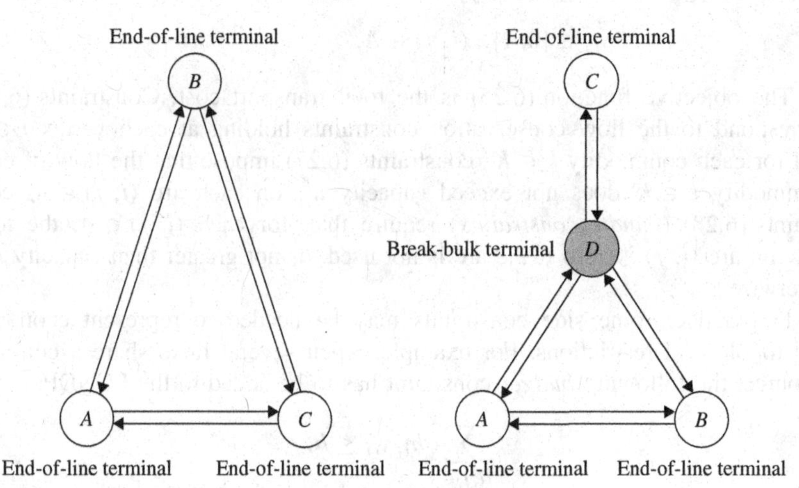

Figure 6.14 Two alternative service networks for a three end-of-line terminal transport system.

In the remainder of this section, the focus is on the basic network design problem, namely, the *fixed charge network design problem* (FCNDP), which can be viewed as a generalization of network flow problems in which a fixed cost f_{ij} has to be paid for using each arc $(i, j) \in A$. Therefore, FCNDPs amount to determining

(a) which arcs have to be employed;

(b) how to transport the commodities on the selected arcs.

Let x_{ij}^k, $(i, j) \in A$, $k \in K$, be the flow of commodity k on arc (i, j), and let y_{ij}, $(i, j) \in A$, be a binary decision variable, equal to 1 if arc (i, j) is used, 0 otherwise. A quite general formulation of the FCNDP is as follows:

$$\text{Minimize} \sum_{k \in K} \sum_{(i,j) \in A} C_{ij}^k(x_{ij}^k) + \sum_{(i,j) \in A} f_{ij} y_{ij} \tag{6.25}$$

subject to

$$\sum_{\{j\in V:(i,j)\in A\}} x_{ij}^k - \sum_{\{j\in V:(j,i)\in A\}} x_{ji}^k = \begin{cases} o_i^k, & \text{if } i \in O(k) \\ -d_i^k, & \text{if } i \in D(k), \\ 0, & \text{if } i \in T(k) \end{cases} \quad i \in V,\ k \in K \tag{6.26}$$

$$x_{ij}^k \leq u_{ij}^k,\ (i,j) \in A,\ k \in K \tag{6.27}$$

$$\sum_{k\in K} x_{ij}^k \leq u_{ij} y_{ij},\ (i,j) \in A \tag{6.28}$$

$$x_{ij}^k \geq 0,\ (i,j) \in A,\ k \in K$$

$$y_{ij} \in \{0,1\},\ (i,j) \in A.$$

The objective function (6.25) is the total transport cost. Constraints (6.26) correspond to the flow conservation constraints holding at each vertex $i \in V$ and for each commodity $k \in K$; constraints (6.27) impose that the flow of each commodity $k \in K$ does not exceed capacity u_{ij}^k on each arc $(i, j) \in A$; constraints (6.28) (*bundle constraints*) require that, for each $(i, j) \in A$, the total flow on arc (i, j) is zero if the arc is not used, or not greater than capacity u_{ij}, otherwise.

In practice, some side constraints may be needed to represent economic and topological restrictions. For example, when several links share a common resource, the following *budget* constraint has to be added to the FCNDP:

$$\sum_{(i,j)\in A} h_{ij} y_{ij} \leq b,$$

where h_{ij}, $(i, j) \in A$ is the consumption of resource made by arc $(i, j) \in A$, and b is the total amount of resource available.

6.3.1 The linear fixed-charge network design model

The *linear fixed-charge network design problem* (LFCNDP) is a particular FCNDP in which the transport costs per flow unit c_{ij}^k are constant (hence the objective function (6.25) is linear).

More formally, the LFCNDP can be formulated as

$$\text{Minimize} \sum_{k\in K} \sum_{(i,j)\in A} c_{ij}^k x_{ij}^k + \sum_{(i,j)\in A} f_{ij} y_{ij}$$

subject to

$$\sum_{\{j\in V:(i,j)\in A\}} x_{ij}^k - \sum_{\{j\in V:(j,i)\in A\}} x_{ji}^k = \begin{cases} o_i^k, & \text{if } i \in O(k) \\ -d_i^k, & \text{if } i \in D(k), \\ 0, & \text{if } i \in T(k) \end{cases} \quad i \in V,\ k \in K$$

$$x_{ij}^k \leq u_{ij}^k,\ (i,j) \in A,\ k \in K \tag{6.29}$$

$$\sum_{k \in K} x_{ij}^k \leq u_{ij} y_{ij}, \ (i, j) \in A$$

$$x_{ij}^k \geq 0, \ (i, j) \in A, \ k \in K$$

$$y_{ij} \in \{0, 1\}, \ (i, j) \in A.$$

FHL is an Austrian fast carrier located in Lienz, whose core business is the transport of palletized load units. The transport service is carried out through five cross-docking facilities by using two types of truck. A customer needs to move pallets from two RDCs to three retailers. The pallets are classified into three categories, corresponding to liquid food, dry food and other goods. The average retailer daily demand in the next trimester is reported in Table 6.4. The pallets are available each day at the RDCS in the quantities indicated in Table 6.4. The transport is generally realized by FHL between each pair of facilities using a single truck with a maximum load of 200 pallets. Only when the terminal is the retailer, FHL uses larger trucks of 250-pallet capacity. The potential transport network is represented by the directed graph $G = (V, A)$ in Figure 6.15. Each arc $(i, j) \in A$ corresponds to a potential transport service from facility i to facility j. If the service is activated, a fixed transport cost (in €) f_{ij} occurs plus variable costs which are proportional to the number of pallets of the three categories transported; c_{ij}^k, $k = 1, 2, 3$, indicates the unit variable transport cost (in €) for each pallet category k. Variable costs and fixed costs are reported in Table 6.5.

Table 6.4 Average daily demand of pallets of the three categories of products from the three retailers and average daily availability at the two RDC in the FHL problem.

	Retailer			RDC	
Category	1	2	3	1	2
Liquid food	80	55	40	80	95
Dry food	74	52	64	90	100
Other	47	42	76	75	90

The model can be formulated as an LFCNDP, whose optimal solution leads to a daily cost of € 34 512 and the activation of the transport services corresponding to the arcs $(1, 3)$, $(1, 4)$, $(2, 4)$, $(2, 5)$, $(3, 6)$, $(4, 6)$, $(4, 9)$, $(5, 7)$, $(6, 8)$ and $(7, 10)$.

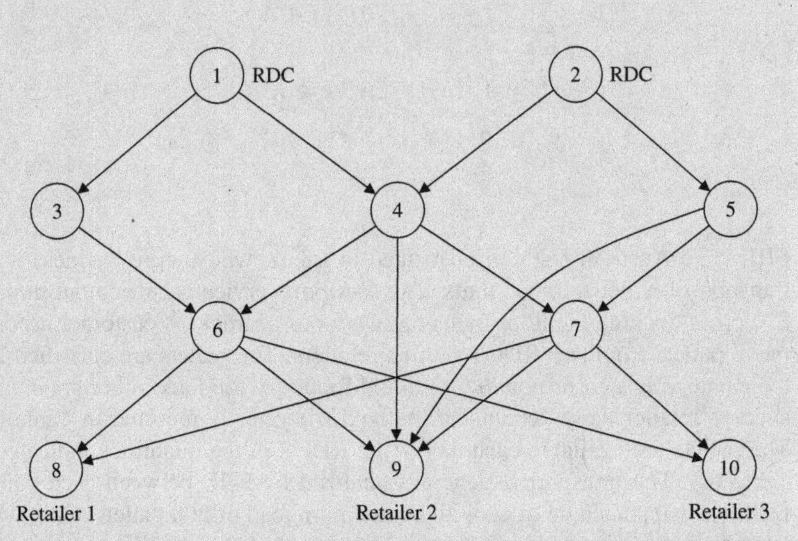

Figure 6.15 Graph representation of the FHL service network design problem.

Table 6.5 Variable costs and fixed costs (in €) in the FHL problem.

	Variable cost			
(i, j)	c_{ij}^1	c_{ij}^2	c_{ij}^3	f_{ij}
(1,3)	22	19	17	450
(1,4)	22	19	17	490
(2,4)	20	21	18	390
(2,5)	20	21	18	435
(3,6)	22	19	18	296
(4,6)	23	22	21	340
(4,7)	26	25	27	685
(4,9)	23	22	21	625
(5,7)	21	19	20	365
(5,9)	25	23	23	655
(6,8)	24	25	25	325
(6,9)	24	25	25	490
(6,10)	24	25	25	725
(7,8)	26	24	27	710
(7,9)	24	24	25	525
(7,10)	23	23	24	410

The LFCNDP is NP-hard, and branch-and-bound algorithms can hardly solve instances with a few hundreds of arcs and tens of commodities. Since instances arising in applications are much larger, heuristics are often used. To evaluate the quality of the solutions provided by heuristics, it is useful, as already observed in Chapter 3, to compute lower bounds on the optimal objective function value z^*_{LFCNDP}. In the following, two distinct continuous relaxations and a simple heuristic are illustrated.

6.3.1.1 The weak continuous relaxation

The weak continuous relaxation is obtained by relaxing the integrity requirement on the decision variables y_{ij}, $(i, j) \in A$:

$$\text{Minimize} \sum_{k\in K}\sum_{(i,j)\in A} c^k_{ij}x^k_{ij} + \sum_{(i,j)\in A} f_{ij}y_{ij} \tag{6.30}$$

subject to

$$\sum_{\{j\in V:(i,j)\in A\}} x^k_{ij} - \sum_{\{j\in V:(j,i)\in A\}} x^k_{ji} = \begin{cases} o^k_i, & \text{if } i \in O(k) \\ -d^k_i, & \text{if } i \in D(k), \\ 0, & \text{if } i \in T(k) \end{cases} \quad i \in V,\ k \in K \tag{6.31}$$

$$x^k_{ij} \leq u^k_{ij},\ (i, j) \in A,\ k \in K \tag{6.32}$$

$$\sum_{k\in K} x^k_{ij} \leq u_{ij}y_{ij},\ (i, j) \in A \tag{6.33}$$

$$x^k_{ij} \geq 0,\ (i, j) \in A,\ k \in K \tag{6.34}$$

$$0 \leq y_{ij} \leq 1,\ (i, j) \in A. \tag{6.35}$$

It is easy to verify that every optimal solution of such a relaxation satisfies each constraint (6.33) as an equality since fixed costs f_{ij}, $(i, j) \in A$, are non-negative. Therefore, decision variables y_{ij}, $(i, j) \in A$, can be expressed as a function of flow variables x^k_{ij}, $(i, j) \in A$, $k \in K$:

$$y_{ij} = \frac{\sum_{k\in K} x^k_{ij}}{u_{ij}},\ (i, j) \in A.$$

Hence, constraints (6.35) can be replaced by the following conditions:

$$\sum_{k\in K} x^k_{ij} \leq u_{ij},\ (i, j) \in A.$$

The relaxed problem (6.30)–(6.35) can be therefore equivalently formulated as:

$$\text{Minimize} \sum_{k\in K}\sum_{(i,j)\in A} \left(c^k_{ij} + \frac{f_{ij}}{u_{ij}}\right) x^k_{ij} \tag{6.36}$$

subject to

$$\sum_{\{j \in V:(i,j) \in A\}} x_{ij}^k - \sum_{\{j \in V:(j,i) \in A\}} x_{ji}^k = \begin{cases} o_i^k, & \text{if } i \in O(k) \\ -d_i^k, & \text{if } i \in D(k), \\ 0, & \text{if } i \in T(k) \end{cases} \quad i \in V, \; k \in K \tag{6.37}$$

$$x_{ij}^k \leq u_{ij}^k, \; (i, j) \in A, \; k \in K \tag{6.38}$$

$$\sum_{k \in K} x_{ij}^k \leq u_{ij}, \; (i, j) \in A \tag{6.39}$$

$$x_{ij}^k \geq 0, \; (i, j) \in A, \; k \in K. \tag{6.40}$$

Model (6.36)–(6.40) is a minimum-cost flow problem with $|K|$ commodities. Let LB_w^* be the lower bound on z_{LFCNDP}^* given by the optimal objective function value of the above relaxation.

6.3.1.2 The strong continuous relaxation

The strong continuous relaxation is obtained by adding the following *valid* inequalities

$$x_{ij}^k \leq u_{ij}^k y_{ij}, \; (i, j) \in A, \; k \in K, \tag{6.41}$$

to the LFCNDP and removing the integrity constraints on the decision variables y_{ij}, $(i, j) \in A$. Taking into account the fact that constraints (6.29) are dominated by constraints (6.41), and can be therefore eliminated, the relaxed problem is

$$\text{Minimize} \sum_{k \in K} \sum_{(i,j) \in A} c_{ij}^k x_{ij}^k + \sum_{(i,j) \in A} f_{ij} y_{ij} \tag{6.42}$$

subject to

$$\sum_{\{j \in V:(i,j) \in A\}} x_{ij}^k - \sum_{\{j \in V:(j,i) \in A\}} x_{ji}^k = \begin{cases} o_i^k, & \text{if } i \in O(k) \\ -d_i^k, & \text{if } i \in D(k), \\ 0, & \text{if } i \in T(k) \end{cases} \quad i \in V, \; k \in K \tag{6.43}$$

$$x_{ij}^k \leq u_{ij}^k y_{ij}, \; (i, j) \in A, \; k \in K \tag{6.44}$$

$$\sum_{k \in K} x_{ij}^k \leq u_{ij} y_{ij}, \; (i, j) \in A \tag{6.45}$$

$$x_{ij}^k \geq 0, \; (i, j) \in A, \; k \in K \tag{6.46}$$

$$0 \leq y_{ij} \leq 1, \; (i, j) \in A. \tag{6.47}$$

Let LB_s^* be the lower bound on z_{LFCNDP}^* given by the optimal objective function value of the relaxation (6.42)–(6.47). Such problem has no special structure and, therefore, can be solved by using any general-purpose LP algorithm. By comparing the two continuous relaxations, it is clear that LB_s^* is always better than, or at least equal to, LB_w^*, that is,

$$\mathrm{LB}_s^* \geq \mathrm{LB}_w^*.$$

This observation leads us to label the former relaxation as *weak*, and the latter as *strong*. Computational experiments have shown that LB_w^* can be as much as 40% lower than LB_s^*.

6.3.1.3 Add-drop heuristics

Add-drop heuristics are simple constructive procedures in which at each step one decides whether a new arc has to be used (*add procedure*) or an arc previously used has to be left out (*drop procedure*). Several criteria can be employed to choose which arc has to be added or dropped. In the following, a very simple drop heuristic is illustrated. In order to describe such a heuristic, it is worth noting that a candidate optimal solution is characterized by the set $A' \subseteq A$ of selected arcs. A solution is feasible if the LMMCFP on the directed graph $G = (V, A')$ induced by A' is feasible. If so, the solution cost is made up of the sum of the fixed costs f_{ij}, $(i, j) \in A'$, plus the optimal solution cost of the LMMCFP. Moreover, it is worth noting that the LFCNDP solution associated to $A' = A$, if feasible, is characterized by a large fixed cost and by a low transport cost. On the other hand, a feasible solution associated to a set A' with a few arcs is expected to be characterized by a low fixed cost and by a high variable cost. Consequently, an improved LFCNDP solution can be obtained by iteratively removing arcs from the set $A' = A$, while the current solution is still feasible and the total cost decreases. The drop heuristic is as follows:

Step 1. Set $h = 0$ and $A^{(h)} = A$. Let $x_{ij}^{k,(h)}$, $(i, j) \in A, k \in K$, be the optimal solution (if any exists) of the LMMCFP on the directed graph $G = (V, A^{(h)})$ and let $z_{\mathrm{LFCNDP}}^{(h)}$ be the cost of the associated LFCNDP. If the LMMCFP is infeasible, *STOP*, the LFCNDP is also infeasible.

Step 2. For each arc $(i, j) \in A^{(h)}$, set $A'^{(h)} = A^{(h)} \setminus \{(i, j)\}$ and solve the LMMCFP on the directed graph $(G = (V, A'^{(h)})$.. If all the LMMCFPs are infeasible, *STOP*, the set of the arcs $A^{(h)}$ and the flow values $x_{ij}^{k,(h)}$, $(i, j) \in A^{(h)}$, $k \in K$, are associated to the best feasible solution found; otherwise, let (v, w) be the arc whose removal from $A^{(h)}$ allows to attain the least cost LFCND feasible solution.

Step 3. Set $A^{(h+1)} = A^{(h)} \setminus \{(v, w)\}$, $h = h + 1$ and go back to Step 2.

The number of iterations of the heuristic is no more than the number of arcs and, at each iteration, Step 2 requires the solution of $O(|A|)$ LMMCFPs.

By applying the drop heuristic to the FHL problem, the following binary decision variables are sequentially set equal to zero:

$$y_{6,10}, y_{78}, y_{59}, y_{79}, y_{69}.$$

The feasible solution leads to a cost of € 34 512, which corresponds to the optimal cost.

6.4 Vehicle allocation problems

VAPs are faced by carriers that generate revenue by transporting full loads over long distances, as in TL trucking and container shipping. Once a vehicle delivers a load, it becomes empty and has to be moved to the pickup point of another load, or has to be repositioned in anticipation of future demands.

The VAP illustrated in this section is a particular version which amounts to deciding the loads to be accepted and the ones to be rejected, as well as repositioning empty vehicles. The problem is modelled as a minimum-cost flow problem on a time-expanded directed graph for the case where all demands are known in advance. For the sake of simplicity, we examine the case where a single vehicle type exists, while the extension to the multi-vehicle type case is left to the reader as an exercise (see Problem 6.9). The case in which demands are random is much more complex and is currently under study by the scientific community.

The planning horizon is supposed to be composed of a finite number $\{1, \ldots, T\}$ of time periods. Let N be the set of points (e.g. cities) where the (full) loads have to picked up and delivered; d_{ijt}, $i \in N$, $j \in N$, $t = 1, \ldots, T$, the number of requested loads to be moved from origin i to destination j at time period t; τ_{ij}, $i \in N$, $j \in N$, the travel time from point i to point j; p_{ij}, $i \in N$, $j \in N$, the profit (revenue minus direct operating costs) derived from moving a load from point i to point j, and c_{ij}, $i \in N$, $j \in N$, the cost of moving an empty vehicle from point i to point j. Moreover, denote by m_{it}, $i \in N$, $t = 1, \ldots, T$, the number of vehicles that enter the system in time period t at point i. The following decision variables are used: x_{ijt}, $i \in N$, $j \in N$, $t = 1, \ldots, T$, representing the number of vehicles that start moving a load from point i to point j at time period t; y_{ijt}, $i \in N$, $j \in N$, $t = 1, \ldots, T$, representing the number of vehicles that start moving empty from point i to point j at time period t. The deterministic single-vehicle VAP can be formulated as follows:

$$\text{Maximize} \sum_{t=1}^{T} \sum_{i \in N} \sum_{j \in N, j \neq i} (p_{ij} x_{ijt} - c_{ij} y_{ijt}) \tag{6.48}$$

subject to

$$\sum_{j\in N}(x_{ij1}+y_{ij1}) = m_{i1},\ i \in N \tag{6.49}$$

$$\sum_{j\in N}(x_{ijt}+y_{ijt}) - \sum_{k\in N, k\neq i: t>\tau_{ki}} (x_{ki(t-\tau_{ki})} + y_{ki(t-\tau_{ki})}) - y_{iit-1} = m_{it},$$
$$i \in N, t \in \{2,\ldots,T\} \tag{6.50}$$

$$x_{ijt} \leq d_{ijt},\ i \in N, j \in N, t \in \{1,\ldots,T\} \tag{6.51}$$

$$x_{ijt} \geq 0,\ i \in N, j \in N, t \in \{1,\ldots,T\}$$

$$y_{ijt} \geq 0,\ i \in N, j \in N, t \in \{1,\ldots,T\}.$$

The objective function (6.48) is the total discounted profit over the planning horizon. Constraints (6.49) and (6.50) impose flow conservation at the beginning of each time period (in this case the flow is given by the number of vehicles). Due to these constraints, the decision variables x_{iit} and y_{iit}, $i \in N$, $j \in N$, $t = 1,\ldots,T$, have integer values implicitly. Constraints (6.51) state that the number of loaded movements of vehicles at each time period $t = 1,\ldots,T$ between each pair origin i destination j, $i \in N$, $j \in N$, is bounded above by the demand. It is worth noting that the $d_{ijt} - x_{ijt}$ differences, $i \in N$, $j \in N$, $t = 1,\ldots,T$, represent the loads that should be rejected, while y_{iit} decision variables, $i \in N$, $t = 1,\ldots,T$, represent vehicles staying idle (the so-called *inventory movements*). It is easy to recognize that VAP can be modelled as a minimum-cost flow problem on a time-expanded directed graph in which vertices are associated to (i, j) pairs, $i \in N$, $j \in N$, and arcs represent loaded, empty and inventory movements. Since there exists a pair of arcs between each pair of nodes, such a network is a directed multigraph.

Murthy is a motor carrier operating in the Andhraachuki region (India). Last July 11, four TL transport requests were made: from Chittoor to Khammam on July 11, from Srikakulam to Ichapur on July 11 and from Ananthapur to Chittoor on July 13 (two loads). On July 11, one vehicle was available in Chittoor and one in Khammam. A further vehicle was currently transporting a previously scheduled shipment and would be available in Chittoor on July, 12. Transport times between terminals are shown in Table 6.6. The revenue provided by a truck carrying a full load is 1.8 times the transport cost of a deadheading truck, estimated equal to 10 000 rupees for each journey day. The problem to solve for Murthy is a VAP in which T = {July 11, July 12, July 13} = {1, 2, 3} and N = {Ananthapur, Chittoor, Ichapur, Khammam, Srikakulam} = {1, 2, 3, 4, 5}. The cost c_{ij} of a journey from city i, $i \in N$, to city j, $j \in N$, is simply obtained by multiplying the cost of each journey day by the journey days τ_{ij} from city i to city j. Consequently, the profit

Table 6.6 Travel times (in days) between terminals in the Murthy problem.

	Ananthapur	Chittoor	Ichapur	Khammam	Srikakulam
Ananthapur	0	1	2	2	2
Chittoor		0	2	2	2
Ichapur			0	2	1
Khammam				0	2
Srikakulam					0

p_{ij}, $i, j \in N$, is equal to $0.8 \times 10\,000 \times \tau_{ij}$. Furthermore, the number of available vehicles m_{it}, $i \in N, t = 1, \ldots, T$, are all zero except the following: $m_{12} = 1$; $m_{21} = 1$; $m_{41} = 1$. The values d_{ijt}, $i \in N$, $j \in N$, $t = 1, \ldots, T$, are also equal to zero, except the following: $d_{241} = 1$; $d_{531} = 1$; $d_{123} = 2$.

The optimal VAP solution is $x^*_{241} = 1$, $x^*_{123} = 1$, $y^*_{441} = 1$, $y^*_{112} = 1$, $y^*_{442} = 1$, $y^*_{443} = 2$, while the values of the remaining decision variables are zero. The corresponding optimal cost is 24 000 rupees. It is worth noting that the requests from Srikakulam to Ichapur on July 11 are not satisfied and from Ananthapur to Chittoor on July 13 are partially satisfied.

6.5 A dynamic driver assignment problem

The *dynamic driver assignment problem* (DDAP) examined in this section arises in TL trucking where full-load trips are assigned to drivers in an on-going fashion. In TL trucking, a trip may take several days (a four-day duration is not unusual both in Europe and in North America) and customer service requests arrive at random. Consequently, each driver is assigned a single trip at a time.

The DDAP can be formulated as a particular single-commodity uncapacitated minimum-cost flow problem (see Figure 6.16) and can be solved efficiently through, for example, the network simplex method.

Let D be the set of drivers waiting to be assigned a task and let L be the current set of transport services to be performed (as each transport service corresponds to a full-load trip and vice versa). Let x_{ij}, $i \in D$, $j \in L$, be a binary decision variable equal to 1 if driver i is assigned to transport service j, 0 otherwise; c_{ij}, $i \in D$, $j \in L$, is the cost of assigning driver i to transport service j. We consider the case in which the number of drivers is greater than the number of transport services required. A possible formulation of the DDAP is the following:

$$\text{Minimize} \sum_{i \in D} \sum_{j \in L} c_{ij} x_{ij}$$

subject to

$$\sum_{j \in L} x_{ij} \leq 1, i \in D$$

$$\sum_{i \in D} x_{ij} = 1, j \in L$$

$$x_{ij} \in \{0, 1\}, \; i \in D, \; j \in L.$$

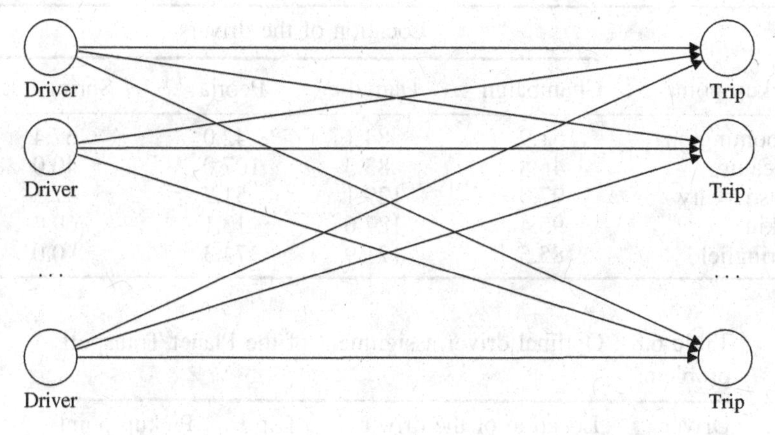

Figure 6.16 An example of a network for a DDAP.

The DDAP model can be easily modified in the case where the number of transport services is greater than or equal to the number of drivers. This modification is left to the reader as an exercise. In practice, the DDAP model is solved as vehicle locations and customer requests are revealed during the planning horizon (this explains the dynamic component of the model). In addition, penalties or bonuses may have added to arc costs c_{ij}, $i \in D$, $j \in L$, to reflect the cost of taking drivers home after a given number of transport services have been performed. A dispatcher who wants to take a driver $i \in D$ home at a given point in time can simply reduce the cost of the assignment of that particular driver to trips $j \in L$ whose delivery points are close to the driver home location. This can be accomplished by subtracting a suitable quantity from the c_{ij} costs.

Planet Transport is an Illinois, US, motor carrier that specializes in TL trucking. Last January 26, the company had to solve the problem of assigning four full-load transport services of each three days long. The pickup points were in Champaign, Danville, Peoria and Springfield. At that time six drivers were available. The first four were located in Bloomington, Decatur, Mason City and Pekin, and the last two in Springfield. The company formulated and solved a DDAP model with $|D| = 6$ and $|L| = 4$. Costs c_{ij}, $i \in D$, $j \in L$, were assumed to be proportional to distances between driver locations i and full-load trip pickup points j (see Table 6.7). The optimal driver assignment is reported in Table 6.8. It is worth noting that the driver located in Mason City and one of the two drivers situated in Springfield were not assigned.

Table 6.7 Distance (in miles) between driver locations and trip pickup points in the Planet Transport problem.

	Location of the drivers			
Pickup point	Champaign	Danville	Peoria	Springfield
Bloomington	51.9	84.1	42.0	67.4
Decatur	46.8	83.3	107.0	40.0
Mason City	97.0	129.1	51.2	47.5
Pekin	95.4	127.6	13.1	69.4
Springfield	85.5	121.9	74.3	0.0

Table 6.8 Optimal driver assignment of the Planet Transport problem.

Driver	Location of the driver	Trip	Pickup point
1	Bloomington	2	Danville
2	Decatur	1	Champaign
3	Mason City	–	–
4	Pekin	3	Peoria
5	Springfield	4	Springfield
6	Springfield	–	–

6.6 Fleet composition

When demand varies over the year, carriers usually cover the baseload of demand through an owned fleet, while using hired vehicles to cover peak periods. In what follows, the least-cost mix of owned and hired vehicles is determined under the assumption that all vehicles are identical. Let n be the number of time periods into which the time horizon of a year is decomposed (for example, $n = 52$ if the time period corresponds to a week); v be the decision variable corresponding to the number of owned vehicles and v_t, $t = 1, \dots, n$, be the required number of vehicles at time period t. Moreover, let c_F and c_V be the fixed and variable cost per time period of an owned vehicle, respectively, and let c_H be the cost per time period of hiring a vehicle (clearly, $c_F + c_V < c_H$). Then, the annual transport cost as a function of the number of owned vehicles, is

$$C(v) = nc_F v + c_V\Big(\sum_{t=1,\dots,n:v_t \le v} v_t + \sum_{t=1,\dots,n:v_t > v} v\Big) + c_H \sum_{t=1,\dots,n:v_t > v} (v_t - v), \tag{6.52}$$

where the right hand side is the sum of the annual fixed cost, the annual variable cost of the owned vehicles and the annual cost of hiring vehicles to cover peak demand. To find the optimal number v^* of owned vehicles, it is sufficient to consider that the cost $C(v)$ corresponding to (6.52) is defined by discrete values of v included in the interval $[0, \bar{v}]$, where:

$$\bar{v} = \max_{t=1,\dots,n} \{v_t\}.$$

As a result, the problem can be easily solved by inspection, calculating the cost $C(v)$ for each $v = 0, \dots, \bar{v}$ and determining v^* as:

$$v^* = \arg \min_{v=0,\dots,\bar{v}} C(v).$$

Fast Courier is a US transport company located in Wichita, Kansas, that specializes in door-to-door deliveries. The company owns a fleet of 14 vans and turns to third parties for hiring vans in case the service demand exceeds the fleet capacity. Last year, the number of vans weekly used for meeting all the transport demand is reported in Table 6.9. For next year, the company has decided to redesign the fleet composition, with the aim of reducing the annual transport cost, assuming that the vehicle requests remain identical to those of the year just ended. Assuming that $c_F = \$350$, $c_V = \$150$

Table 6.9 Weekly number of vans used by Fast Courier last year.

t	v_t	t	v_t	t	v_t	t	v_t
1	12	14	18	27	23	40	25
2	15	15	17	28	22	41	25
3	16	16	16	29	24	42	24
4	17	17	14	30	26	43	22
5	17	18	13	31	27	44	22
6	18	19	13	32	28	45	19
7	20	20	14	33	30	46	20
8	20	21	15	34	32	47	18
9	21	22	16	35	32	48	17
10	22	23	17	36	30	49	16
11	24	24	19	37	29	50	16
12	22	25	21	38	28	51	14
13	20	26	22	39	26	52	13

and $c_H = \$ \, 800$, the transport costs $C(v)$ calculated by 6.52 when v varies in the interval $[0, 32]$ are shown in Table 6.10. As a result, the optimal number of owned vehicles is equal to $v^* = 19$ (or, equivalently, 20), with an annual transport cost equal to $C(v^*) = \$ \, 606\,600$. With respect to the currently adopted solution with 14 owned vehicles used, the saving the company will obtain by adopting this new solution is, therefore, equal to $638\,450 - 606\,600 = \$ \, 31\,850$.

Table 6.10 Annual transport costs (in $) at variation of the number of owned vehicles of Fast Courier.

v	$C(v)$	v	$C(v)$	v	$C(v)$
0	853 600	11	682 000	22	613 100
1	838 000	12	666 400	23	620 900
2	822 400	13	651 450	24	629 350
3	806 800	14	638 450	25	639 750
4	791 200	15	627 400	26	651 450
5	775 600	16	617 650	27	664 450
6	760 000	17	611 150	28	678 100
7	744 400	18	607 900	29	693 050
8	728 800	19	606 600	30	708 650
9	713 200	20	606 600	31	725 550
10	697 600	21	609 200	32	742 450

6.7 Shipment consolidation

In this section, we examine a consolidation and dispatching problem often faced by companies that carry out their own transport services. Here, the company has to choose the best way of delivering a set of orders to its customers over a planning horizon made up of T days. The company must decide

- the best mode of transport for each shipment;
- how orders have to be consolidated;
- the features of owned vehicle schedules (start times, intermediate stops (if any), the order in which stops are visited etc.).

Each order $k \in K$ is characterized by a destination i_k, a weight $w_k \geq 0$, a release day l_k (the day in which order k is ready for shipment; we assume, for simplicity, that order k can be shipped at least one day after the release day) and a deadline day d_k (the day within which order k must be delivered to i_k). The company may transport its products by using its owned vehicles, following predetermined routes, or by using a carrier. An owned truck may follow any route

r of a preestablished set R. With each route $r \in R$ are associated a set of stops S_r (visited in a given order), a (fixed) cost f_r, and a capacity q_r (the maximum weight that the vehicle operating route r can carry). Moreover, let τ_{kr}, $k \in K$, $r \in R$, be the number of travel days to deliver order k on route r ($\tau_{kr} = 0$ means same-day delivery). Transporting order k, $k \in K$, to its destination by a carrier costs g_k and takes τ'_k days. The decision variables are all binary: x_{krt}, $k \in K$, $r \in R$, $t = 1, \ldots, T$, having a value equal to 1 if order k is assigned to route r starting on day t, 0 otherwise; y_{rt}, $r \in R$, $t = 1, \ldots, T$, equal to 1 if route r is operated on day t, 0 otherwise; z_k, equal to 1 if order k is transported by a carrier, 0 otherwise (such a decision variable is defined only if $l_k + \tau'_k \leq d_k$).

The problem in then

$$\text{Minimize} \sum_{r \in R} \sum_{t=1}^{T} f_r y_{rt} + \sum_{k \in K} g_k z_k \tag{6.53}$$

subject to

$$\sum_{k: l_k \leq t \leq d_k - \tau_{kr}, i_k \in S_r} w_k x_{krt} \leq q_r y_{rt}, \quad r \in R, t = 1, \ldots, T \tag{6.54}$$

$$\sum_{r: i_k \in S_r} \sum_{t: l_k \leq t \leq d_k - \tau_{kr}} x_{krt} + z_k = 1, \quad k \in K \tag{6.55}$$

$$x_{krt} \in \{0, 1\}, \quad k \in K, r \in R, t = 1, \ldots, T \tag{6.56}$$

$$y_{rt} \in \{0, 1\}, \quad r \in R, t = 1, \ldots, T \tag{6.57}$$

$$z_k \in \{0, 1\}, \quad k \in K. \tag{6.58}$$

The objective function (6.53) is the total cost paid to transport orders. Constraints (6.54) state that, for each route $r \in R$ and for each day $t = 1, \ldots, T$, the total weight carried on route r, on day t must not exceed capacity q_r if y_{rt} is equal to 1, and is equal to zero otherwise. Constraints (6.55) impose that each order is assigned to a route operated by a owned truck or to a carrier. It is easy to show that formulation (6.53)–(6.58) can be transformed into a network design model on a time-expanded directed graph.

Oxximet manufactures semi-finished chemical products in a plant located close to Milan, Italy. The main customers are located close to Lausanne, Switzerland, and Lyon and Marseille, France. Last June 11, 15 customer orders were waiting to be satisfied (see Table 6.11).

Oxximet could use its owned trucks, according to the set of routes reported in Table 6.12, visiting one or more customers. Each of the truck routes has a duration of one day at most and can be scheduled in every day of the planning horizon. Truck capacity is 260 quintals. The company can also use carriers, whose cost and delivery time for each order are shown in the last two columns of Table 6.11.

Table 6.11 Details of the orders received on June 11 by Oxximet.

Order	Customer location	Release day	Deadline day	Weight (q)	Carrier shipping cost (€)	Carrier delivery days (days)
1	Marseille	June 11	June 15	207.34	600	1
2	Marseille	June 11	June 16	19.05	300	0
3	Lyon	June 11	June 16	19.59	300	0
4	Marseille	June 11	June 16	35.23	300	1
5	Lausanne	June 11	June 16	61.54	300	1
6	Lausanne	June 11	June 16	38.31	300	1
7	Lyon	June 13	June 16	100.46	600	0
8	Marseille	June 13	June 16	15.44	500	1
9	Lyon	June 13	June 16	56.89	500	0
10	Marseille	June 13	June 15	39.55	500	1
11	Lausanne	June 14	June 15	242.65	350	0
12	Lausanne	June 14	June 16	102.65	300	0
13	Marseille	June 14	June 15	154.79	550	0
14	Marseille	June 14	June 16	78.53	500	1
15	Marseille	June 14	June 16	45.42	500	0

Table 6.12 Routes of the owned vehicles and related fixed costs (in €) of the Oxximet problem.

ID	Route	Cost
1	Milan–Lausanne	800
2	Milan–Lyon	750
3	Milan–Marseille	800
4	Milan–Lausanne–Lyon	830
5	Milan–Lyon–Marseille	870
6	Milan–Lausanne–Marseille	890

The problem can be formulated by using the model defined by (6.53)–(6.58), with a scheduling time horizon $T = 5$, from June 12 to June 16 included, $|K| = 15$ and $|R| = 6$. The parameters l_k, d_k, w_k, g_k and τ'_k related to the orders $k \in K$ are shown in Table 6.11.

By observing the carrier delivery times, we find that

$$l_k + \tau'_k \leq d_k, k \in K,$$

that is, the decision variable z_k exists for each $k \in K$. As observed before, $\tau_{kr} = 0$, $k \in K, r \in R$.

The optimal solution provides for delivery scheduling owned truck routes shown in Table 6.13. In addition, orders 1 and 11 should be delivered by using carriers ($z_1^* = z_{11}^* = 1$), with a shipping day from June 12 to June 14 for order 1, and in June 15 for order 11. The overall optimal cost is € 3 450.

Table 6.13 Optimal solution of the owned vehicle routes for the Oxximet problem.

Day	Route	Orders
June 12	–	–
June 13	–	–
June 14	–	–
June 15	Milan–Marseille	2, 4, 10, 13
June 15	Milan–Lausanne–Lyon	5, 6, 9, 12
June 16	Milan–Lyon–Marseille	3, 7, 8, 14, 15

6.8 Vehicle routing problems

VRPs amount to finding optimal delivery or collection routes from one or several depots to a number of users. VRPs can be defined on a mixed graph $G = (V, A, E)$, where V is a set of vertices, A is a set of arcs and E is a set of edges. A vertex 0 represents the depot at which m vehicles are based, while a subset $U \subseteq V$ of *required vertices* and a subset $R \subseteq A \cup E$ of *required arcs* and *required edges* represent the users. VRPs consist of determining a least cost set of m circuits based at a depot, and including the required vertices, arcs and edges.

In this graph representation, arcs and edges correspond to road segments, and vertices correspond to road intersections. In addition, isolated users are represented by required vertices, whereas subsets of customers distributed almost continuously along a set of customers are modelled as required arcs or edges (this is often the case of mail delivery and solid waste collection in urban areas) . See Figures 6.17 and 6.18 for an example. If $R = \emptyset$, the VRP is called a *node routing problem* (NRP), while if $U = \emptyset$ it is called an *arc routing problem* (ARP). NRPs have been studied more extensively than ARPs and are usually referred to simply as VRPs. However, for the sake of clarity, in this textbook we use the appellation NRPs. If $m = 1$ and there are no side constraints, the NRP is the classical *travelling salesman problem* which consists of determining a single circuit spanning the vertices of G, whereas the ARP is the *rural postman problem* (RPP) which amounts to designing a single circuit including the arcs and edges of R. The RPP reduces to the *Chinese postman problem* (CPP) if every arc and edge have to be serviced ($R = A \cup E$).

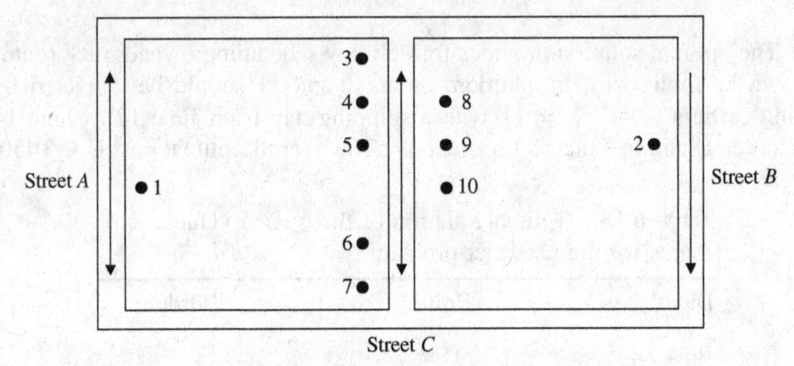

Figure 6.17 A road network where 10 customers (represented by black dots) are to be served. Streets A and C are two-way. Street B is one-way.

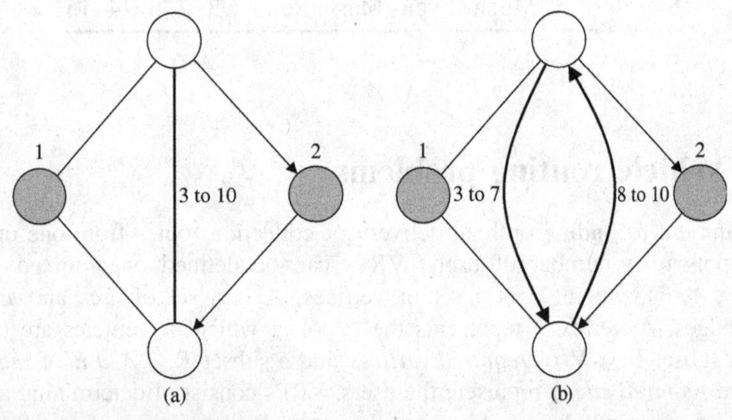

Figure 6.18 (a) A graph representation of a vehicle, traversing road B of Figure 6.17, can serve the customers on both sides; and (b) a graph representation of a vehicle, traversing road B, can serve the customers on a single side. Service vertices are in grey; service arcs and edges are in bold.

Operational constraints The most common operational constraints are

- the number of vehicles m can be fixed or can be a decision variable, possibly subject to an upper bound constraint;
- the total demand transported by a vehicle at any time must not exceed its capacity;
- the duration of any route must not exceed a work shift duration;
- customers must be served within preestablished time windows;
- some customers must be served by specific vehicles;

- the service of a customer must be performed by a single vehicle or may be shared by several vehicles;
- customers are subject to precedence relations.

When customers impose service time windows or when travel times vary during the day, time issues have to be considered explicitly in the design of vehicle routes, in which case VRPs are often referred to as *vehicle routing and scheduling problems* (VRSPs).

Precedence constraints arise naturally whenever some goods have to be transported between specified pairs of pickup and delivery points. In such problems, a pickup and delivery pair is to be serviced by the same vehicle (no transshipment is allowed) and each pickup point must visited before the associated delivery point. Another kind of precedence relations has to be imposed whenever vehicles have first to perform a set of deliveries (*linehaul customers*) and then a set of pickups (*backhaul customers*), as is customary in some industries (VRPs with backhauls, see Figure 6.19).

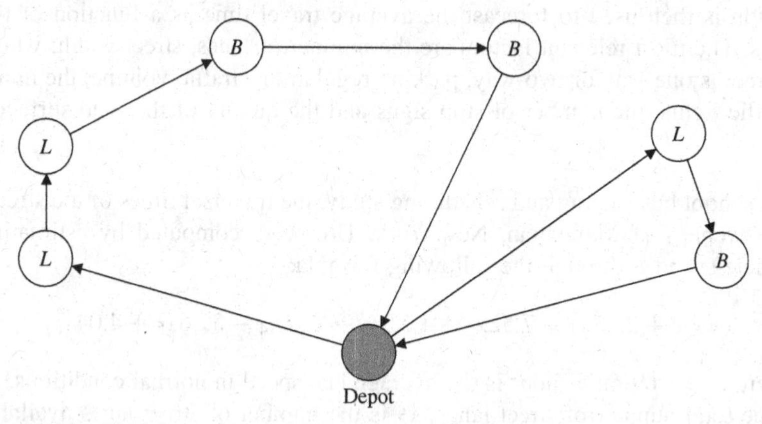

Figure 6.19 Vehicle routing with backhauls (L: linehaul customer, B: backhaul customers).

Objective With each arc and edge $(i, j) \in A \cup E$ are associated a travel time t_{ij} and a travel cost c_{ij}. In addition, with each vehicle may be associated a fixed cost. The most common objective is to minimize the cost of traversing the arcs and edges of the graph plus the sum of the fixed costs associated to using the vehicles.

Travel time estimate While the computation of the distances between the nodes of a road network is straightforward, the accurate estimation of the travel times is often difficult. The reason is twofold. Firstly, the average speed depends on the period of the day as well as on the day of the week and the occurrence of holidays. In particular, in most urban areas a rush hour (or peak hour) happens twice a day, once in the morning and once in the evening, the time intervals when most people commute. Moreover, in some areas the traffic is lighter on weekends than between Monday and Friday. Secondly, remarkable fluctuations around the average speed are caused by the weather conditions, accidents, strikes disrupting the public transport system and events like sport matches, concerts, political protests and so on. In the last few decades, there has been a proliferation of online travel time information systems which provide an estimate of the current traversal times through a suitable processing of data coming from a number of sensing technologies (including inductive loops placed in the roadbed, video vehicle detection and, more recently, GPS-based mobile phones). At the moment, such systems cover only a portion of the whole road network, but it is expected that they will be more and more widespread in the coming years. The data collected by the travel time information systems can be used to make forecasts on the future average speeds by using the techniques illustrated in Chapter 2. Whenever such data are not available, one can devise a travel time estimate on the basis of the features of the road by using a regression analysis. To this end, the factors affecting travel time along a street are identified, and a regression equation is then used to forecast the average travel time as a function of these factors. The most relevant factors are the number of lanes, street width, whether the street is one-way or two-way, parking regulations, traffic volume, the number of traffic lights, the number of stop signs and the quality of the road surface.

In a school bus routing and scheduling study, the traversal times of the streets and avenues of Manhattan, New York, US, were computed by estimating vehicle speed v through the following formula:

$$v = \bar{v} + 2.07x_1 + 7.52x_2 + 1.52x_3 + 1.36x_4 - 3.26x_5 + 4.04x_6,$$

where $\bar{v} = 7.69$ miles/hour is the average bus speed in normal conditions, x_1 is the total number of street lanes, x_2 is the number of street lanes available for buses, x_3 is a binary constant equal to 1 in case of a one-way street and 0 otherwise, x_4 is equal to 1 in case of bad road surface conditions and 2 in case of good road surface conditions, x_5 takes into account the traffic volume (1 = low, 2 = medium, 3 = high) and x_6 is the time fraction of green lights. Variable coefficients were estimated through regression analysis.

6.8.1 The travelling salesman problem

In the absence of operational constraints, there always exists an optimal NRP solution in which a single vehicle is used (see Problem 6.17). Hence, the NRP reduces to a TSP which consists of finding a least-cost (often called tour) circuit including all the required vertices and the depot. In any TSP feasible solution on graph G, each vertex of $U \cup \{0\}$ appears at least once and two successive vertices of $U \cup \{0\}$ are linked by a least-cost path. As a consequence, the TSP can be reformulated on an auxiliary complete directed graph $G' = (V', A')$, where $V' = U \cup \{0\}$ is the vertex set and A' is the arc set. With each arc $(i, j) \in A'$ is associated a cost c_{ij} equal to that of a least-cost path from i to j in G. These costs satisfy the *triangle inequality*:

$$c_{ij} \leq c_{ik} + c_{kj}, \forall (i, j) \in A', \forall k \in V', k \neq i, j.$$

Because of this property, there exists a TSP optimal solution which is a Hamiltonian tour in G', that is, a cycle in which each vertex in V' appears exactly once. In what follows, the search for an optimal or suboptimal TSP solution is restricted to Hamiltonian tours.

If $c_{ij} = c_{ji}$ for each pair of distinct vertices $i, j \in V'$, the TSP is said to be *symmetric* (STSP), otherwise it is called *asymmetric* (ATSP). The STSP is suitable for inter-city transport, while the ATSP is recommended in urban settings because of one-way streets. Of course, the solution techniques developed for the ATSP can also be applied to the STSP. This method could, however, be very inefficient, as explained later. It is therefore customary to deal with the two cases separately.

6.8.1.1 The asymmetric travelling salesman problem

Let x_{ij}, $(i, j) \in A'$, be a binary decision variable equal to 1 if arc (i, j) is part of the solution, 0 otherwise. The ATSP can then be formulated as follows:

$$\text{Minimize} \sum_{(i,j) \in A'} c_{ij} x_{ij}$$

subject to

$$\sum_{i \in V' \setminus \{j\}} x_{ij} = 1, \quad j \in V' \tag{6.59}$$

$$\sum_{j \in V' \setminus \{i\}} x_{ij} = 1, \quad i \in V' \tag{6.60}$$

$$\sum_{i \in S} \sum_{j \notin S} x_{ij} \geq 1, \quad S \subset V', \ |S| \geq 2 \tag{6.61}$$

$$x_{ij} \in \{0, 1\}, \quad (i, j) \in A'.$$

Equations (6.59) and (6.60) are referred to as *degree constraints*. Constraints (6.59) mean that a unique arc enters each vertex $j \in V'$. Similarly, constraints (6.60) state that a single arc exits each vertex $i \in V'$. Constraints (6.61) guarantee that the tour has at least one arc coming out from each proper and a non-empty subset S of vertices in V' (*connectivity constraints*). They are redundant for $|S| = 1$ because of constraints (6.60). It is worth noting that the number of constraints (6.61) is $2^{|V'|} - |V'| - 2$. Such constraints can be formulated in an alternative way, algebraically equivalent (see Problem 6.19):

$$\sum_{i \in S} \sum_{j \in S} x_{ij} \leq |S| - 1, \;\; S \subset V', \; |S| \geq 2. \tag{6.62}$$

Inequalities (6.62) prevent the formation of subtours containing less than $|S|$ vertices (*subtour elimination constraints*).

A lower bound The ATSP has been shown to be NP-hard. A good lower bound on the ATSP optimal solution cost z^*_{ATSP} can be obtained by removing constraints (6.61) from ATSP formulation. The optimal solution of the relaxed problem can be obtained by solving the following *assignment problem* (AP):

$$\text{Minimize} \sum_{i \in V'} \sum_{j \in V'} c_{ij} x_{ij} \tag{6.63}$$

subject to

$$\sum_{i \in V'} x_{ij} = 1, \;\; j \in V' \tag{6.64}$$

$$\sum_{j \in V'} x_{ij} = 1, \;\; i \in V' \tag{6.65}$$

$$x_{ij} \in \{0, 1\}, \;\; i, j \in V', \tag{6.66}$$

where, in the objective function (6.63), $c_{ii} = \infty$, $i \in V'$, in order to force $x^*_{ii} = 0$, for all $i \in V'$.

Note that, due to the particular structure of the constraints (6.64) and (6.65), the relations (6.66) can be replaced with

$$x_{ij} \geq 0, \;\; i, j \in V'.$$

The optimal AP solution x^*_{AP} corresponds to a collection of p directed subtours $C_1, \ldots, C_p$, spanning all vertices of the directed graph G'. If $p = 1$, the AP solution is feasible (and hence optimal) for the ATSP.

As a rule, z^*_{AP} is a good lower bound on z^*_{ATSP} if the cost matrix is strongly asymmetric (in this case, it has been empirically demonstrated that the deviation $(z^*_{\text{ATSP}} - z^*_{\text{AP}})/z^*_{\text{AP}}$ from the optimal solution cost is often less than 1%). In contrast, in the case of symmetric costs, the deviation is typically 30% or more. The reason of this behavior can be explained by the fact that for symmetric costs, if the AP solution contains arc $(i, j) \in A'$, then the AP optimal solution is likely to include arc $(j, i) \in A'$ too. As a result, the optimal AP solution usually shows several small subtours of only two vertices and is quite different from the ATSP optimal solution.

Bontur is a pastry producer that was founded in Prague, Czech Republic, in the 19th century. The firm currently operates, in addition to four modern plants, a workshop in Gorazdova street where the founder started working. The workshop serves Prague and its surroundings. Every day at 6:30 AM, a fleet of vans carries the pastries from the workshop to several retail outlets (small shops, supermarkets and hotels). In particular, all outlets of the Vltava River district are usually served by a single vehicle. For the sake of simplicity, arc transport costs are assumed to be proportional to arc lengths. In Figure 6.20 the road network is modelled as a mixed graph $G = (V, A)$, where a length d_{ij} is associated with each arc or edge (i, j). The workshop and the vehicle depot are located in vertex 0. Last March 23, seven shops (located at vertices 1, 3, 9, 18, 20 and 22) needed to be supplied. The problem can be formulated as an ATSP on a complete directed graph $G' = (V', A')$, where V' is formed by the seven vertices associated to the customers and by vertex 0. With each arc $(i, j) \in A'$ is associated a cost c_{ij} corresponding to the length of the shortest path from i to j on G (see Table 6.14). The optimal AP solution x^*_{AP} is made up of the following three subtours (see Figure 6.21):

$$C_1 = \{(1, 4), (4, 3), (3, 9), (9, 1)\},$$

of cost equal to 11.0 km;

$$C_2 = \{(0, 18), (18, 0)\},$$

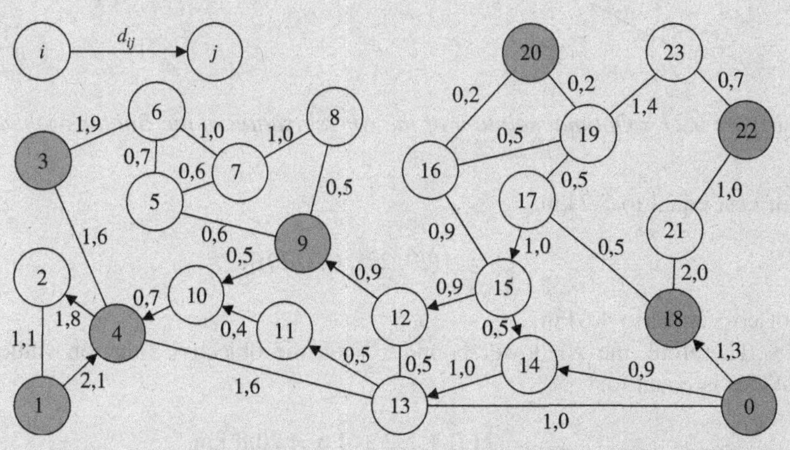

Figure 6.20 A graph representation of a Bontur distribution problem (one-way street segments are represented by arcs, while two-way street segments are modelled through edges). The depot and the service vertices are in grey.

Table 6.14 Shortest path length (in km) from i to j, $i, j \in V'$, in the Bontur problem.

	0	1	3	4	9	18	20	22
0	0.0	5.5	4.2	2.6	2.4	1.3	2.5	4.3
1	4.7	0.0	3.7	2.1	5.1	6.0	7.2	9.0
3	4.2	4.5	0.0	1.6	3.2	5.5	6.7	8.5
4	2.6	2.9	1.6	0.0	3.0	3.9	5.1	6.9
9	3.8	4.1	2.8	1.2	0.0	5.1	6.3	8.1
18	3.9	7.4	6.1	4.5	3.3	0.0	1.2	3.0
20	3.5	7.0	5.7	4.1	2.9	1.2	0.0	2.3
22	5.8	9.3	8.0	6.4	5.2	3.0	2.3	0.0

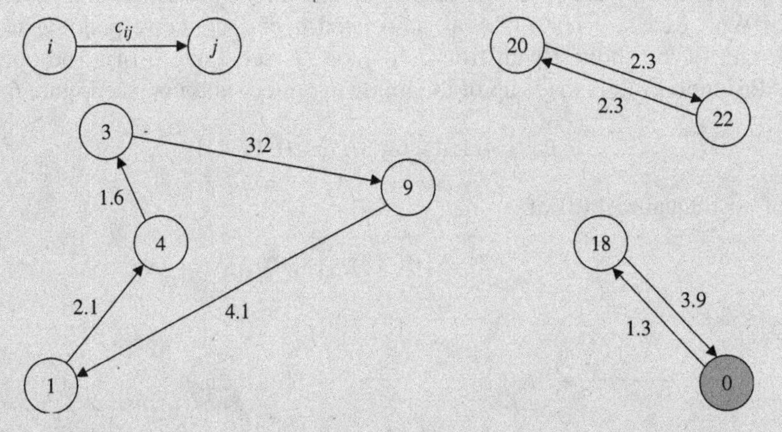

Figure 6.21 Optimal solution of the AP relaxation of the Bontur problem.

of cost equal to 5.2 km;

$$C_3 = \{(20, 22), (22, 20)\},$$

of cost equal to 4.6 km.

Therefore, the AP lower bound z^*_{AP} on the objective function value of ATSP is equal to:

$$z^*_{AP} = 11.0 + 5.2 + 4.6 = 20.8 \text{ km}.$$

Patching heuristic The patching heuristic works as follows. First, the AP relaxation is solved. If a single tour is obtained, the procedure stops (the AP solution is the optimal ATSP solution). Otherwise, a feasible ATSP solution is constructed by merging the subtours of the AP solution. When merging two subtours, one arc is removed from each subtour and two new arcs are added in such a way that a single connected subtour is obtained.

Step 0. *Initialization.* Let $C = \{C_1, \ldots, C_p\}$ be the set of the p subtours in the AP optimal solution. If $p = 1$, *STOP*. The AP solution is feasible (and hence optimal) for the ATSP.

Step 1. Identify the two subtours $C_h, C_k \in C$ with the largest number of vertices.

Step 2. Merge C_h and C_k in such a way that the cost increase is kept at minimum. Update C and let $p = p - 1$. If $p = 1$, *STOP*, an ATSP feasible solution has been determined, otherwise go back to Step 1.

In order to find a feasible solution $\bar{x}_{\text{ATSP}}$ to the Bontur distribution problem, the patching heuristic is applied to the AP solution shown in Figure 6.21. At the first iteration, C_1 and C_2 are selected to be merged (alternatively, C_3 could be used instead of C_2). By merging C_1 and C_2 at minimum cost (through the removal of arcs $(3, 9)$ and $(18, 0)$ and the insertion of arcs $(3, 0)$ and $(18, 9)$), the following subtour (having length equal to 16.6 km) is obtained (see Figure 6.22):

$$C_4 = \{(0, 18), (18, 9), (9, 1), (1, 4), (4, 3), (3, 0)\}.$$

The partial solution, formed by the two subtours C_3 and C_4, is depicted in Figure 6.22. The total length increases by 0.4 km with respect to the initial solution. At the end of the second iteration, the two subtours in Figure 6.22 are merged at the minimum cost increase of 0.3 km through

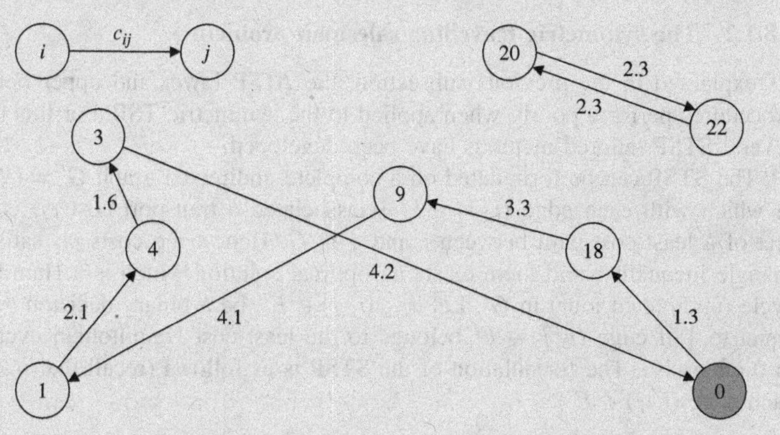

Figure 6.22 Partial solution at the end of the first iteration of the patching heuristic in the Bontur problem.

the removal of arcs (18, 9) and (20, 22) and the insertion of arcs (18, 22) and (20, 9).

This way, a feasible ATSP solution of cost $\bar{z}_{ATSP} = 21.5$ km is obtained (see Figure 6.23). In order to evaluate the quality of the heuristic solution, the following deviation from the AP lower bound can be computed:

$$\frac{\bar{z}_{ATSP} - z^*_{AP}}{z^*_{AP}} = \frac{21.5 - 20.8}{20.8} = 0.0337,$$

which corresponds to a percentage deviation of 3.37%.

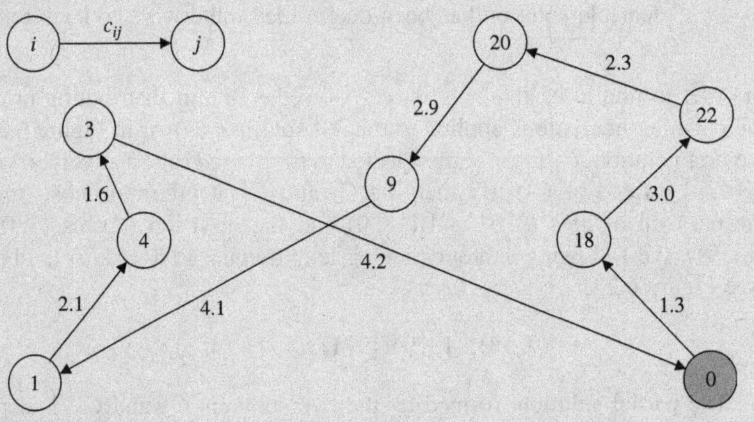

Figure 6.23 Feasible Hamiltonian cycle (obtained by the patching heuristic) of the ATSP associated to the Bontur problem.

6.8.1.2 The symmetric travelling salesman problem

As explained in the previous subsection, the ATSP lower and upper bounding procedures perform poorly when applied to the symmetric TSP. For this reason, several STSP tailored methods have been developed.

The STSP can be formulated on a complete undirected graph $G' = (V', E')$, in which with each edge $(i, j) \in E'$ is associated a transport cost c_{ij} equal to that of a least-cost path between i and j in G. Hence, the costs c_{ij} satisfy the triangle inequality, and there exists an optimal solution which is a Hamiltonian cycle (undirected tour) in G'. Let x_{ij}, $(i, j) \in E'$, be a binary decision variable equal to 1 if edge $(i, j) \in E'$ belongs to the least-cost Hamiltonian cycle, and to 0 otherwise. The formulation of the STSP is as follows (recall that $i < j$ for each edge $(i, j) \in E'$):

$$\text{Minimize} \sum_{(i,j)\in E'} c_{ij}x_{ij}$$

subject to

$$\sum_{i \in V':(i,j) \in E'} x_{ij} + \sum_{i \in V':(j,i) \in E'} x_{ji} = 2, \ j \in V' \tag{6.67}$$

$$\sum_{(i,j) \in E':i \in S, j \notin S} x_{ij} + \sum_{(j,i) \in E':i \in S, j \notin S} x_{ji} \geq 2, \ S \subset V', \ 2 \leq |S| \leq \lceil |V'|/2 \rceil \tag{6.68}$$

$$x_{ij} \in \{0, 1\}, \ (i, j) \in E'.$$

Equations (6.67) mean that exactly two edges must be incident to every vertex $j \in V'$ (*degree constraints*). Inequalities (6.68) state that, for every vertex subset S, there exist at least two edges with one endpoint in $S \subset V'$ and the other endpoint in $V' \setminus S$ (*connectivity constraints*). Since the connectivity constraints of a subset S and that of its complement $V' \setminus S$ are equivalent, one has to consider only inequalities (6.68) associated to subsets $S \subset V'$ such that $|S| \leq \lceil |V'|/2 \rceil$. Constraints (6.68) are redundant if $|S| = 1$ because of (6.67). Alternatively, the connectivity constraints (6.68) can be replaced with the following equivalent *subtour elimination constraints*:

$$\sum_{(i,j) \in E':i,j \in S} x_{ij} + \sum_{(j,i) \in E':j,i \in S} x_{ji} \leq |S| - 2, \ S \subset V', \ 2 \leq |S| \leq \lceil |V'|/2 \rceil.$$

A lower bound The STSP is an NP-hard problem. A lower bound on the optimal solution cost z^*_{STSP} can be obtained by solving the following problem (see Problem 6.22):

$$\text{Minimize} \sum_{(i,j) \in E'} c_{ij} x_{ij} \tag{6.69}$$

subject to

$$\sum_{i \in V':(i,r) \in E'} x_{ir} + \sum_{i \in V':(r,i) \in E'} x_{ri} = 2 \tag{6.70}$$

$$\sum_{(i,j) \in E':i \in S, i \neq r, j \notin S, j \neq r} x_{ij} + \sum_{(j,i) \in E':i \in S, i \neq r, j \notin S, j \neq r} x_{ji} \geq 1,$$

$$S \subset V', \ 1 \leq |S| \leq \lceil |V'|/2 \rceil \tag{6.71}$$

$$x_{ij} \in \{0, 1\}, \ (i, j) \in E', \tag{6.72}$$

where $r \in V'$ is arbitrarily chosen (*root vertex*). Model (6.69)–(6.72) corresponds to a minimum spanning r-tree problem (MSrTP), for which the optimal solution is a least-cost connected subgraph spanning G' and such that vertex $r \in V'$ has degree two. The MSrTP can be solved in $O(|V'|^2)$ steps with the following procedure:

Step 1. Determine a minimum-cost tree $T^* = (V' \setminus \{r\}, E_T)$ spanning $V' \setminus \{r\}$.

Step 2. Insert in T^* vertex r as well as the two least-cost edges incident to vertex r.

Saint-Martin distributes fresh fishing products in Normandy, France. Last June 7, the company received seven orders from sales points all located in northern Normandy. It was decided to serve the seven requests by means of a single vehicle sited in Betteville. The problem can be formulated as an STSP on a complete graph $G' = (V', E')$, where V' is composed of eight vertices corresponding to the sales points and of vertex 0 associated with the depot. With each edge $(i, j) \in E'$ is associated a cost c_{ij} equal to the shortest distance between vertices i and j (see Table 6.15). The minimum spanning r-tree is depicted in Figure 6.24, to which corresponds a cost $z^*_{\text{MS}r\text{TP}} = 225.8$ km.

Table 6.15 Distances (in km) between terminals in the Saint-Martin problem.

	Betteville	Bolbec	Dieppe	Fécamp	Le Havre	Luneray	Rouen	Valmont
Betteville	0.0	27.9	54.6	42.0	56.5	37.0	30.9	34.1
Bolbec		0.0	67.2	25.6	28.8	48.4	57.4	21.6
Dieppe			0.0	60.5	95.8	18.8	60.4	52.1
Fécamp				0.0	39.4	43.1	70.2	12.2
Le Havre					0.0	77.2	84.5	44.4
Luneray						0.0	51.6	34.0
Rouen							0.0	59.3
Valmont								0.0

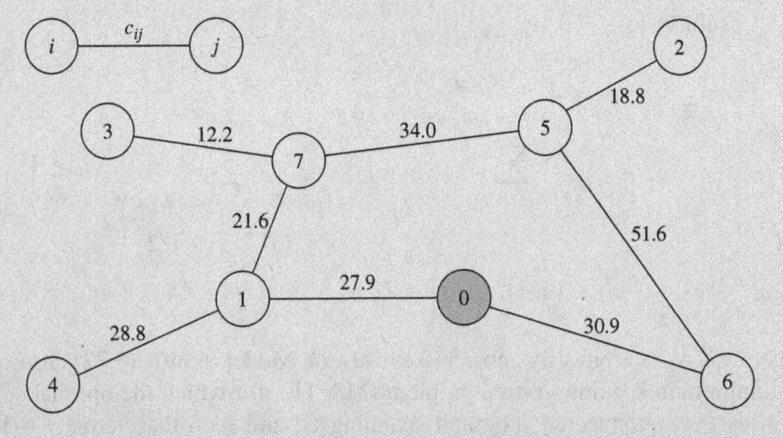

Figure 6.24 Minimum-cost spanning r-tree in the Saint-Martin problem.

The MSrTP lower bound can be improved in two ways. In the former method, the MSrTP relaxation is solved for more choices of the root $r \in V'$ and then the largest MSrTP lower bound is selected. In the latter method, $r \in V'$ is fixed but each constraint (6.67) with the only exception of $j = r$ is relaxed in a Lagrangian fashion. Let λ_j, $j \in V' \setminus \{r\}$, be the Lagrangian multiplier attached to vertex $j \in V' \setminus \{r\}$. A Lagrangian relaxation of the STSP is as follows:

$$\text{Minimize} \sum_{(i,j)\in E'} c_{ij}x_{ij} + \sum_{j\in V'\setminus\{r\}} \lambda_j\Big(\sum_{i\in V':(i,j)\in E'} x_{ij} + \sum_{i\in V':(j,i)\in E'} x_{ji} - 2\Big) \tag{6.73}$$

subject to

$$\sum_{i\in V':(i,r)\in E'} x_{ir} + \sum_{i\in V':(r,i)\in E'} x_{ri} = 2 \tag{6.74}$$

$$\sum_{(i,j)\in E':i\in S, i\neq r, j\notin S, j\neq r} x_{ij} + \sum_{(j,i)\in E':i\in S, i\neq r, j\notin S, j\neq r} x_{ji} \geq 1,$$

$$S \subset V',\ 1 \leq |S| \leq \lceil |V'|/2 \rceil \tag{6.75}$$

$$x_{ij} \in \{0, 1\},\ (i, j) \in E'. \tag{6.76}$$

Setting arbitrarily $\lambda_r = 0$, the objective function (6.73) can be rewritten as

$$\sum_{(i,j)\in E'} \left(c_{ij} + \lambda_i + \lambda_j\right) x_{ij} - 2\sum_{j\in V'} \lambda_j. \tag{6.77}$$

To determine the optimal Lagrangian multipliers (or at least a set of good multipliers), a suitable variant of the subgradient method illustrated in Section 3.3.2.1 can be used. In particular, at the hth iteration the updating formula of the Lagrangian multipliers is the following:

$$\lambda_j^{(h+1)} = \lambda_j^{(h)} + \beta^{(h)} s_j^{(h)},\ j \in V' \setminus \{r\},$$

where:

$$s_j^{(h)} = \sum_{i\in V':(i,j)\in E'} x_{ij}^{(h)} + \sum_{i\in V':(j,i)\in E'} x_{ji}^{(h)} - 2,\ j \in V' \setminus \{r\},$$

$x_{ij}^{(h)}$, $(i, j) \in E'$, is the optimal solution of the Lagrangian relaxation MSrTP(λ) (6.77), (6.74)–(6.76) at the hth iteration, and $\beta^{(h)}$ can be set equal to

$$\beta^{(h)} = \frac{1}{h},\ h = 1, \ldots$$

The results of the first three iterations of the subgradient method in the Saint-Martin problem ($r = 0$) are

$$\lambda_j^{(1)} = 0, j \in V' \setminus \{r\};$$

$$z^*_{\text{MS}r\text{TP}(\lambda^{(1)})} = 225.8;$$

$$s^{(1)} = [1; -1; -1; -1; 1; 0; 1]^{\text{T}};$$

$$\beta^{(1)} = 1;$$

$$\lambda^{(2)} = [1; -1; -1; -1; 1; 0; 1]^{\text{T}};$$

$$z^*_{\text{MS}r\text{TP}(\lambda^{(2)})} = 231.8;$$

$$s^{(2)} = [1; -1; -1; -1; 1; 0; 1]^{\text{T}};$$

$$\beta^{(2)} = 0.5;$$

$$\lambda^{(3)} = [3/2; -3/2; -3/2; -3/2; 3/2; 0; 3/2]^{\text{T}};$$

$$z^*_{\text{MS}r\text{TP}(\lambda^{(3)})} = 234.8.$$

Nearest neighbour heuristic The nearest neighbour heuristic is a simple constructive procedure that builds a Hamiltonian path by iteratively linking the vertex inserted at the previous iteration to its nearest unrouted neighbour. Finally, a Hamiltonian cycle is obtained by connecting the two endpoints of the path. The nearest neighbour heuristic often provides low-quality solutions, since the edges added in the final iterations may be very costly.

Step 0. Set $C = \{r\}$, where $r \in V'$ is a vertex chosen arbitrarily, and set $h = r$.

Step 1. Identify the vertex $k \in V' \setminus S$ such that $c_{hk} = \min_{j \in V' \setminus C} \{c_{hj}, c_{jh}\}$. Add k at the end of C.

Step 2. If $|C| = |V'|$, add r at the end of C, *STOP* (C corresponds to a Hamiltonian cycle), otherwise let $h = k$ and go back to Step 1.

In order to find a feasible solution $\bar{x}_{\text{STSP}}$ to the Saint-Martin distribution problem, the nearest neighbour heuristic is applied ($r = 0$), and the following Hamiltonian cycle is obtained (see Figure 6.25):

$$C = \{(0, 1), (1, 7), (3, 7), (3, 4), (4, 5), (2, 5), (2, 6), (0, 6)\},$$

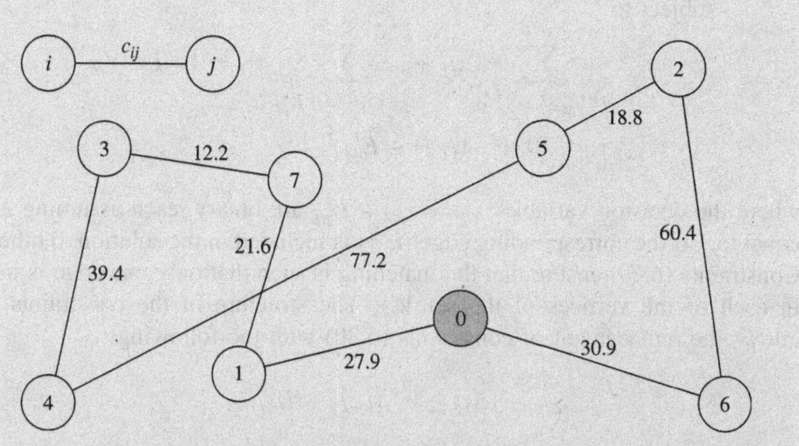

Figure 6.25 Feasible Hamiltonian cycle generated by the nearest neighbour heuristic in the Saint-Martin problem.

whose cost is 288.4 km. The deviation of this solution cost from the available lower bound LB = 234.8 km is

$$\frac{\bar{z}_{\text{STSP}} - \text{LB}}{\text{LB}} = \frac{288.4 - 234.8}{234.8} = 0.2283,$$

which corresponds to a percentage deviation of 22.83%.

Christofides heuristic The Christofides heuristic is a constructive procedure that works as follows:

Step 1. Compute a minimum-cost tree $T = (V', E'_T)$ spanning the vertices of $G' = (V', E')$. Let z^*_T be the cost of this tree.

Step 2. Compute a least-cost perfect-matching $M = (V'_M, E'_M)$ among the vertices of odd degree in the tree T ($|V'_M|$ is always an even number). Let z^*_M be the optimal matching cost. Let $U = (V', E'_U)$ be the Eulerian subgraph (or multigraph) of G' induced by the union of the edges of T and M (i.e. $E'_U = E'_T \cup E'_M$).

Step 3. Determine a Eulerian cycle C_E on $U = (V', E'_U)$. Observe that $z(C_E) = z^*_T + z^*_M$.

Step 4. Extract a Hamiltonian cycle C_H from C_E.

The least-cost perfect-matching problem (Step 2 of the Christofides heuristic) corresponds to solve the following problem:

$$\text{Minimize} \sum_{(i,j)\in E'_M} c_{ij}x_{ij} \tag{6.78}$$

subject to

$$\sum_{i \in V'_M:(i,j) \in E'_M} x_{ij} + \sum_{i \in V'_M:(j,i) \in E'_M} x_{ji} = 1, \ j \in V'_M \tag{6.79}$$

$$x_{ij} \in \{0, 1\}, \ (i, j) \in E'_M, \tag{6.80}$$

where the decision variables x_{ij}, $(i, j) \in E'_M$ are binary, each assuming a value equal to 1 if the corresponding edge (i, j) is included in the solution, 0 otherwise. Constraints (6.79) ensure that the matching is such that only one edge is incident in each of the vertices of the set V'_M. The structure of the constraints (6.79) allows the replacement of constraints (6.80) with the following:

$$x_{ij} \geq 0, \ (i, j) \in E'_M. \tag{6.81}$$

The problem (6.78), (6.79), (6.81) can be solved exactly in polynomial time (see Problem 6.23) with an algorithm whose a complexity is $O(|V'_M|)^3$.

A Eulerian cycle on a Eulerian (multi)graph $U = (V', E'_U)$ (Step 3 of the Christofides heuristic) can be obtained by using the following *end-pairing procedure*

Step 1. Determine a spanning of the edges E'_U of U, defined by the set $C = \{C_1, \ldots, C_p\}$, $p \geq 1$, of cycles; in each of them every edge is crossed exactly once.

Step 2. If $|C| = 1$, *STOP*, the cycle obtained is Eulerian.

Step 3. Identify two cycles in C which contain at least one common vertex. Merge the two cycles, so as to obtain a single cycle in which each edge of the two cycles is crossed exactly once; update C and go back to Step 2.

At Step 1 of this procedure, to find a spanning of the edges E'_U of U, it is sufficient to visit the (multi)graph (in width or in depth) until a vertex already visited is reached; in this way, an element of the set C is obtained and, removing such edges from E'_U, once again a (multi)graph is obtained, eventually not connected, in which all the vertices are again of even degree. It is possible to repeat the visit of the (multi)graph to determine a cycle, until the spanning of the whole (multi)graph is obtained. At Step 3, the vertex in common between the two cycles acts as a pivot. The merge of the two cycles into one can be obtained by visiting the edges of the first cycle up to the pivot, traversing all the edges of the second cycle, returning to the pivot and continuing the visit with the remaining edges of the first cycle.

At Step 4 of the Christofides heuristic, a Hamiltonian cycle C_H can be easily obtained from the Eulerian cycle C_E by removing any repeated vertices from C_E. In particular, let $C_E = \{i_0, i_1, \ldots, i_{m-1}, i_m = i_0\}$ and suppose that there is at least one repeated vertex, that is, $i_h = i_k$, $h < k$. In this case, the vertex i_k

can be removed from C_E, which means, in the sequence of edges that form C_E, the pair of edges (i_{k-1}, i_k), (i_k, i_{k+1}) can be replaced by the edge (i_{k-1}, i_{k+1}) (*shortcut*). This technique typically involves a cost reduction (or at least not an increase), since the triangle inequality holds. For this reason, $z(C_H) \leq z(C_E)$.

Moreover, note that, by simply reversing the order in which the vertices appear in C_E, in some cases, a second Hamiltonian cycle can be obtained whose cost can be compared to that of the first cycle. Finally, it can be shown that the cost of the Christofides solution is at most 50% higher than the optimal solution cost (see Problem 6.24).

The Saint-Martin distribution problem is solved by means of the Christofides heuristic. The minimum spanning tree is made up of edges $\{(0, 1), (0, 6), (1, 4), (1, 7), (2, 5), (3, 7), (5, 7)\}$, and has a cost of 174.2 km. The optimal matching of the odd-degree vertices (1, 2, 3, 4, 6 and 7) is composed of edges $(1, 4)$, $(2, 6)$ and $(3, 7)$, and has a cost of 101.4 km. The Eulerian multigraph generated at the end of Step 3 of the Christofides heuristic is shown in Figure 6.26. At Step 4 of the heuristic, the following Eulerian cycle is found:

$$C_E = \{0, 1, 4, 1, 7, 3, 7, 5, 2, 6, 0\},$$

with cost $z(C_E) = 174.2 + 101.4 = 275.6$ km. Starting from C_E, the following two Hamiltonian cycles are found:

$$C_{H^{(1)}} = \{0, 1, 4, 7, 3, 5, 2, 6, 0\};$$
$$C_{H^{(2)}} = \{0, 6, 2, 5, 7, 3, 1, 4, 0\},$$

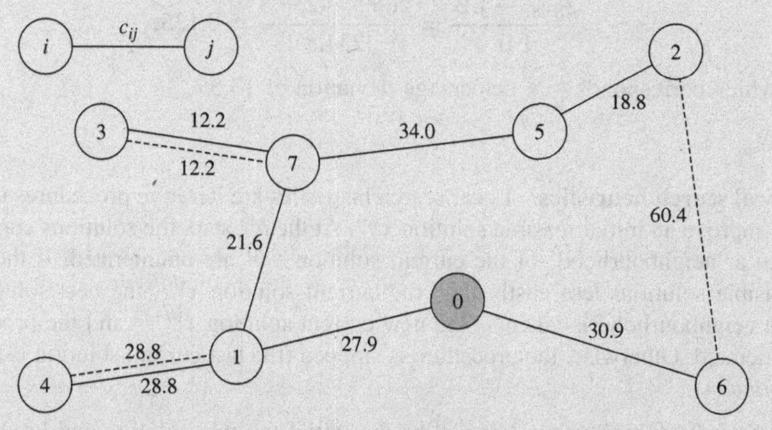

Figure 6.26 Eulerian multigraph generated at Step 2 of the Christofides heuristic in the Saint-Martin proble (edges of the minimum-cost spanning tree are full lines and minimum-cost matching edges are dashed lines).

with $C_{H^{(2)}}$ obtained by removing the vertices repeated from the sequence of vertices $\{0, 6, 2, 5, 7, 3, 7, 1, 4, 1, 0\}$, corresponding to C_E reversed. We will obtain

$$z(C_{H^{(1)}}) = 266.5 \text{ km};$$

$$z(C_{H^{(2)}}) = 267.2 \text{ km}.$$

The chosen Hamiltonian cycle is, therefore, $C_{H^{(1)}}$ (see Figure 6.27).

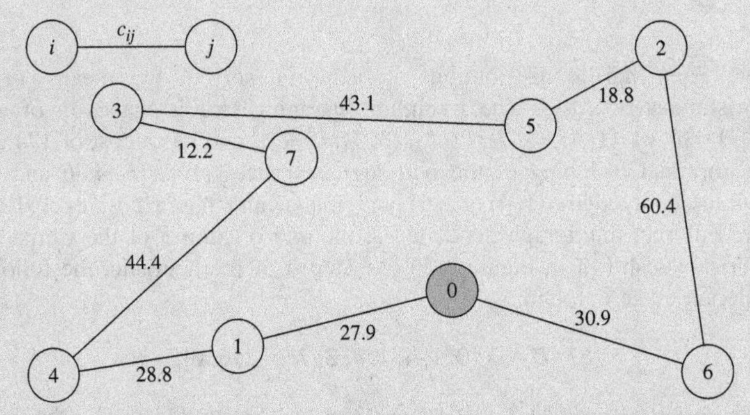

Figure 6.27 Hamiltonian cycle provided by the Christofides heuristic in the Saint-Martin problem.

In order to evaluate the quality of the heuristic solution, the following deviation from the LB can be computed:

$$\frac{z_{STSP} - LB}{LB} = \frac{266.5 - 234.8}{234.8} = 0.135,$$

which corresponds to a percentage deviation of 13.5%.

Local search heuristics Local search heuristics are iterative procedures that try to improve an initial feasible solution $x^{(0)}$. At the hth step, the solutions contained into a 'neighbourhood' of the current solution $x^{(h)}$ are enumerated. If there are feasible solutions less costly than the current solution $x^{(h)}$, the best solution of the neighbourhood is taken as the new current solution $x^{(h+1)}$ and the procedure is iterated. Otherwise, the procedure is stopped (the last current solution is a *local optimum*).

Step 0. *Initialization.* Let $x^{(0)}$ be the initial feasible solution and let $N(x^{(0)})$ be its neighbourhood. Set $h = 0$.

Step 1. Enumerate the feasible solutions belonging to $N(x^{(h)})$. Select the best feasible solution $x^{(h+1)} \in N(x^{(h)})$.

Step 2. If the cost of $x^{(h+1)}$ is less than that of $x^{(h)}$, set $h = h + 1$ and go back to Step 1; otherwise, *STOP*, $x^{(h)}$ is the best solution found.

For the STSP, $N(x^{(h)})$ is commonly defined as the set of all Hamiltonian cycles that can be obtained by substituting k edges $(2 \le k \le |V'|)$ of $x^{(h)}$ for k other edges in E' (*k-exchange*) (see Figure 6.28).

Step 0. Let $C^{(0)}$ be the initial Hamiltonian cycle, and let $z^{(0)}_{\text{STSP}}$ be the cost of $C^{(0)}$. Set $h = 0$.

Step 1. Identify the best feasible solution $C^{(h+1)}$ that can be obtained through a k-exchange. If $z^{(h+1)}_{\text{STSP}} \ge z^{(h)}_{\text{STSP}}$, *STOP*, $C^{(h)}$ is a Hamiltonian cycle for the STSP.

Step 2. Let $h = h + 1$, and go back to Step 1.

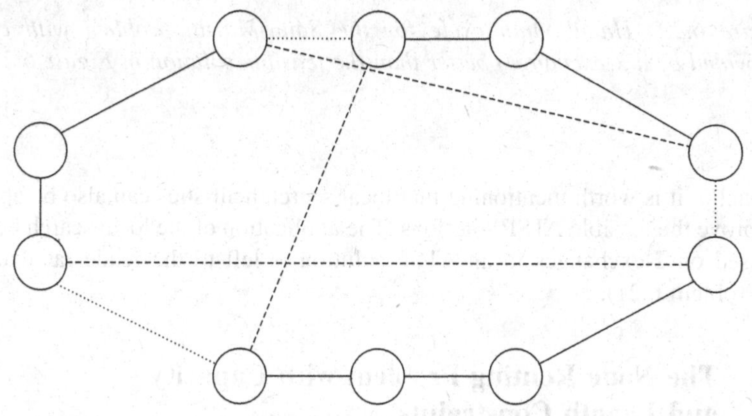

Figure 6.28 A feasible 3-exchange (dotted edges are removed, dashed edges are inserted).

In a local search heuristic based on k-exchanges, k can be constant or can be dynamically increased in order to intensify the search when improvements are likely to occur. If k is constant, each execution of Step 1 requires $O(|V'|^k)$ operations. In general, k is set equal to 2 or 3 at most, in order to limit the computational effort.

If a 2-exchange procedure is applied to the solution provided by the Christofides heuristic in the Saint-Martin problem, a less costly Hamiltonian cycle (see Figure 6.29) is obtained at the first iteration by replacing edges (3, 5) and (4, 7) with edges (3, 4) and (5, 7). As a consequence, the solution cost decreases by 14.1 km.

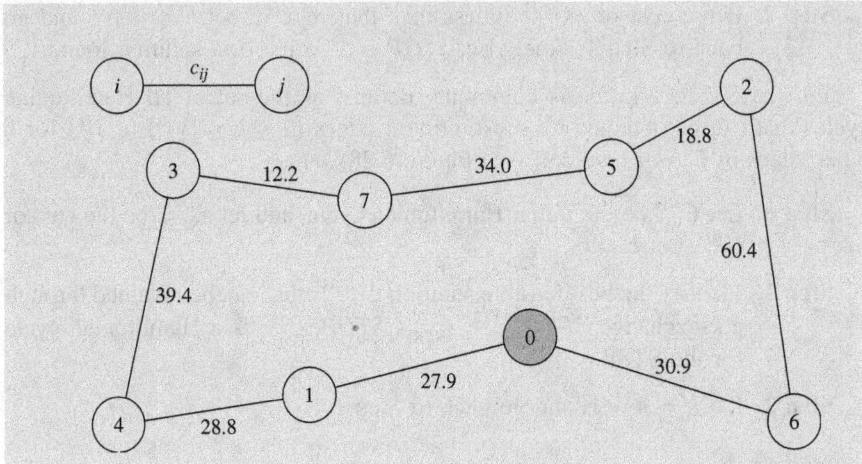

Figure 6.29 Hamiltonian cycle for the Saint-Martin problem with cost (provided by a 2-exchange) better than the feasible solution in Figure 6.27.

Finally, it is worth mentioning that local search heuristics can also be applied to improve the feasible ATSP solutions. The application of the local search heuristic based on 2-exchanges to an ATSP solution is left to the reader as exercise (see Problem 6.21).

6.8.2 The Node Routing Problem with Capacity and Length Constraints

As previously illustrated, in several settings, operational constraints come into play when designing vehicle routes. These restrictions lead to a large number of variants, and the algorithms described in literature are often dependent on the type of constraints. For this reason, in the remainder of this subsection the most important constrained NRPs are examined and a limited number of techniques, representative of the most used methods, are described. As usual, the interested reader is referred to the relevant scientific literature for further information.

The *Node Routing Problem with Capacity and Length Constraints* (NRPCL) consists of designing a set of m least-cost vehicle routes starting and ending at the depot, such that:

- each customer is visited exactly once;
- a demand p_i is associated to each customer i, $i \in U$ (demands are either collected or delivered, but not both); then the total demand of each vehicle cannot exceed a given vehicle capacity q (vehicles are assumed to be identical) (*capacity constraints*);

- a service time s_i is associated to each customer i, $i \in U$; then the total duration of each route, including service and travel times, may not exceed a given work shift duration T.

This problem can be formulated on a complete directed graph $G' = (V', A')$ or on a complete undirected graph $G' = (V', E')$ depending on whether the cost matrix is asymmetric or symmetric. In both cases, the vertex set V' is composed of the depot 0 and the customers in U. In what follows, the focus is on the symmetric version of the problem.

The NRPCL can be formulated by suitably modifying the STSP model:

$$\text{Minimize} \sum_{(i,j)\in E'} c_{ij} x_{ij}$$

subject to

$$\sum_{i\in V':(i,j)\in E'} x_{ij} + \sum_{i\in V':(j,i)\in E'} x_{ji} = 2, \ j \in U \tag{6.82}$$

$$\sum_{i\in V':(0,i)\in E'} x_{0i} = 2m \tag{6.83}$$

$$\sum_{(i,j)\in E':i\in S, j\notin S} x_{ij} + \sum_{(j,i)\in E':i\in S, j\notin S} x_{ji} \geq 2\alpha(S), \ S \subseteq U, |S| \geq 2 \tag{6.84}$$

$$\sum_{(i,j)\in E':i\in S, j\notin S} x_{ij} + \sum_{(j,i)\in E':i\in S, j\notin S} x_{ji} \geq 4, \ S \subseteq U, \ |S| \geq 2,$$

$$t^*_{\text{STSP}}(S) > T \tag{6.85}$$

$$x_{ij} \in \{0, 1\}, \ (i, j) \in E'.$$

Constraints (6.82) state that two edges are incident to each customer $j \in U$ (*customer degree constraints*). Similarly, constraints (6.83) guarantee that $2m$ edges are incident to vertex 0 (*depot degree constraints*). Capacity constraints (6.84) impose that the number of vehicles serving customers in S is at least twice the lower bound $\alpha(S)$ on the optimal solution value of a 1-BP problem (see Section 5.4.2.1) with $|S|$ items having weights p_i, $i \in S$ and bins of capacity q. In practice, it is common to use $\alpha(S) = \left\lceil \left(\sum_{i\in S} p_i\right) / q \right\rceil$. Length constraints (6.85) state that a single route is not sufficient to serve all the customers in S whenever the duration $t^*_{\text{STSP}}(S)$ of a least-cost Hamiltonian cycle spanning $S \cup \{0\}$ exceeds T.

6.8.2.1 Set partitioning and covering formulations

An alternative formulation of the NRPCL can be obtained as follows. Let K be the set of routes in G' satisfying the capacity and length constraints and let c_k,

$k \in K$, be the cost of route k. Define a_{ik}, $i \in U$, $k \in K$, as a binary constant equal to 1 if vertex i is included into route k, and to 0 otherwise. Let y_k, $k \in K$, be a binary decision variable equal to 1 if route k is used in an optimal solution, and to 0 otherwise. The NRPCL can be reformulated as a *setpartitioning problem* (NRPSP) in the following way:

$$\text{Minimize} \sum_{k\in K} c_k y_k$$

subject to

$$\sum_{k\in K} a_{ik} y_k = 1 \ , i \in U \tag{6.86}$$

$$y_k \in \{0, 1\}, \ k \in K.$$

Constraints (6.86) establish that each customer $i \in U$ must be served. Moreover, the following constraint could be considered:

$$\sum_{k\in K} y_k = m,$$

which limits the number of routes that can be activated to m.

The NRPSP is very flexible as it can be easily modified in order to include additional operational constraints. Its main weakness is the large number of variables, especially for 'weakly constrained' problems. For example, if $p_i = 1$, $i \in U$, and the length constraints are not binding, $|K| = O\left(\sum_{h=1}^{q} \binom{|V'|}{h}\right)$. Consequently, even if $|U| = 50$ and $q = 10$, there can be several billion decision variables. However, in some applications the characteristics of the operational constraints can considerably reduce $|K|$. This happens very often, for example, in fuel distribution where the demand of a user $i \in U$ (a gas pump) is customarily a small part (usually a half or a third) of a vehicle capacity. Therefore, the number of customers visited in each route is typically at most three. As a consequence, $|K| = O\left(\binom{|V'|}{3} + \binom{|V'|}{2} + \binom{|V'|}{1}\right) = O(|V'|^3)$.

It is easy to show (see Problem 6.26) that constraints (6.86) can be replaced with the following relations:

$$\sum_{k\in K} a_{ik} y_k \geq 1, \ i \in U,$$

in which case an easier-to-solve *setcovering* formulation (NRPSC) is obtained.

Bengalur Oil manufactures and distributes fuel to filling stations in the Karnataka region of India. Last July 2, the firm received five orders (see Table 6.16). The distances between the gas stations and the firm's depot are reported in Table 6.17. The vehicles have a capacity of 160 hectolitres.

Table 6.16 Orders (in hectolitres) received by Bengalur Oil.

Gas station	Order
1	50
2	75
3	50
4	50
5	75

Table 6.17 Distance (in km) between the gas stations and the firm's depot in the Bengalur Oil problem (depot corresponds to vertex 0).

		Gas station				
	0	1	2	3	4	5
0	0	90	100	90	80	80
1		0	10	20	10	30
2			0	10	20	40
3				0	10	30
4					0	20
5						0

In order to formulate the problem as a NRPSC, the feasible routes are enumerated (see Table 6.18; note that, since the problem is symmetric, the routes in which the customers appear in reverse order with respect to those reported in the table are not considered).

Table 6.18 Feasible routes in the Bengalur Oil problem.

ID	Route	Cost (km)	ID	Route	Cost (km)
1	{0, 1, 0}	180	10	{0, 2, 3, 0}	200
2	{0, 2, 0}	200	11	{0, 2, 4, 0}	200
3	{0, 3, 0}	180	12	{0, 2, 5, 0}	220
4	{0, 4, 0}	160	13	{0, 3, 4, 0}	180
5	{0, 5, 0}	160	14	{0, 3, 5, 0}	200
6	{0, 1, 2, 0}	200	15	{0, 4, 5, 0}	180
7	{0, 1, 3, 0}	200	16	{0, 1, 3, 4, 0}	200
8	{0, 1, 4, 0}	180	17	{0, 1, 4, 3, 0}	200
9	{0, 1, 5, 0}	200	18	{0, 3, 1, 4, 0}	200

The NRPSC model is

$$\begin{aligned}\text{Minimize } & 180y_1 + 200y_2 + 180y_3 + 160y_4 + 160y_5 + 200y_6 + 200y_7 \\ & +180y_8 + 200y_9 + 200y_{10} + 200y_{11} + 220y_{12} + 180y_{13} \\ & +200y_{14} + 180y_{15} + 200y_{16} + 200y_{17} + 200y_{18}\end{aligned}$$

subject to

$$y_1 + y_6 + y_7 + y_8 + y_9 + y_{16} + y_{17} + y_{18} \geq 1$$

$$y_2 + y_6 + y_{10} + y_{11} + y_{12} \geq 1$$

$$y_3 + y_7 + y_{10} + y_{13} + y_{14} + y_{16} + y_{17} + y_{18} \geq 1$$

$$y_4 + y_8 + y_{11} + y_{13} + y_{15} + y_{16} + y_{17} + y_{18} \geq 1$$

$$y_5 + y_9 + y_{12} + y_{14} + y_{15} \geq 1$$

$$y_1,\ y_2,\ y_3,\ y_4,\ y_5,\ y_6,\ y_7,\ y_8,\ y_9,$$

$$y_{10},\ y_{11},\ y_{12},\ y_{13},\ y_{14},\ y_{15},\ y_{16},\ y_{17},\ y_{18} \in \{0, 1\}.$$

A feasible solution to the problem can be determined by applying the heuristic procedure illustrated in Section 3.3.5. Considering that $c_k > 0$, $k = 1, \ldots, 18$, and that the variables y_k, $k = 1, \ldots, 18$, appear in all the constraints with unit coefficient, the following ratios are calculated:

$$\begin{array}{ll}
c_1/n_1 = 180/1 = 180; & c_2/n_2 = 200/1 = 200; \\
c_3/n_3 = 180/1 = 180; & c_4/n_4 = 160/1 = 160; \\
c_5/n_5 = 160/1 = 160; & c_6/n_6 = 200/2 = 100; \\
c_7/n_7 = 200/2 = 100; & c_8/n_8 = 180/2 = 90; \\
c_9/n_9 = 200/2 = 100; & c_{10}/n_{10} = 200/2 = 100; \\
c_{11}/n_{11} = 200/2 = 100; & c_{12}/n_{12} = 220/2 = 110; \\
c_{13}/n_{13} = 180/2 = 90; & c_{14}/n_{14} = 200/2 = 100; \\
c_{15}/n_{15} = 180/2 = 90; & c_{16}/n_{16} = 200/3 = 66.66; \\
c_{17}/n_{17} = 200/3 = 66.66; & c_{18}/n_{18} = 200/3 = 66.66.
\end{array}$$

Recall that n_k, $k = 1, \ldots, 18$, represents the number of constraints in which the corresponding decision variable y_k appears with a unit coefficient. The ratios with the lowest values are obtained for $k = 16$, 17 and 18; $k = 16$ is chosen arbitrarily, $\bar{y}_{16} = 1$ is set, and constraints 1, 3 and 4 are removed from the original model, obtaining the following problem:

$$\begin{aligned}\text{Minimize } & 180y_1 + 200y_2 + 180y_3 + 160y_4 + 160y_5 + 200y_6 + 200y_7 \\ & +180y_8 + 200y_9 + 200y_{10} + 200y_{11} + 220y_{12} + 180y_{13} + 200y_{14} \\ & +180y_{15} + 200y_{17} + 200y_{18}\end{aligned}$$

subject to

$$y_2 + y_6 + y_{10} + y_{11} + y_{12} \geq 1$$

$$y_5 + y_9 + y_{12} + y_{14} + y_{15} \geq 1$$

$$y_1,\ y_2,\ y_3,\ y_4,\ y_5,\ y_6,\ y_7,\ y_8,\ y_9, y_{10},$$

$$y_{11},\ y_{12},\ y_{13},\ y_{14},\ y_{15},\ y_{17},\ y_{18} \in \{0, 1\}.$$

Coefficients c_1, c_3, c_4, c_7, c_8, c_{13}, c_{17} and c_{18} are greater than zero and the corresponding variables do not appear in any constraint, so we can set $\bar{y}_1 = \bar{y}_3 = \bar{y}_4 = \bar{y}_7 = \bar{y}_8 = \bar{y}_{13} = \bar{y}_{17} = \bar{y}_{18} = 0$. The following ratios are then calculated:

$$c_2/n_2 = 200/1 = 200; \quad c_5/n_5 = 160/1 = 160;$$
$$c_6/n_6 = 200/1 = 200; \quad c_9/n_9 = 200/1 = 200;$$
$$c_{10}/n_{10} = 200/1 = 200; \quad c_{11}/n_{11} = 200/1 = 200;$$
$$c_{12}/n_{12} = 220/2 = 110; \quad c_{14}/n_{14} = 200/1 = 200;$$
$$c_{15}/n_{15} = 180/1 = 180.$$

The ratio with the lowest value is obtained for $k = 12$. For this reason, we set $\bar{y}_{12} = 1$; hence, constraints 1 and 2 are removed. The problem becomes

$$\text{Minimize } 200y_2 + 160y_5 + 200y_6 + 200y_9 + 200y_{10}$$
$$+200y_{11} + 200y_{14} + 180y_{15}$$

subject to

$$y_2,\ y_5,\ y_6,\ y_9,\ y_{10},\ y_{11},\ y_{14},\ y_{15} \in \{0, 1\}.$$

This means that $\bar{y}_2 = \bar{y}_5 = \bar{y}_6 = \bar{y}_9 = \bar{y}_{10} = \bar{y}_{11} = \bar{y}_{14} = \bar{y}_{15} = 0$ are set. Hence, the feasible solution found provides for the activation of routes 12 and 16, with a total cost of 420 km (see Figure 6.30), which, in this case, corresponds to the optimal solution of the problem.

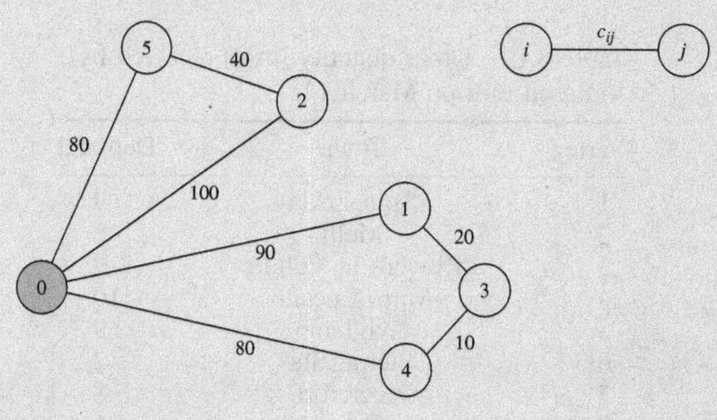

Figure 6.30 Optimal solution of the Bengalur Oil problem.

The setcovering formulation can also be used to find a suboptimal solution to the NRPCL. Let K' a subset of K of feasible routes, with the condition that each customer $i \in U$ belongs to at least one route $k \in K'$. The following relaxed problem is obtained:

$$\text{Minimize} \sum_{k \in K'} c_k y_k \tag{6.87}$$

subject to

$$\sum_{k \in K'} a_{ik} y_k \geq 1 \ , i \in U \tag{6.88}$$

$$y_k \in \{0, 1\} \, , k \in K', \tag{6.89}$$

whose optimal solution y^*, since $K' \subset K$, could provide for some customer $i \in U$ that $\sum_{k \in K'} a_{ik} y_k = p > 1$, in which case it is sufficient to remove customer i from $p - 1$ routes where he is present (this operation does not prejudice the feasibility of the solution). The solution y^* obtained in this way, since the problem (6.87)–(6.89) is a relaxation of the NRPCL, is just a feasible solution of NRPCL, and its quality clearly depends on the choice of the subset K' of feasible routes.

Last March 14, the Italian oil mill Viola had to plan, starting from the production plant located in Venosa, the delivery of extra-virgin oil drums to its customers situated in 10 towns: Spinazzola, Melfi, Rionero in Vulture, Muro Lucano, Avigliano, Pietragalla, Acerenza, Tolve, Genzano di Lucania and Palazzo S. Gervasio. A vehicle with a capacity of 35 quintals is used for the deliveries. The amounts to be delivered are reported in Table 6.19, whereas Tables 6.20 and 6.21 indicate the shortest distances between each pair of towns involved.

Table 6.19 Order quantity (in q) received by Viola oil mill on March 14.

Vertex	Town	Demand
1	Spinazzola	9
2	Melfi	8
3	Rionero in Vulture	9
4	Muro Lucano	10
5	Avigliano	9
6	Pietragalla	12
7	Acerenza	8
8	Tolve	6
9	Genzano di Lucania	7
10	Palazzo San Gervasio	5

Table 6.20 Part I–Shortest distance (in km) between towns in the Viola problem.

	Venosa	Spinazzola	Melfi	Rionero in Vulture	Muro Lucano	Avigliano
Venosa	0	27	22	18	63	48
Spinazzola		0	49	45	90	67
Melfi			0	13	58	42
Rionero in Vulture				0	42	31
Muro Lucano					0	34
Avigliano						0

Table 6.21 Part II–Shortest distance (in km) between towns in the Viola problem.

	Pietragalla	Acerenza	Tolve	Genzano di Lucania	Palazzo S. Gervasio
Venosa	52	38	56	33	19
Spinazzola	52	40	41	19	12
Melfi	47	52	67	55	41
Rionero in Vulture	35	40	56	49	36
Muro Lucano	62	75	67	85	81
Avigliano	21	34	39	50	55
Pietragalla	0	16	29	30	48
Acerenza		0	25	15	32
Tolve			0	21	38
Genzano di Lucania				0	15
Palazzo S. Gervasio					0

To find a feasible solution to the routing problem, a subset of 16 feasible routes is generated (reported in Table 6.22). Assigning a binary decision variable y_k, $k = 1, \dots, 16$, to each of the feasible routes reported in Table 6.22, the following relaxed NRPCL problem is obtained:

$$\text{Minimize } 98y_1 + 53y_2 + 145y_3 + 121y_4 + 119y_5 + 110y_6 + 67y_7 + 58y_8 + 107y_9 + 170y_{10} + 140y_{11} + 80y_{12} + 118y_{13} + 161y_{14} + 138y_{15} + 158y_{16}$$

subject to

$$y_1 + y_8 + y_9 + y_{12} + y_{15} \geq 1$$

$$y_1 + y_2 + y_9 + y_{11} + y_{16} \geq 1$$

$$y_2 + y_9 + y_{11} \geq 1$$
$$y_3 + y_{10} + y_{11} \geq 1$$
$$y_3 + y_4 + y_{10} + y_{14} + y_{16} \geq 1$$
$$y_4 + y_{10} + y_{14} \geq 1$$
$$y_5 + y_{13} + y_{14} + y_{15} \geq 1$$
$$y_5 + y_6 + y_{13} + y_{14} + y_{15} + y_{16} \geq 1$$
$$y_6 + y_7 + y_{12} + y_{13} + y_{15} + y_{16} \geq 1$$
$$y_7 + y_8 + y_{12} + y_{13} + y_{15} + y_{16} \geq 1$$
$$y_1,\ y_2,\ y_3,\ y_4,\ y_5,\ y_6,\ y_7,\ y_8,\ y_9,$$
$$y_{10},\ y_{11},\ y_{12},\ y_{13},\ y_{14},\ y_{15},\ y_{16} \in \{0, 1\},$$

Table 6.22 Feasible routes of the Viola problem.

ID	Route	Cost (km)	ID	Route	Cost (km)
1	{0, 1, 2, 0}	98	9	{0, 1, 2, 3, 0}	107
2	{0, 2, 3, 0}	53	10	{0, 4, 5, 6, 0}	170
3	{0, 4, 5, 0}	145	11	{0, 2, 3, 4, 0}	140
4	{0, 5, 6, 0}	121	12	{0, 1, 9, 10, 0}	80
5	{0, 7, 8, 0}	119	13	{0, 7, 8, 9, 10, 0}	118
6	{0, 8, 9, 0}	110	14	{0, 5, 6, 8, 7, 0}	161
7	{0, 9, 10, 0}	67	15	{0, 7, 8, 9, 10, 1, 0}	138
8	{0, 1, 10, 0}	58	16	{0, 2, 5, 8, 9, 10, 0}	158

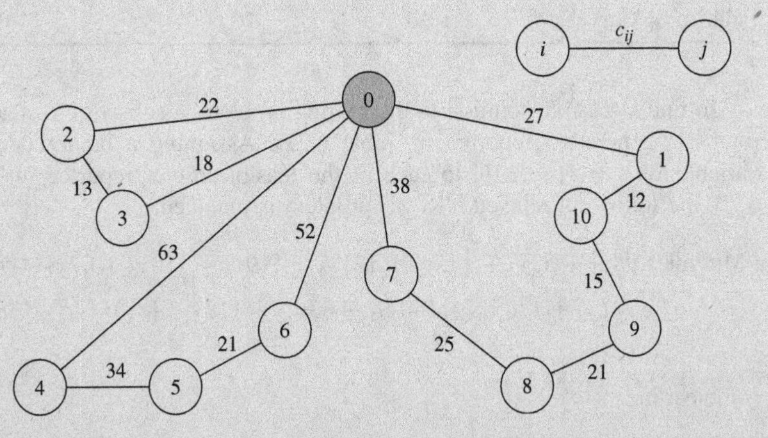

Figure 6.31 Feasible solution of the Viola problem.

whose optimal solution (see Figure 6.31) is the following: $y_2^* = 1$, $y_{10}^* = 1$, $y_{15}^* = 1$ (the value of the remaining decision variables is zero), with $z(y^*) =$ 361 km. Note that with this solution (which, as noted above, may not be optimal for the routing problem considered), all the constraints of the model are satisfied for equality (i.e. each customer is served by one activated route).

6.8.2.2 Constructive heuristics

In the sequel, some constructive heuristics procedures for the NRPCL are illustrated.

'Cluster first, route second' heuristics 'Cluster first, route second' heuristics attempt to determine a good NRPCL solution in two steps. First, customers are partitioned into subsets $U_k \subset U$, each of which is associated to a vehicle $k = 1, \ldots, m$. Second, for each vehicle $k = 1, \ldots, m$, the STSP on the complete subgraph induced by $U_k \cup \{0\}$ is solved (exactly or heuristically). The partitioning of the customer set can be made visually or by more formalized procedures (as the one proposed by Fisher and Jaikumar). Readers interested in a deeper examination of this subject are referred to the relevant scientific literature.

In the Bengalur Oil problem, customers can be partitioned into two clusters:

$$U_1 = \{2, 5\},$$
$$U_2 = \{1, 3, 4\}.$$

Then, two STSPs are solved on the complete subgraphs induced by $U_1 \cup \{0\}$ and $U_2 \cup \{0\}$, respectively. The result of the 'cluster first, route second' heuristic is already shown in Figure 6.30 and corresponds to the optimal solution.

'Route first, cluster second' heuristics 'Route first, cluster second' heuristics attempt to determine an NRPCL solution in two stages. Firstly, a single Hamiltonian cycle (generally infeasible for the NRPCL) is generated through an exact or heuristic STSP algorithm. Then, the cycle is decomposed into m feasible routes, originating and terminating at the depot. The route decomposition can be performed visually or by means of formalized procedures, like the one proposed by Beasley. Readers interested in this method should consult the relevant scientific literature.

Applying a 'route first, cluster second' heuristic to the Bengalur Oil problem, a Hamiltonian cycle, having cost equal to 240 km, is generated (Figure 6.32). At the second stage, the cycle is decomposed into two feasible routes, which are the same as those illustrated in Figure 6.30, corresponding also in this case to the optimal solution.

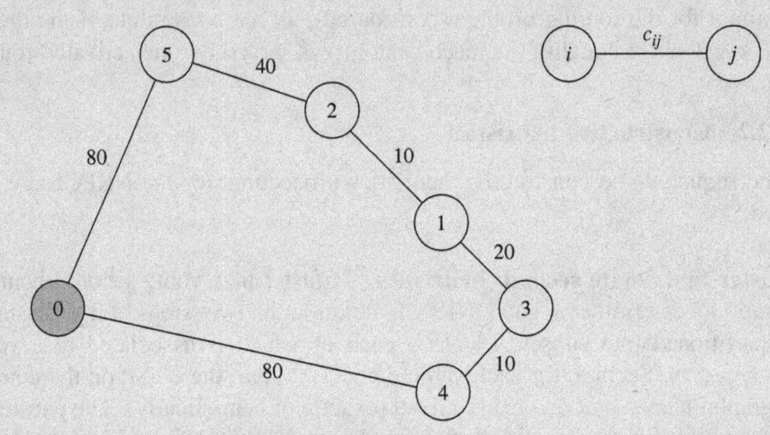

Figure 6.32 Infeasible Hamiltonian cycle generated at the end of the first phas of the 'route first, cluster second' heuristic in the Bengalur Oil problem.

Savings heuristic The savings heuristic is an iterative procedure that initially generates $|U|$ distinct routes each of which serves a single customer. At each subsequent iteration, the heuristic attempts to merge a pair of routes in order to obtain a cost reduction (a *saving*). The cost saving s_{ij} achieved when servicing customers i and j, $i, j \in U$ on one route, as opposed to servicing them individually is (see Figure 6.33):

$$s_{ij} = c_{0i} + c_{0j} - c_{ij}, \ \ i, j \in U, \ \ i \neq j. \tag{6.90}$$

It is worth noting that s_{ij}, $i, j \in U$, are non-negative since the triangle inequality holds for all costs c_{ij}, $i, j \in E'$. The savings formula still holds if $i \in U$ is the last customer of the first route involved in a merge, and $j \in U$ is the first customer of the second route. The heuristic stops when it is no more possible to merge feasibly a pair of routes.

Step 0. *Initialization.* Let C be the set of $|U|$ initial routes $C_i = \{0, i, 0\}$, $i \in U$. For each pair of vertices $i, j \in U$, $i \neq j$, compute the saving s_{ij} by using Equations (6.90). Let L be the list of savings sorted in a non-increasing fashion (since $s_{ij} = s_{ji}$, $i, j \in U$, $i \neq j$, then the list L contains only one saving value for each pair of different customers).

Step 1. Extract from the top of list L a saving s_{ij}. If vertices i and j belong to two separate routes of C in which i and j are directly linked to the depot (see e.g. Figure 6.34), and if the route obtained by replacing

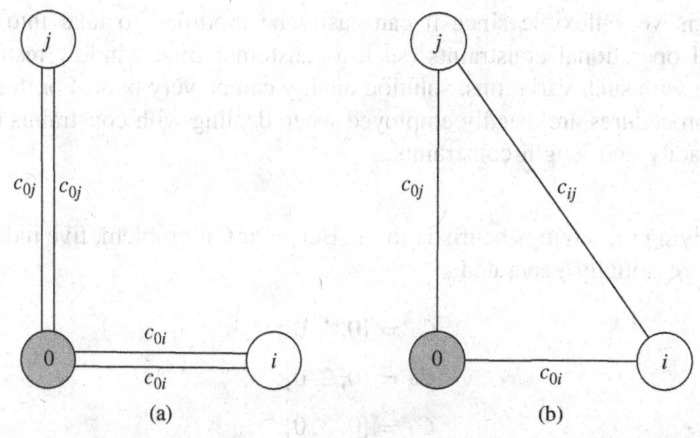

Figure 6.33 Computation of saving s_{ij}: (a) the cost of the two distinct routes is $2c_{0i} + 2c_{0j}$; (b) the cost of the merged route is $c_{0i} + c_{0j} - c_{ij}$.

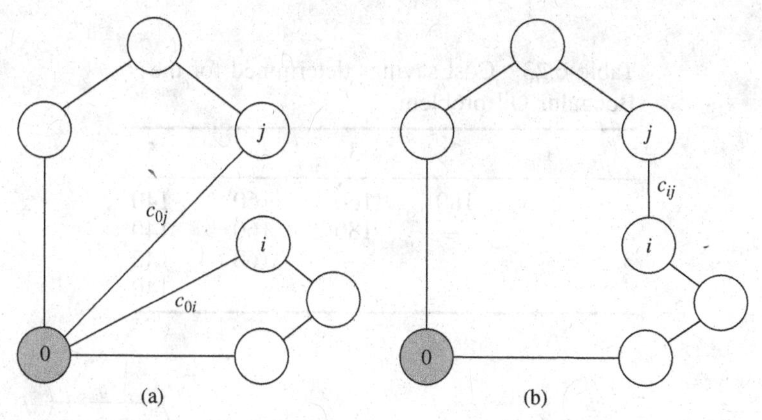

Figure 6.34 Merging two routes in a single route (the saving $s_{ij} = c_{0i} + c_{0j} - c_{ij}$).

edges $(0, i)$ and $(0, j)$ with edge (i, j) is feasible, then merge the two routes and update C.

Step 2. If $L = \emptyset$, *STOP*, it is not possible to merge further pairs of routes. If $L \neq \emptyset$, go back to Step 1.

The computational complexity of the heuristic is determined by the saving sorting phase and is therefore $O(|V'|^2 \log |V'|^2) = O(|V'|^2 \log |V'|)$. In practice, the heuristic is very fast as it takes less than a second on the most common computers to solve a problem with hundreds of customers. However, the quality of the solutions can be poor. According to extensive computational experiments, the error made by the savings heuristic is typically in the 5–20% range. The savings heuristic is very flexible since it can easily be modified to take into account additional operational constraints (such as customer time window restrictions). However, with such variations, solution quality can be very poor. For this reason, tailored procedures are usually employed when dealing with constraints different from capacity and length constraints.

By applying the savings heuristic to the Bengalur Oil problem, five individual routes are initially generated:

$$C_1 = \{0, 1, 0\};$$

$$C_2 = \{0, 2, 0\};$$

$$C_3 = \{0, 3, 0\};$$

$$C_4 = \{0, 4, 0\};$$

$$C_5 = \{0, 5, 0\}.$$

Table 6.23 Cost savings determined for the Bengalur Oil problem.

	1	2	3	4	5
1	–	180	160	160	140
2		–	180	160	140
3			–	160	140
4				–	140

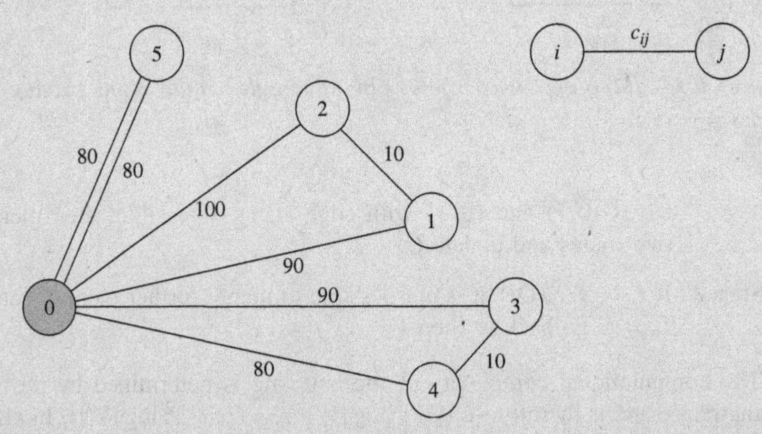

Figure 6.35 Solution provided by the savings heuristic for the Bengalur Oil problem.

Then savings s_{ij}, $i, j \in U$, $i \neq j$ are calculated (see Table 6.23) and list L is initialized:

$$L = \{s_{12}, s_{23}, s_{13}, s_{14}, s_{24}, s_{34}, s_{15}, s_{25}, s_{35}, s_{45}\}.$$

Subsequently, routes C_1 and C_2 are merged while savings s_{23}, s_{13}, s_{14}, s_{24} are discarded. Then, saving s_{34} is implemented by merging routes C_3 and C_4. At this stage there are no further feasible route merges. The final solution has a cost of 540 km (see Figure 6.35).

6.8.3 The Node Routing and Scheduling Problem with Time Windows

In several settings, customers need to be serviced within specified time windows. This is the case, for example, of retail outlets that cannot be replenished during busy periods. In the simplest version of *Node Routing and Scheduling Problem with Time Windows* (NRSPTW), each customer specifies a single time window, while in other variants each customer can set multiple time windows (e.g. a time window in the morning and one in the afternoon). Let e_i, $i \in U$, be the *earliest time* at which service can start at customer i, and let l_i, $i \in U$, be the *latest time* (or *deadline*) at which service must start at customer i. Similarly, let e_0 be the earliest time at which vehicles can leave the depot, and let l_0 be the deadline within which vehicles must return to the depot. In the NRSPTW, the service starting time at each customer $i \in U$ is a decision variable b_i. If a vehicle arrives too early at a customer $j \in U$, it has to wait. Therefore, b_j, $j \in U$, is given by

$$b_j = \max\{e_j, b_i + s_i + t_{ij}\}, j \in U,$$

where i is the customer visited just before j, t_{ij} is the quickest travel time between customers i and j and s_i is the service time of customer i.

It is worth noting, that even though travel costs and times are symmetric, a solution is made up of a set of directed cycles, because of the time windows that do not allow reversals of route orientations.

In what follows, a constructive heuristic is illustrated for the NRSPTW and a tabu search procedure capable of providing high-quality solutions to a large number of constrained NRPs is described.

6.8.3.1 Insertion heuristics

Insertion-type heuristics are among the most efficient for the NRSPTW. Among them, the I1 Solomon heuristic is described. The heuristic builds a feasible solution by constructing one route at a time. At each iteration it is decided which new customer $u^* \in U$ has to be inserted in the current solution, and between which adjacent customers $i(u^*)$ and $j(u^*)$ the new customer u^* has to be inserted on the current route. When choosing u^*, the heuristic takes into account both the cost increase associated with the insertion of u^*, and the delay in service time at customers following u^* on the route.

Step 0. *Initialization.* The first route is initially $C_1 = \{0, \bar{i}, 0\}$, where $\bar{i}$ is the customer with the earliest deadline. Set $k = 1$.

Step 1. Let $C_k = \{i_0, i_1, \ldots, i_m\}$ be the current route, where $i_0 = i_m = 0$. Set

$$f_1(i_{p-1}, u, i_p) = \alpha(c_{i_{p-1}u} + c_{ui_p} - \mu c_{i_{p-1}i_p}) + (1 - \alpha)(b^u_{i_p} - b_{i_p}), \tag{6.91}$$

where $0 \leq \alpha \leq 1$, $\mu \geq 0$ and $b^u_{i_p}$ is the time when service begins at customer i_p provided that customer u is inserted between i_{p-1} and i_p. For each unrouted customer u, compute its best feasible insertion position in route C_k as

$$f_1(i(u), u, j(u)) = \min_{p=1,\ldots,m} f_1(i_{p-1}, u, i_p),$$

where $i(u)$ and $j(u)$ are the two adjacent vertices of the current route between which u should to be inserted. Determine the best unrouted customer u^* to be inserted, yielding

$$f_2(i(u^*), u^*, j(u^*)) = \max_u\{f_2(i(u), u, j(u)\},$$

where

$$f_2(i(u), u, j(u)) = \lambda c_{0u} - f_1(i(u), u, j(u)) \tag{6.92}$$

with $\lambda \geq 0$.

Step 2. Insert customer u^* in route C_k between $i(u^*)$ and $j(u^*)$ and go back to Step 1. If u^* does not exist, but there are still unrouted customers, set $k = k + 1$, initialize a new route C_k (as in Step 0) and go back to Step 1. Otherwise, *STOP*, a feasible solution of the NRSPTW has been found.

The insertion heuristic tries to maximize the benefit obtained when servicing a customer on the current route rather than on a direct route. For example, when $\mu = \alpha = \lambda = 1$, Equation (6.92) corresponds to the saving in distance from servicing customer u on the same route as customers i and j rather than using an individual route. The best feasible insertion place of an unrouted customer is determined by minimizing a measure, defined by Equation (6.91), of the extra distance and the extra time required to visit it. Different values of the parameters μ, α and λ lead to different possible criteria for selecting the customer to be inserted and its best position in the current route.

McNish is a chain of supermarkets located in Scotland. Last October 13, the warehouse situated in Aberdeen was required to serve 12 sales points located in Banchory, Clova, Cornhill, Dufftown, Fyvie, Huntly, Newbyth, Newmill, Peterhead, Strichen, Towie and Turriff. The number of requested pallets and the time windows within which service was allowed are reported in Table 6.24, whereas distances and travel times on the fastest routes between the cities are reported in Tables 6.25–6.28. Each vehicle has a capacity of 30 pallets. It leaves the warehouse at 9:00 AM and must return to the warehouse by 2:00 PM. Service time (time needed for unloading a vehicle) can be assumed to be 15 min on average for every sales point, regardless of demand. For the sake of simplicity, time is computed in minutes starting at 9:00 AM (e.g. 11:00 AM corresponds to 120 min). The I1 insertion heuristic (with parameters $\alpha = 0.9$, $\mu = \lambda = 1$) provides the following results. Initially,

$$C_1 = \{0, 9, 0\}.$$

Table 6.24 Number of pallets and time windows for the sales point in the McNish problem.

Vertex	Sales point	Number of pallets	Time window
1	Banchory	9	9:00 AM–11:30 AM
2	Clova	7	9:00 AM–12:30 PM
3	Cornhill	5	9:00 AM–12:30 PM
4	Dufftown	4	11:00 AM– 2:30 PM
5	Fyvie	8	9:00 AM–12:30 PM
6	Huntly	8	11:00 AM– 1:30 PM
7	Newbyth	7	9:00 AM–12:30 PM
8	Newmill	6	10:00 AM–12:30 PM
9	Peterhead	6	9:00 AM–10:30 AM
10	Strichen	6	9:00 AM–11:15 AM
11	Towie	4	10:00 AM–12:45 PM
12	Turriff	6	10:00 AM–12:30 PM

Table 6.25 Part I–Distances (in km) between cities computed on the fastest route in the McNish problem.

	Aberdeen	Banchory	Clova	Cornhill	Dufftown	Fyvie	Huntly
Aberdeen	0.0	28.4	58.1	84.7	83.4	41.1	63.0
Banchory		0.0	48.1	89.7	76.1	52.7	67.9
Clova			0.0	46.3	34.1	48.6	24.5
Cornhill				0.0	36.8	34.9	23.2
Dufftown					0.0	53.8	21.8
Fyvie						0.0	33.4
Huntly							0.0

Table 6.26 Part II–Distances (in km) between cities computed on the fastest route in the McNish problem.

	Newbyth	Newmill	Peterhead	Strichen	Towie	Turriff
Aberdeen	59.5	31.8	51.5	56.7	62.1	55.1
Banchory	75.3	36.8	82.5	87.8	47.5	66.7
Clova	66.1	48.6	95.3	80.0	13.0	51.9
Cornhill	35.1	55.2	70.0	47.0	53.3	21.0
Dufftown	63.4	64.4	100.4	77.3	41.0	49.2
Fyvie	22.7	21.0	50.2	34.8	55.5	14.2
Huntly	42.9	43.9	79.9	56.8	31.4	28.7
Newbyth	0.0	43.5	39.8	16.9	72.9	15.1
Newmill		0.0	50.3	43.9	52.7	34.9
Peterhead			0.0	24.7	100.6	47.8
Strichen				0.0	86.9	28.4
Towie					0.0	58.8
Turriff						0.0

Table 6.27 Part I–Travel times (in minutes) on the fastest route between cities in the McNish problem.

	Aberdeen	Banchory	Clova	Cornhill	Dufftown	Fyvie	Huntly
Aberdeen	0	34	74	89	87	51	65
Banchory		0	58	96	87	63	72
Clova			0	56	42	70	31
Cornhill				0	43	42	27
Dufftown					0	65	26
Fyvie						0	43
Huntly							0

At the first iteration, for each unrouted customer the values shown in Table 6.29 are obtained, from which we have that $C_1 = \{0, 9, 10, 0\}$.

At the second iteration, the values shown in Table 6.30, and $C_1 = \{0, 9, 10, 7, 0\}$, are obtained. At the third iteration, we obtain the values shown in Table 6.31, and the route $C_1 = \{0, 9, 10, 7, 12, 0\}$, with a cost of 163.3 km.

At the end of the third iteration, the feasibility constraints are not satisfied for any unrouted customer, so a new route, $C_2 = \{0, 1, 0\}$, is initialized. After another three iterations, $C_2 = \{0, 1, 8, 5, 3, 0\}$ is obtained, with a cost of 205.8 km. Finally, the last route will be $C_3 = \{0, 11, 2, 4, 6, 0\}$ with a cost of 194 km. The final solution covers 563.1 km and corresponds to the scheduling shown in Tables 6.32–6.34.

Table 6.28 Part II–Travel times (in minutes) on the fastest route between cities in the McNish problem.

	Newbyth	Newmill	Peterhead	Strichen	Towie	Turriff
Aberdeen	70	37	58	62	67	67
Banchory	90	44	88	92	54	79
Clova	80	61	117	97	19	64
Cornhill	38	68	80	52	61	27
Dufftown	74	67	113	91	47	58
Fyvie	27	28	63	43	75	16
Huntly	51	44	90	68	36	35
Newbyth	0	54	49	21	84	18
Newmill		0	63	57	63	44
Peterhead			0	30	118	59
Strichen				0	102	35
Towie					0	69
Turriff						0

Table 6.29 Best feasible insertion on route C_1 for each unrouted customer in the McNish problem at the first iteration of the I1 heuristic.

u	$f_1(i, u, j)$	$i(u)$	$j(u)$	$f_2(i, u, j)$
2	134.5	9	0	−60.5
3	112.5	9	0	−23.5
4	143.5	9	0	−56.5
5	57.5	9	0	−6.5
6	98.5	9	0	−33.5
7	62.5	9	0	7.5
8	43.5	9	0	−6.5
10	35.5	9	0	26.5
11	128.5	9	0	−61.5
12	69.5	9	0	−2.5

Table 6.30 Best feasible insertion on route C_1 for each unrouted customer in the McNish problem at the second iteration of the I1 heuristic.

u	$f_1(i, u, j)$	$i(u)$	$j(u)$	$f_2(i, u, j)$
3	80.5	10	0	8.5
5	33.5	10	0	17.5
6	72.5	10	0	−7.5
7	30.5	10	0	39.5
8	33.5	10	0	3.5
12	41.5	10	0	25.5

Table 6.31 Best feasible insertion on route C_1 for each unrouted customer in the McNish problem at the third iteration of the I1 heuristic.

u	$f_1(i, u, j)$	$i(u)$	$j(u)$	$f_2(i, u, j)$
3	58.5	7	0	30.5
5	9.5	7	0	41.5
6	47.5	7	0	17.5
8	22.5	7	0	14.5
12	16.5	7	0	50.5

Table 6.32 Schedule of the first route in the McNish problem.

City	Arrival (hh:mm)	Departure (hh:mm)	Cumulated load (pallets)
Aberdeen	–	9:00 AM	0
Peterhead	9:58 AM	10:13 AM	6
Strichen	10:43 AM	10:58 AM	12
Newbyth	11:19 AM	11:34 AM	19
Turriff	11:52 AM	12:07 PM	25
Aberdeen	1:14 PM	–	

Table 6.33 Schedule of the second route in the McNish problem.

City	Arrival (hh:mm)	Departure (hh:mm)	Cumulated load (pallets)
Aberdeen	–	9:00 AM	0
Banchory	9:34 AM	9:49 AM	9
Newmill	10:33 AM	10:48 AM	15
Fyvie	11:16 AM	11:31 AM	23
Cornhill	12:13 AM	12:28 PM	28
Aberdeen	1:57 PM	–	

Table 6.34 Schedule of the third route in the McNish problem.

City	Arrival (hh:mm)	Departure (hh:mm)	Cumulated load (pallets)
Aberdeen	–	9:00 AM	0
Towie	10:07 AM	10:22 AM	4
Clova	10:41 AM	10:56 AM	11
Dufftown	11:38 AM	11:53 AM	15
Huntly	12:19 PM	12:34 PM	23
Aberdeen	1:39 PM	–	

6.8.3.2 Tabu Search heuristic

The field of heuristics has been transformed by the development of metaheuristics. One representative method of this class of heuristics is *Tabu Search* (TS). This is essentially a local search heuristic that generates a sequence of solutions in the hope of generating better local optima. TS differs from classical methods in that the successive solutions it examines do not necessarily improve upon each other. A key concept at the heart of TS is that of neighbourhood. The neighbourhood $N(s)$ of a solution s is the set of all solutions that can be reached from s by performing a simple operation. For example, in the context of the NRP, two common neighbourhood structures are obtained by moving a customer from its current route to another route or by swapping two customers between two different routes. The standard TS mechanism is to move from s to the best neighbour in $N(s)$. This way of proceeding may, however, induce cycling. For example, s' may be the best neighbour of s which, in turn, is the best neighbour of s'. To avoid cycling the search process is prevented from returning to solutions processing some attributes of solutions recently considered. Such solutions are declared 'tabu' for a number of iterations. For example, if a customer v is moved from route r to route r' at iteration t, then moving v back to route r will be declared tabu until iteration $t + \theta$, where θ is called the length of the tabu tenure (typically θ is chosen between 5 and 10). When the tabu tenure has expired, v may be moved back to route r at which time the risk of cycling will most likely have disappeared because of changes that have occurred elsewhere in the solution.

Not only is it possible to accept deteriorating solutions in TS, but also it may be interesting to consider infeasible solutions. For example, in the sequence of solutions s, s', s'', both s and s'' may be feasible while s' is infeasible. If s'' cannot be reached directly from s, but only from s', and if it improves upon s, then it pays to go through the infeasible solution s'. This can occur if, for example, s' contains a route r that violates vehicle capacity due to the inclusion of a new customer in that route. Feasibility may be restored at the next iteration if a customer is removed from route s'. A practical way of handling infeasible solutions in TS is to work with a penalized objective function. If $f(s)$ is the actual cost of solution s, then the penalized objective is defined as:

$$f'(s) = f(s) + \alpha Q(s) + \beta D(s) + \gamma W(s),$$

where $Q(s)$, $D(s)$ and $W(s)$ are the total violations of the vehicle capacity constraints, route duration constraints and time window constraints, respectively. Other types of constraints can of course be handled in the same way. The parameters α, β and γ are positive weights associated with constraint violations. These parameters are initially set equal to 1 and self-adjust during the course of the search to produce a mix of feasible and infeasible solutions. For example, if at a given iteration s is feasible with respect to the vehicle capacity constraint, then dividing α by a factor $(1 + \delta)$ (where $\delta > 0$) will increase the likelihood of generating an infeasible solution at the next iteration. Conversely, if s is infeasible, multiplying α by $(1 + \delta)$ will help the search move to a feasible solution. A good choice of δ is typically 0.5. The same principle applies to β and γ.

TS repeatedly performs these operations starting from an initial solution which may be infeasible. It stops after a preset number of iterations. As is common in TS, this number can be very large (i.e. several thousands).

The application of TS to NRPs with various kinds of constraints has proved the absolute validity of this heuristic over the years.

6.8.4 Arc routing problems

As previously defined, an arc routing problem consists of designing a least-cost set of vehicle routes in a graph $G = (V, A, E)$, such that each arc and edge in a subset $R \subseteq A \cup E$ should be visited. A required arc is a one-way street containing customers or a side of a two-way street whose customers cannot be served simultaneously with the customers of the other side (a typical example is solid waste collection along high-speed two-way streets). A required edge is a two-way street for which customers service can take place driving along the street in any direction (a typical example is solid waste collection along streets with light traffic).

Unlike NRPs (that are formulated and solved on an auxiliary complete graph G'), ARPs are generally modelled directly on G.

In this subsection, unconstrained ARPs, namely the CPP and the RPP, are examined. Constrained ARPs can be approached using the algorithmic ideas employed for the constrained NRPs along with the solution procedures for the CPP and the RPP. For example, the ARP with capacity and length constraints, whose applications arise in garbage collection and mail delivery, can be solved using a 'cluster first, route second' heuristic: in a first stage, required arcs and edges are divided into clusters, each of which is assigned to a vehicle, while at a second stage an ARP is solved for each cluster. In this textbook, constrained ARPs are not tackled for the sake of brevity.

6.8.4.1 The Chinese postman problem

The Chinese postman problem is to determine a minimum-cost route traversing all arcs and all edges of a graph at least once. Its main applications arise in garbage collection, mail delivery, network maintenance, snow removal and meter reading in urban areas.

The CPP is related to the problem of determining whether a graph $G = (V, A, E)$ is Eulerian, that is, whether it contains a tour traversing each arc and each edge of the graph exactly once. Obviously, in a Eulerian graph with non-negative arc and edge costs, each Eulerian tour constitutes an optimal CPP solution. In a non-Eulerian graph, an optimal CPP solution must traverse at least one arc or edge twice.

Necessary and sufficient conditions for the existence of a Eulerian tour depend on the type of graph G considered (directed, undirected or mixed), as stated in the following propositions whose proofs are omitted for brevity.

Property. A directed and strongly connected graph G is Eulerian if and only if it is *symmetric*, that is, for any vertex the number of incoming arcs (*incoming degree*) is equal to the number of outgoing arcs (*outgoing degree*) (*symmetric vertex*).

Property. An undirected and connected graph G is Eulerian if and only if it is *even*, that is, each vertex has an even degree (*even vertex*).

Property. A mixed and strongly connected graph G is Eulerian if and only if:

(a) the total number of arcs and of edges incident to any vertex is even (*even graph*);

(b) for each set S of vertices ($S \subset V$ and $S \neq \emptyset$), the difference between the number of the arcs traversing the cut $(S, V \setminus S)$ in the two directions is less than or equal to the number of edges of the cut (*balanced graph*).

Furthermore, since an even and symmetric graph is balanced, the following proposition holds.

Property. A mixed, strongly connected, even and symmetric graph G is Eulerian.

This condition is sufficient, but not necessary, as illustrated in the example reported in Figure 6.36.

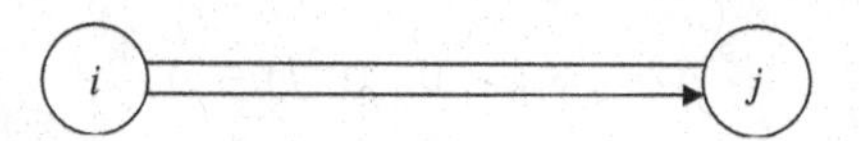

Figure 6.36 A mixed Eulerian graph which is not symmetric.

The solution of the CPP can be decomposed in two steps.

Step 1. Define a least-cost set of arcs $A^{(a)}$ and of edges $E^{(a)}$ such that the multigraph $G^{(a)} = (V, A \cup A^{(a)}, E \cup E^{(a)})$ is Eulerian (if G is itself Eulerian, then $A^{(a)} = \emptyset$ and $E^{(a)} = \emptyset$ and, therefore, $G = G^{(a)}$).

Step 2. Determine a Eulerian tour in $G^{(a)}$.

The first step of the procedure can be executed in polynomial time if G is a directed or undirected graph, while it results in an NP-hard procedure if G is mixed. The second step can be performed in $O(|A \cup E|)$ time with the end-pairing procedure illustrated above.

In the sequel it is shown how to determine $G^{(a)}$ in case G is directed or undirected.

Directed Chinese postman problem When G is directed, in an optimal solution the arcs in $A^{(a)}$ form a set of least-cost paths connecting the asymmetric vertices. Therefore $A^{(a)}$ can be obtained by solving a minimum-cost flow problem namely, a transportation problem) on a bipartite directed graph suitably defined. Let V^+ and V^- be the subsets of V whose vertices have a positive and a negative difference between the incoming degree and the outgoing degree, respectively. The bipartite directed graph is $G_T = (V^+ \cup V^-, A_T)$, where $A_T = \{(i, j) : i \in V^+,\ j \in V^-\}$.

With each arc $(i, j) \in A_T$ is associated a cost w_{ij} equal to that of a least-cost path in G from vertex i to vertex j. Let also $o_i\ (> 0),\ i \in V^+$, be the supply of the vertex i, equal to the difference between its incoming degree and its outgoing degree. Similarly, let $d_i\ (> 0),\ i \in V^-$, be the demand of vertex i, equal to the difference between its outgoing degree and its incoming degree. Furthermore, let s_{ij}, $(i, j) \in A_T$, be the decision variable associated to the flow along arc (i, j).

The transportation problem is

$$\text{Minimize} \sum_{(i,j)\in A_T} w_{ij} s_{ij} \tag{6.93}$$

subject to

$$\sum_{j\in V^-} s_{ij} = o_i,\ i \in V^+ \tag{6.94}$$

$$\sum_{i\in V^+} s_{ij} = d_j,\ j \in V^- \tag{6.95}$$

$$s_{ij} \geq 0,\ (i, j) \in A_T. \tag{6.96}$$

Of course, $\sum_{i\in V^+} o_i = \sum_{j\in V^-} d_j$, so that problem (6.93)–(6.96) is feasible. Let s^*_{ij}, $(i, j) \in A_T$, be an optimal (integer) solution of the transportation problem. $A^{(a)}$ is formed by the arcs $(r, s) \in A$ belonging to the least-cost paths associated to the arcs $(i, j) \in A_T$ such that $s^*_{ij} > 0$ ((r, s) is taken s_{ij} times).

In the directed graph $G = (V, A)$ shown in Figure 6.37, the differences between the incoming and outgoing degrees of vertices 0, 1, 2, 3, 4 and 5 are -1, 0, 1, 0, 1 and -1, respectively. The least-cost paths from vertex 2 to vertex 0 and from vertex 2 to vertex 5 are given by the sequences of arcs $\{(2, 3), (3, 4), (4, 5), (5, 0)\}$ (of cost equal to 109) and $\{(2, 3), (3, 4), (4, 5)\}$ (of cost equal to 86), respectively. Similarly, the least-cost path from vertex 4 to vertex 0 is formed by $\{(4, 5), (5, 0)\}$, of cost 51, while the least-cost path from vertex 4 to vertex 5 is given by arc $(4, 5)$ whose cost is equal to 28. We can therefore formulate the transportation problem on the bipartite directed graph $G_T = (V^+ \cup V^-, A_T)$ represented in Figure 6.38.

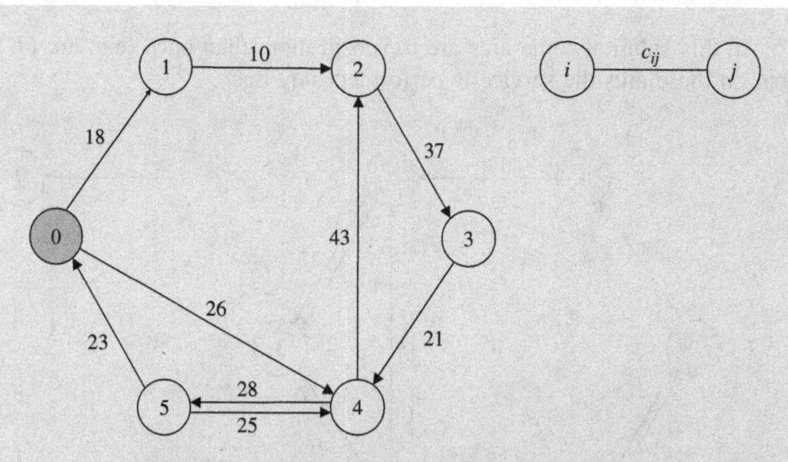

Figure 6.37 A directed graph $G = (V, A)$.

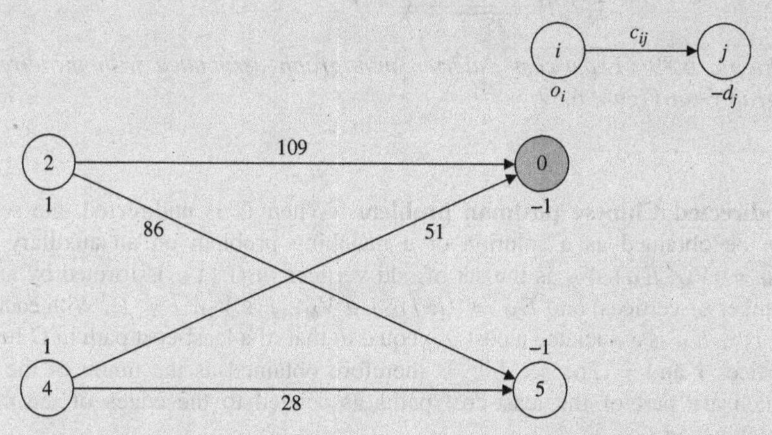

Figure 6.38 Bipartite directed graph $G_T = (V^+ \cup V^- A_T)$ associated with graph G in Figure 6.37.

The optimal solution to the transportation problem is

$$s^*_{20} = 1, s^*_{25} = 0, s^*_{40} = 0, s^*_{45} = 1,$$

of cost equal to 137. Therefore, set $A^{(a)}$ is formed by the arcs of the least-cost paths from vertex 2 to vertex 0 and from vertex 4 to vertex 5. Adding such arcs to the directed graph G, a least-cost Eulerian multigraph is obtained (Figure 6.39). Hence, an optimal CPP solution of cost 368 is defined by the following arcs:

$$\{(0, 1), (1, 2), (2, 3), (3, 4), (4, 5), (5, 0), (0, 4),$$
$$(4, 2), (2, 3), (3, 4), (4, 5), (5, 4), (4, 5), (5, 0)\}.$$

In this solution some arcs are traversed more than once (e.g. arc (4, 5)), but on these arcs the service is performed only once.

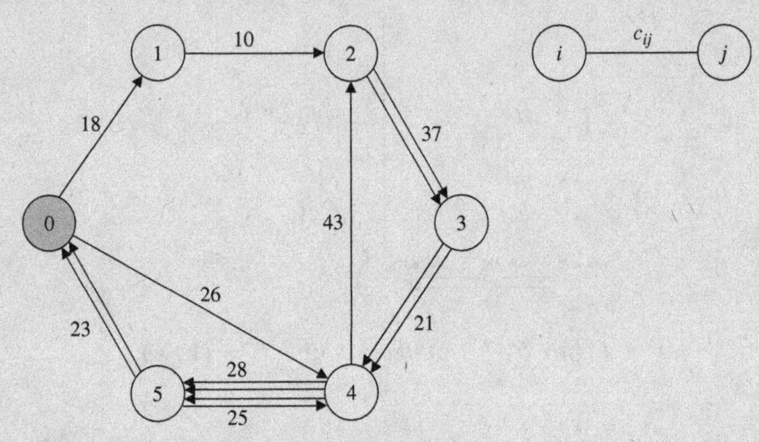

Figure 6.39 Least-cost Eulerian multigraph associated with the directed grap G in Figure 6.37.

Undirected Chinese postman problem When G is undirected, the set $E^{(a)}$ can be obtained as a solution of a matching problem on an auxiliary graph $G_M = (V_M, E_M)$. V_M is the set of odd vertices in G (V_M is formed by an even number of vertices) and $E_M = \{(i, j) : i \in V_M,\ j \in V_M,\ i \neq j\}$. With each edge $(i, j) \in E_M$ is associated a cost w_{ij} equal to that of a least-cost path in G between vertices i and j. The set $E^{(a)}$ is therefore obtained as the union of the edges which are part of the least-cost paths associated to the edges of the optimal matching on G_M.

Welles is in charge of maintaining the road network of Wales in the United Kingdom. Among other things, the company has to monitor periodically the roads in order to locate crack and potholes in the asphalt. To this purpose, the road network has been divided into about 10 subnetworks, each of which has to be visited every 15 days by a dedicated vehicle. The graph representing one such subnetwork is shown in Figure 6.40. In order to determine the optimal CPP solution, a minimum-cost matching problem between the odd-degree vertices (vertices 2, 3, 6, 7, 9 and 11) is solved. The optimal matching is: 2–3, 6–7 and 9–11. The associated set of paths in G is (2, 3), (6, 7) and (9, 11) (total cost is 7.5 km). Adding these edges to G, the Eulerian multigraph in Figure 6.41 is obtained. Finally, by using the end-pairing procedure, the

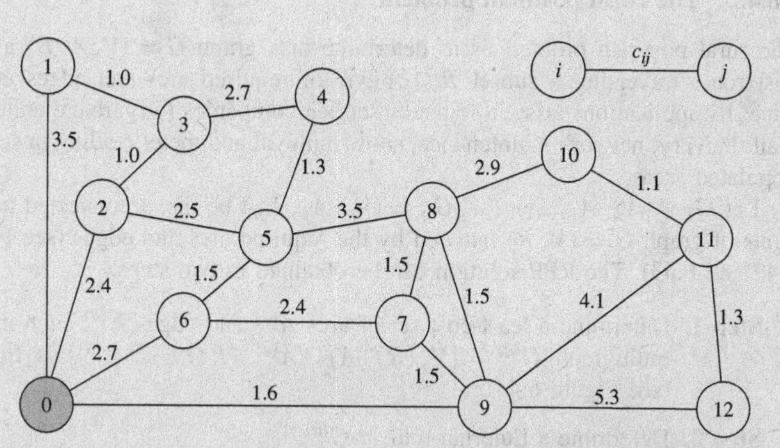

Figure 6.40 Graph representation used in the Welles problem. Costs c_{ij}, $(i, j) \in E$ are in km.

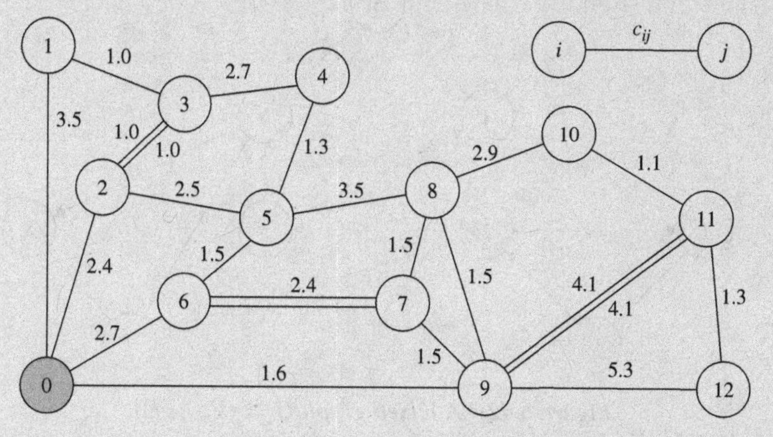

Figure 6.41 Least-cost Eulerian multigraph in the Welles problem.

optimal CPP solution is obtained:

$$\{(0, 1), (1, 3), (3, 2), (2, 5), (5, 4), (4, 3), (3, 2), (2, 0),$$
$$(0, 6), (6, 7), (7, 8), (8, 5), (5, 6), (6, 7), (7, 9), (9, 11),$$
$$(11, 10), (10, 8), (8, 9), (9, 11), (11, 12), (12, 9), (9, 0)\},$$

whose cost is 52.8 km, that is, the cost of the arcs of G, plus 7.5 km. In this solution, edges (2, 3), (6, 7) and (9, 11) are traversed twice.

6.8.4.2 The rural postman problem

The rural postman problem is to determine in a graph $G = (V, A, E)$ a least-cost route traversing a subset $R \subseteq A \cup E$ of required arcs and edges at least once. Its applications arise in logistics service companies for garbage collection, mail delivery, network maintenance, snow removal and meter reading in scarcely populated areas.

Let $G_1 = (V_1, A_1, E_1), \ldots, G_p = (V_p, A_p, E_p)$ be the p connected components of graph $G = (V, R)$ induced by the required arcs and edges (see Figures 6.42 and 6.43). The RPP solution can be obtained in two steps.

Step 1. Determine a least-cost set of arcs $A^{(a)}$ and edges $E^{(a)}$ such that the multigraph $G^{(a)} = \left(V, (R \cap A) \cup A^{(a)}, (R \cap E) \cup E^{(a)}\right)$ is Eulerian (see Figure 6.44).

Step 2. Determine a Eulerian tour in $G^{(a)}$.

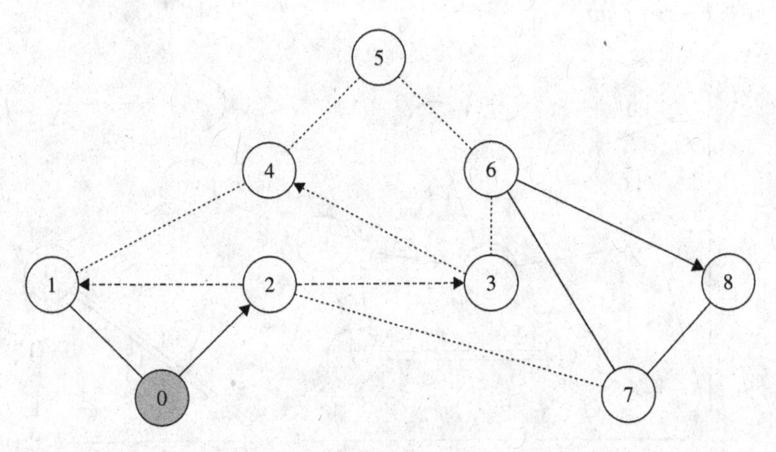

Figure 6.42 A mixed graph $G = (V, A, E)$.

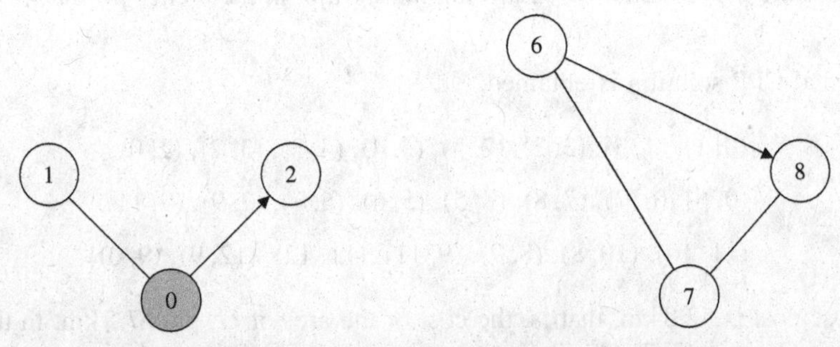

Figure 6.43 Connected components induced by the required arcs and edges of graph G in Figure 6.42.

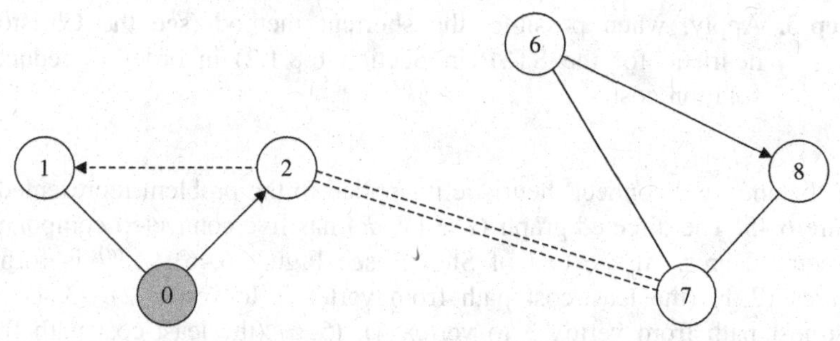

Figure 6.44 Least-cost Eulerian multigraph associated with graph $G = (V, R)$ of Figure 6.42.

The first step is NP-hard even for directed and undirected graphs, if $p > 1$. For $p = 1$, the RPP can be reduced to a CPP. The second step can be solved in polynomial time with the end-pairing procedure. In what follows, the first stage of two constructive heuristics is illustrated for directed and undirected graphs.

Directed rural postman problem A sub-optimal solution to the directed RPP can be obtained through the 'balance and connect' heuristic.

Step 1. Using the procedure employed for the directed CPP, construct a directed symmetric graph $G'^{(a)} = (V, R \cup A'^{(a)})$, by adding to $G = (V, R)$ a suitable set of least-cost paths between nonsymmetric vertices. If $G'^{(a)} = (V, R \cup A'^{(a)})$ is connected, *STOP*, $G'^{(a)} = G^{(a)}$ is Eulerian.

Step 2. Let p' $(1 < p' \leq p)$ be the number of connected components of $G'^{(a)} = (V, R \cup A'^{(a)})$. Construct an auxiliary undirected graph $G^{(c)} = (V^{(c)}, E^{(c)})$, in which there is a vertex $h \in V^{(c)}$ for each connected component of $G'^{(a)}$, and, between each pair of vertices $h, k \in V^{(c)}$, $h \neq k$, there is an edge $(h, k) \in E^{(c)}$. With edge (h, k) is associated a cost g_{hk} equal to:

$$g_{hk} = \min_{i \in V_h, j \in V_k} \{w_{ij} + w_{ji}\},$$

where w_{ij} and w_{ji} are the costs of the least-cost paths from vertex j to vertex j and from vertex j to vertex j in G, respectively. Compute the minimum-cost tree $T^{(c)} = (V^{(c)}, E_T^{(c)})$ spanning the vertices of graph $G^{(c)}$. Construct a symmetric, connected and directed graph $G^{(a)} = (V, R \cup A'^{(a)} \cup A''^{(a)})$ by adding to $G'^{(a)} = (V, R \cup A'^{(a)})$ the set of arcs $A''^{(a)}$ belonging to the least-cost paths corresponding to the edges $E_T^{(c)}$ of the tree $T^{(c)}$.

Step 3. Apply, when possible, the shortcut method (see the Christofides heuristic for the STSP in Section 6.8.1.2) in order to reduce the solution cost.

The 'balance and connect' heuristic is applied to the problem represented in Figure 6.45. The directed graph $G = (V, R)$ has five connected components of required arcs. At the end of Step 1 (see Figure 6.46), $A'^{(a)}$ is formed by arcs $(2, 1)$ (the least-cost path from vertex 1 to vertex 2), $(3, 4)$ (the least-cost path from vertex 3 to vertex 4), $(5, 6)$ (the least-cost path from vertex 5 to vertex 6), $(8, 9)$ and $(9, 11)$ (the least-cost path from vertex 8 to vertex 11) and $(10, 7)$ (the least-cost path from vertex 10 to vertex 7).

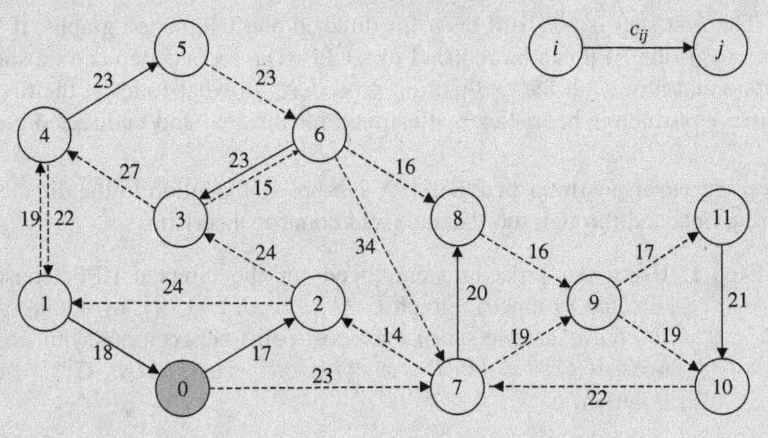

Figure 6.45 Directed graph $G = (V, A)$.

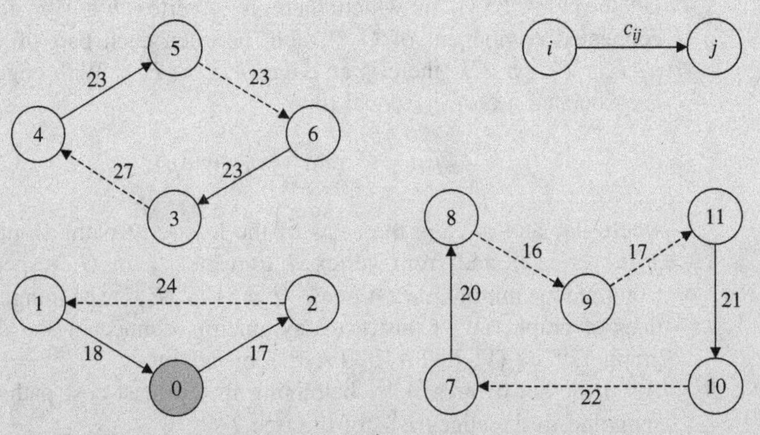

Figure 6.46 Graph $G'^{(a)} = (V, R \cup A'^{(a)})$ obtained at the end of Step 1 of the 'balance and connect' heuristic.

At Step 2, $V^{(c)} = \{1, 2, 3\}$, and the least-cost paths from vertex 1 to vertex 4 (arc $(1, 4)$) and vice versa (arc $(4, 1)$), and from vertex 6 to vertex 7 (arc $(6, 7)$) and vice versa (arcs $(7, 2)$, $(2, 3)$ and $(3, 6)$), are added to the partial solution (see Figure 6.47). Finally, using the end-pairing procedure, the following tour of cost 379 is obtained:

$$\{(0, 2), (2, 1), (1, 4), (4, 5), (5, 6), (6, 7), (7, 8), (8, 9), (9, 11),$$
$$(11, 10), (10, 7), (7, 2), (2, 3), (3, 6), (6, 3), (3, 4), (4, 1), (1, 0)\}.$$

It can be shown that in this case, the 'balance and connect' solution is optimal.

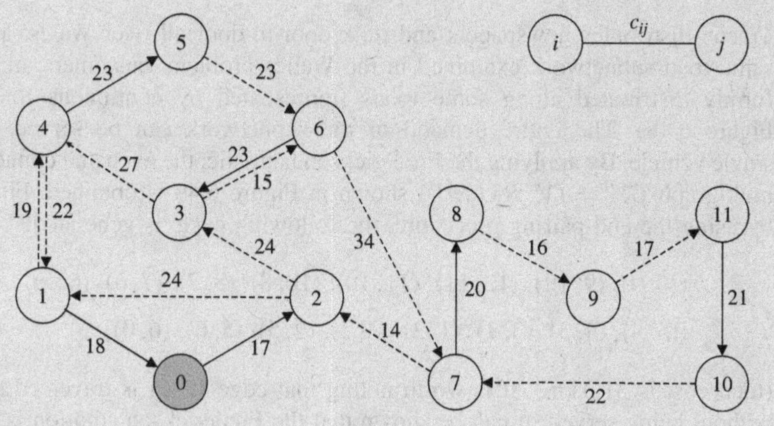

Figure 6.47 Symmetric and connected directed graph $G^{(a)} = (V, R \cup A^{'(a)} \cup A^{''(a)})$ *obtained at the end of Step 2 of the 'balance and connect' heuristic.*

Undirected rural postman problem A suboptimal solution to the undirected RPP can be obtained through the Frederickson heuristic.

Step 1. Using the matching procedure illustrated for the undirected CPP, construct an even graph $G^{'(a)} = (V, R \cup E^{'(a)})$. If $G^{'(a)} = (V, R \cup E^{'(a)})$ is also connected, *STOP*, $G^{'(a)} = G^{(a)}$ is Eulerian.

Step 2. Let p' $(1 < p' \leq p)$ be the number of connected components of $G^{'(a)} = (V, R \cup E^{'(a)})$. Construct an auxiliary undirected graph $G^{(c)} = (V^{(c)}, E^{(c)})$, in which there is a vertex $h \in V^{(c)}$ for each connected component of $G^{'(a)}$, and between each couple of vertices $h,\ k \in V^{(c)}$, $h \neq k$, there is an edge $(h, k) \in E^{(c)}$. With each edge

(h, k) is associated a cost g_{hk} equal to

$$g_{hk} = \min_{i \in V_h, j \in V_k} \{w_{ij}\},$$

where w_{ij} is the cost of the least-cost path between vertices i and j in G. Compute a minimum-cost tree $T^{(c)} = (V^{(c)}, E_T^{(c)})$ spanning the vertices of graph $G^{(c)}$. Construct an even and connected graph $G^{(a)} = (V, R \cup E'^{(a)} \cup E''^{(a)})$ by adding to $R \cup E'^{(a)}$ the set of edges $E''^{(a)}$ (each of which taken twice) belonging to the least-cost paths corresponding to the edges $E_T^{(c)}$ of tree $T^{(c)}$.

Step 3. Apply, if possible, the shortcuts method (see the Christofides heuristic for the STSP in Section 6.8.1.2) in order to reduce the solution cost.

Tracon distributes newspapers and milk door-to-door all over Wales. In the same road subnetwork examined in the Welles problem, customers are uniformly distributed along some roads (represented by continuous lines in Figure 6.48). The entire demand of the subnetwork can be served by a single vehicle. By applying the Frederickson heuristic, the even and connected multigraph $G^{(a)} = (V, R \cup E^{(a)})$ shown in Figure 6.49 is obtained. Finally, by using the end-pairing procedure, the following cycle is generated:

$$\{(0,9), (9,12), (12,11), (11,10), (10,8), (8,7), (7,6), (6,5),$$
$$(5,4), (4,3), (3,1), (1,3), (3,2), (2,5), (5,6), (6,0)\}$$

(total cost is 31.3 km). It is worth noting that edge $(5, 6)$ is traversed twice without being served. It can be shown that the Frederickson solution is optimal.

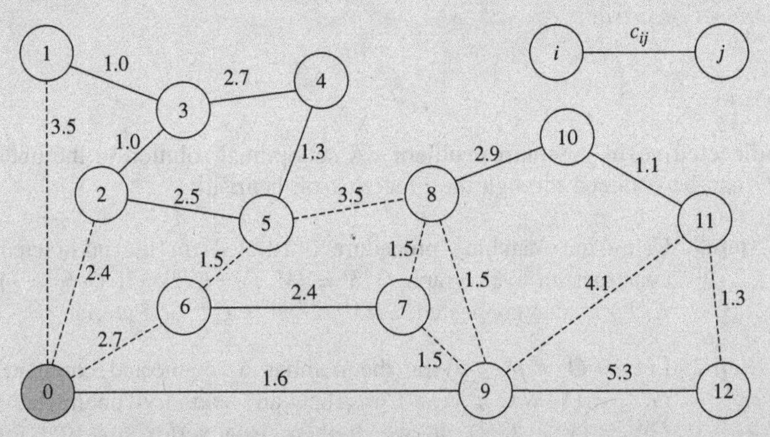

Figure 6.48 Graph $G = (V, E)$ associated to the Tracon problem (costs are in km).

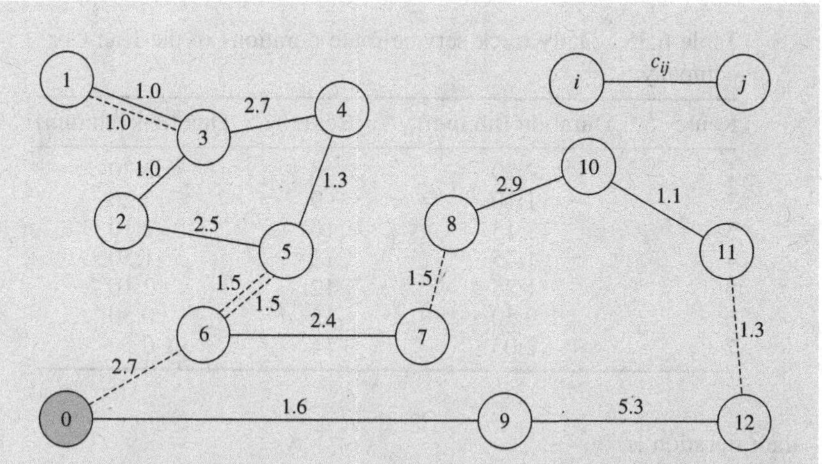

Figure 6.49 Even and connected multigraph $G^{(a)} = (V, R \cup E^{(a)})$, obtained at the end of Step 2 of the Frederickson heuristic applied to the Tracon problem.

6.8.5 Route sequencing

It is worth observing that in the VRPs illustrated before, each vehicle is used only once during the planning period. This situation is not representative of real cases, in which, typically, the vehicles could be used for more journeys on the same working shift. This happens, in particular, when each route in the VRP solution has a rather short duration. In these cases it is possible to reduce the number of vehicles used by allocating more routes to the same vehicle, in respect of the total time length constraint. The problem, known as the *vehicle routing problem with multiple trips* (VRPMT), can be treated as a special case of 1-BP (see Section 5.4.2.1), in which the routes to allocate to the vehicles correspond to the objects to be loaded, each of them having a 'weight' equal to the time duration, whereas the vehicles correspond to bins of capacity equal to T, where T is the time duration of the working shift.

Han-Cor, a public Korean carrier, has solved a daily routing problem which corresponds to the truck service routes shown in Table 6.35. The daily duration of the transport service is six hours and 20 minutes (or, equivalently, 380 minutes). To allocate the routes to vehicles, the BFD heuristic (see Section 5.4.2) can be applied to the 1-BP problem with 14 objects, each of them with 'weight' equal to d_i, $i = 1, \ldots, 14$ (see the second and fourth columns of Table 6.35). The list S of the routes ordered by non-increasing values of

Table 6.35 Daily truck service route durations of the Han-Cor company.

Route	Duration (hh:mm)	Route	Duration (hh:mm)
1	2:30	8	3:20
2	1:10	9	2:20
3	3:15	10	2:10
4	1:45	11	1:50
5	1:25	12	2:10
6	0:50	13	0:50
7	2:05	14	0:50

their duration is:

$$S = \{8, 3, 1, 9, 10, 12, 7, 11, 4, 5, 2, 6, 13, 14\}.$$

At the end of the application of the heuristic, it is found that five vehicles ($v_1, \ldots, v_5$) are needed; to each of them is allocated one of the following routes:

$$v_1 = \{8, 1\}, \text{ used for 350 minutes overall;}$$

$$v_2 = \{3, 9\}, \text{ used for 335 minutes overall;}$$

$$v_3 = \{10, 12, 11\}, \text{ used for 370 minutes overall;}$$

$$v_4 = \{7, 4, 5, 6\}, \text{ used for 365 minutes overall;}$$

$$v_5 = \{2, 13, 14\}, \text{ used for 170 minutes overall.}$$

6.9 Real-time vehicle routing problems

As pointed out in Section 6.8, several important short-haul transport problems must be solved in real time. In this section, the main features of such problems are illustrated.

In *real-time* VRDPs, uncertain data are gradually revealed during the operational interval, and routes are constructed in an on-going fashion as new data arrive. The *events* that lead to route modifications can be: (a) the arrival of new user requests, (b) the arrival of a vehicle at a destination or (c) the update of travel times.

Every event must be processed according to the policies set by the company or organization operating the fleet of vehicles. As a rule, when a new request is received, one must decide whether it can be serviced on the same day, or whether it must be delayed or rejected. If the request is accepted, it is assigned temporarily to a position in a vehicle route. The request is effectively serviced as planned if no other event occurs in the meantime. Otherwise, it can be assigned to a different position of the same vehicle route, or even dispatched to a different vehicle.

It is worth noting that at any time each driver just needs to know his next stop. Hence, when a vehicle reaches a destination it has to be assigned a new destination. Because of the difficulty of estimating the current position of a moving vehicle, reassignments could not easily made until quite recently. However, due to advances in vehicle positioning and communication technologies, route diversions and reassignments are now a feasible option and should take place if this results in a cost saving or in an improved service level.

Finally, if an improved estimation of vehicle travel times is available, it may be useful to modify the current routes or even the decision of accepting a request or not. For example, if an unexpected traffic jam occurs, some user services can be deferred. If the demand rate is low, it is somewhat useful to relocate idle vehicles in order to anticipate future demands or to escape a forecasted traffic congestion.

Real-time problems possess a number of particular features, some of which have just been described. In the following, the remaining characteristics are outlined.

Quick response. Algorithms for solving real-time VRDPs must provide a quick response so that route modifications can be transmitted in a timely manner to the fleet. To this end, two methods can be used: simple policies (like the 'first come, first served policy'), or more involved algorithms running on parallel hardware. The choice between them depends mainly on the objective, the degree of dynamism and the demand rate.

Denied or deferred service. In some applications it is valid to deny service to some users, or to leave them to a competitor, in order to avoid excessive delays or unacceptable costs. For instance, requests that cannot be serviced within a given time windows are rejected.

Congestion. If the demand rate exceeds a given threshold, the system becomes saturated, that is, the expected waiting time of a request goes to infinity.

The degree of dynamism. Designing an algorithm for solving real-time VRDPs depends to a large extent on how much the problem is dynamic. To quantify this concept, the *degree of dynamism* of a problem has been defined. Let $[0, T]$ be the operational interval and let n_s and n_d be the number of static and dynamic requests, respectively ($n_s + n_d = |U|$). Moreover, let $t_i \in [0, T]$ be the *occurrence time* of service request of customer $i \in U$. Static requests are such that $t_i = 0$, $i \in U$, while dynamic ones have $t_i \in [0, T]$, $i \in U$. The degree of dynamism δ can be simply defined as:

$$\delta = \frac{n_d}{n_s + n_d}$$

and may vary between 0 and 1. Its meaning is straightforward. For instance, if δ is equal to 0.3, then three customers out of 10 are dynamic. Such definition can be generalized in order to take into account both dynamic request occurrence times and possible time windows. For a given δ value, a problem is more dynamic if immediate requests occur at the end of the operational interval $[0, T]$. As a result, the measure of dynamism can be generalized as follows:

$$\delta' = \frac{\sum_{i=1}^{n_d} t_i/T}{n_s + n_d}.$$

Again δ' ranges between 0 and 1. It is equal to 0 if all user requests are known in advance, while it is equal to 1 if all user requests occur at time T. Finally, the definition of δ' can be modified to take into account possible time windows on user service time. Let a_i and b_i be the *ready time* and *deadline* of customer $i \in U$, respectively. Then,

$$\delta'' = \frac{\sum_{i=1}^{n_d} [T - (b_i - t_i)]/T}{n_s + n_d}.$$

It can be shown that δ'' also varies between 0 and 1. Moreover, it is worth noting that if no time windows are imposed (i.e. $a_i = t_i$ and $b_i = T$ for each customer $i \in U$), then $\delta'' = \delta'$. As a rule, vendor-based distribution logistics systems (such as those distributing heating oil) are weakly dynamic. Problems faced by long-distance couriers and appliance repair service companies are moderately dynamic. Finally, emergency services exhibit a strong dynamic behaviour.

Objectives. In real-time VRDPs, the objective to be optimized is often a combination of different measures. In weakly dynamic systems, the focus is on minimizing routing cost but, when operating a strongly dynamic system, minimizing the expected *response time* (i.e. the expected time lag between the instant a user service begins and its occurrence time) becomes a key issue. Another meaningful criterion which is often considered (alone or combined with other measures) is throughput optimization, that is, the maximization of the expected number of requests serviced within a given period of time.

6.10 Integrated location and routing problems

Facility location and vehicle routing are two of the most fundamental decisions in logistics. Location decisions that are very costly and difficult to change are said to be strategic, for example those involving major installations such as factories, airports, fixed transport links and so on. Others are said to be tactical because while still being relatively costly, they can still be modified after several years. Warehouse and store locations fall in that category. Finally, operational location decisions involve easily movable facilities such as parking areas, mail boxes and the like.

Once facilities are located, a routing plan must be put in place to link them together on a regular basis. All too often, facilities are first located without sufficient consideration of transport costs, which may result in systemic inefficiencies. When planning to locate facilities, it is preferable to integrate in the analysis the routing cost that these will generate. This applies equally well to strategic, tactical and operational decisions. A strategic location-routing decision is the location of airline hubs whose choice bears on routing costs. The location of depots and warehouses in a logistics system is a tactical decision influencing delivery costs to customers. A simple example arising at the operational level is mail box location. Locating a large number of mail boxes in a city will improve customer convenience since average walking distance to a mail box will be reduced. At the same time, the cost of emptying a larger number of mail boxes on a regular basis will be higher.

Unfortunately, integrated location-routing optimization models combining these two aspects will often contain too many integer decision variables and constraints to be solvable optimally. Heuristics based on a decomposition principle are often used instead. Facilities are first located, predecessor and successor nodes are assigned to facilities and routing is then performed. These three decisions are then iteratively updated until no significant improvement can be reached.

3L Multimedia is a company that distributes newspapers and magazines throughout France. The distribution system, set up in 2008, includes 54 hubs which serve 9 542 sales points (supermarkets, newsstands and bookshops) in France. On the basis of the historical data available about the deliveries carried out by using the different fleets of vehicles available, a forecasting study was carried out of the total logistics costs in the next three years, evaluating 10 different potential locations for the hubs.

The data resulting from this study are reported in Table 6.36. Note that the disposal costs are referred to existing hubs which are no longer foreseen

Table 6.36 Total logistics costs (in €) in correspondence to the 10 location alternatives for the 3L Multimedia problem.

Alternative	Number of hubs	Sales cost	Triennial location cost	Triennial transport cost	Total cost
1	47	140 000	534 000	856 000	1 530 000
2	55	90 000	610 000	790 000	1 490 000
3	44	122 000	491 000	882 000	1 495 000
4	50	80 000	565 000	844 000	1 489 000
5	38	180 000	443 000	964 000	1 587 000
6	40	75 000	446 000	933 000	1 454 000
7	58	22 000	658 000	781 000	1 461 000
8	60	210 000	663 000	762 000	1 635 000
9	32	180 000	357 000	1 081 000	1 618 000
10	35	55 000	397 000	1 033 000	1 485 000

in the locations where they are situated. The transport costs are calculated, simulating the daily delivery plans for the planning horizon (three years), by solving the vehicle routing problem with capacity and length constraints.

The best simulated alternative proves to be the sixth, providing for 40 hubs.

6.11 Vendor-managed inventory routing

IRPs are combinatorial problems in which inventory control and routing decisions have to be made in an integrated fashion and simultaneously. With respect to the traditional method, where the supplier receives orders from its own customers and plans the deliveries in relation to the choices made by the customers themselves, in IRP, the vendor decides the best replenishment policy of the customers, avoiding stockout and making sure that the quantities delivered are compatible with the storage capacity of the warehouses and with the capacity of the vehicles available for deliveries. The aim is to minimize the total inventory and transport cost over a given time horizon.

The greater difficulty in dealing with IRPs than with VRPs arises from the fact that such problems integrate a level of tactical decisions, related to the coordinated inventory replenishment of a set of customers distributed over a limited geographical area, with operational decisions regarding the best scheduling of the deliveries over some time periods of the time horizon. Inventory management comes into play whenever all the customers consume products with a fixed or variable rate within a limited time interval, and have to manage the quantities of commodity to be supplied in relation to the storage capacity of their warehouses. This consideration naturally leads to the introduction of the time dimension in the context of classical routing, where the customers' demand is known a priori.

Different variants of the IRP have been defined in scientific literature and, for each of them, many solution methods have been proposed. In this textbook, a classical single-commodity formulation on a complete graph $G' = (V', E')$ is illustrated, in which $V' = U \cup \{0\}$, where U is the set of customers to be served, 0 represents the supplier and a homogeneous fleet K of vehicles with capacity Q is available for the deliveries. To each edge $(i, j) \in E'$ is associated a non-negative cost c_{ij}, corresponding to the cost of a shortest path from i to j (so that the triangular inequality holds). Consumption of a commodity by customer $i \in U$ is represented by q_i^t or μ_i^t, according to whether the planning horizon T is a discrete set or a continuous interval, respectively. In particular, if $T = 1, \dots, H$, q_i^t is the consumption of the commodity by customer $i \in U$ in the time period $t \in T$; whenever $T = [1, H]$, μ_i^t represents, instead, the commodity consumption rate of customer i at time period t. To the supplier, the quantity of commodity available in time period $t \in T$ is represented by p_0^t or, more simply, p^t. I_0^0 represents the initial inventory level of the commodity at the supplier, whereas I_i^0 is the initial stock level of the commodity at customer $i \in U$. The inventory levels of the commodity at the supplier and the customer $i \in U$ at the beginning (or at the end) of time period t, are denoted by I_0^t and I_i^t, respectively. The maximum storage capacities at the supplier and the customer $i \in U$ are indicated by L_0 and L_i, respectively. Finally, the inventory cost per product unit at the supplier and the customer $i \in U$ are represented by h_0 and h_i.

To understand how difficult it is to solve even small instances of IRPs, we provide a simple example.

Consider an IRP with a discrete planning horizon (days), formulated on the graph in Figure 6.50. Table 6.37 shows the maximum level of inventory L_i and the periodic consumption q_i by each customer $i \in U$. Assume that $I_i^0 = L_i$, $i \in U$, and that the supplier has a fleet with an unlimited number of vehicles of capacity $Q = 5\,000$ units of commodity. Assume, moreover, that the supplier's commodity availability is unlimited ($L_0 = \infty$), and that the inventory cost is negligible with respect to the transport cost.

The aim is to find the best resupply policy of the four customers, by avoiding stockouts and fulfilling the constraints related to the maximum storage and the vehicles' capacities.

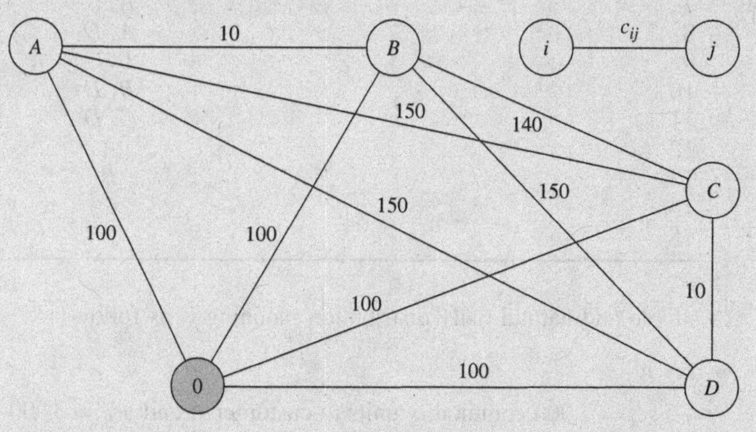

Figure 6.50 Complete undirected graph representative of a small instance of the considered IRP. Vertex 0 indicates the supplier, while the edge costs are expressed in €.

Table 6.37 Maximum inventory level L_i and periodic consumption q_i (in units of commodity) of each customer $i \in U$ in the example of the IRP.

Customer	Inventory level	Consumption
A	5 000	1 000
B	3 000	3 000
C	2 000	2 000
D	4 000	1 500

The periodicity of the policy implies that the customers inventory levels are equal at the beginning and the end of the time horizon.

This allows the assumption that the time horizon is infinite. There are 15 different ways to select the combinations of customers to be served, as shown in Table 6.38, but it is difficult to enumerate all feasible delivery amounts (it is an infinite number for a continuous problem).

Table 6.38 Possible delivery plans for the four customers of the IRP example (the different routes for the first 11 plans with more than one customer are not reported).

Delivery plan	Customers served
1	A, B, C, D
2	A, B, C
3	A, B, D
4	A, C, D
5	B, C, D
6	A, B
7	A, C
8	A, D
9	B, C
10	B, D
11	C, D
12	A
13	B
14	C
15	D

A simple and natural daily distribution planning is as follows:

- day t:
  - $w_A^t = 1\,000$ commodity units to customer A and $w_B^t = 3\,000$ commodity units to customer B, with a transport cost of € 210;
  - $w_C^t = 2\,000$ commodity units to customer C and $w_D^t = 1\,500$ commodity units to customer D, with a transport cost of € 210,

  having a total daily cost of € 420.

A better solution, instead, provides for different deliveries on successive days:

- day t:
  - $w_B^t = 3\,000$ commodity units to customer B and $w_C^t = 2\,000$ commodity units to customer C, with a transport cost of € 340;

- day $t+1$:
  - $w_A^{(t+1)} = 2\,000$ commodity units to customer A and $w_B^{(t+1)} = 3\,000$ commodity units to customer B, with a transport cost of € 210;
  - $w_C^{(t+1)} = 2\,000$ commodity units to customer C and $w_D^{(t+1)} = 3\,000$ commodity units to customer D, with a transport cost of € 210,

  with an average daily cost of € 380.

To define a MIP model for the IRP on a discrete time horizon T, the following decision variables are defined I_i^t, $i \in U$, $t \in T'$, each of which represents the customer inventory level i at the beginning of time period t, where $T' = T \cup \{H+1\}$ is the extended time horizon where time period $H+1$ is defined in order to evaluate the impact of the decisions made at the end of T; x_{ik}^t, $i \in U$, $k \in K$, $t \in T$, each of which represents the quantity of commodity delivered in the time period t to customer i with vehicle k; y_{ijk}^t, $(i,j) \in E'$, $k \in K$, $t \in T$, each of which represents the number of times that the vehicle k traverses the edge (i,j) in time period t, and z_{ik}^t, $i \in V'$, $k \in K$, $t \in T$, each of which is binary, assuming value 1 if vehicle k visits the node (customer or supplier) i in time period t, 0 otherwise. The model is as follows:

$$\text{Minimize} \sum_{i \in V'} \sum_{t \in T'} h_i I_i^t + \sum_{t \in T} \sum_{k \in K} \sum_{(i,j) \in E'} c_{ij} y_{ijk}^t \tag{6.97}$$

subject to

$$I_0^t = I_0^{t-1} + p^{t-1} - \sum_{i \in U} \sum_{k \in K} x_{ik}^{t-1}, \quad t \in T' \tag{6.98}$$

$$I_i^t = I_i^{t-1} + \sum_{k \in K} x_{ik}^{t-1} - q_i^{t-1}, \quad i \in U, \ t \in T' \tag{6.99}$$

$$I_0^t \geq \sum_{i \in U} \sum_{k \in K} x_{ik}^t, \quad t \in T \tag{6.100}$$

$$\sum_{k \in K} x_{ik}^t \leq L_i - I_i^t, \quad i \in U, \ t \in T \tag{6.101}$$

$$x_{ik}^t \leq L_i z_{ik}^t, \quad i \in U, \ k \in K, \ t \in T \tag{6.102}$$

$$\sum_{i \in U} x_{ik}^t \leq Q, \quad k \in K, \ t \in T \tag{6.103}$$

$$\sum_{i \in V' : (i,j) \in E'} y_{ijk}^t + \sum_{i \in V' : (j,i) \in E'} y_{jik}^t = 2 z_{jk}^t, \quad j \in V', \ k \in K, \ t \in T \tag{6.104}$$

$$\sum_{(i,j)\in E':i\in S,j\notin S} y^t_{ijk} + \sum_{(j,i)\in E':i\in S,j\notin S} y^t_{jik} \geq 2z^t_{hk},$$

$$S \subset V', 2 \leq |S| \leq \lceil |V'|/2 \rceil, \; h \in S, \; k \in K, \; t \in T, \quad (6.105)$$

$$I^t_i \geq 0, \; i \in U, \; t \in T' \quad (6.106)$$

$$x^t_{ik} \geq 0, \; i \in U, \; k \in K, \; t \in T \quad (6.107)$$

$$y^t_{ijk} \in \{0, 1\}, \; (i, j) \in E', \; k \in K, \; t \in T \quad (6.108)$$

$$y^t_{0ik} \in \{0, 1, 2\}, \; (0, i) \in E', \; k \in K, \; t \in T \quad (6.109)$$

$$z^t_{ik} \in \{0, 1\}, \; i \in V', \; k \in K, \; t \in T. \quad (6.110)$$

The objective function (6.97) represents the total inventory and transport cost over the time horizon. Constraints (6.98) mean that the inventory level of the supplier in time period $t \in T'$ is given by the sum of the inventory level at the beginning of the previous time period, plus the product quantity $p^{(t-1)}$ available in the time period $t - 1$, minus the quantities of the commodity delivered to customers in time period $t - 1$. Observe that $p^0 = 0$ and $x^0_{ik} = 0$, for each $i \in U$ and for each $k \in K$. Constraints (6.99) guarantee that the customer inventory level i, $i \in U$, at the beginning of each time period $t \in T'$ is equal to the sum of level at the beginning of the previous time period, plus the quantity of commodity delivered, minus the quantity consumed in $t - 1$. Note that $q^0_i = 0$, for each $i \in U$. Constraints (6.100) ensure that the supplier is always able to satisfy the customer requests, whereas constraints (6.101) ensure that the quantities delivered do not exceed the customers' storage capacity in each time period $t \in T$. Constraints (6.102) require that each customer is visited every time that a quantity of commodity greater than zero is delivered. Constraints (6.103) guarantee that the quantities delivered by each vehicle are compatible with the capacity of those vehicles that begin the delivery route starting from the supplier. Constraints (6.104) and (6.105) are classic routing constraints. They ensure, respectively, that the number of edges incident in a node and traversed by vehicle $k \in K$ is even, and that there are no subtours.

Finally, note that model (6.97)–(6.110) allows that, in a given time period $t \in T$, the quantities to be supplied to the same customer $i \in U$ can be delivered using several vehicles (split delivery).

Split is a Finnish company supplying each week four customers of a special mixture which is used as a basic component for preparing a drink sold at retail stores. The product has a consumption demand recorded every Monday, Wednesday and Friday. Split intends to adopt an integrated management strategy for the optimal planning of the production and distribution of this mixture. The geographic coordinates of the Split DC and of the four customers are shown in Table 6.39. The Split distribution problem can be formulated by using model (6.97)–(6.110) on a

Table 6.39 Cartesian coordinates of the supplier (vertex 0) and of the four Split customers.

Vertex	Abscissa	Ordinate
0	154	417
1	172	334
2	267	87
3	148	433
4	355	444

complete graph $G' = (V', E')$, with $V' = \{0, 1, 2, 3, 4\}$ and $|E'| = 10$. Vertex 0 represents the Split DC, from which at most two vehicles, each having a capacity of 120 tons, can operate each day. The transport costs between any pair of vertices are calculated in proportion to the Euclidean distances and are expressed in € (the coefficient of proportionality, for simplicity, is set equal to 1). The data relative to the initial inventory levels and to the maximum storage capacity at the Split DC and of its customers are shown in Table 6.40.

Table 6.40 Initial inventory level at the first period and maximum storage capacity (in tons) at the Split DC and at its customers.

Vertex	Inventory level	Storage capacity
0	510	–
1	130	195
2	70	105
3	58	116
4	48	73

The quantities of product (in tons) available to the Split DC and the quantities of product (in tons) consumed per day by the customers are assumed constant throughout the planning horizon $T = \{1, 2, 3\}$. In particular: $p^t = 193$, $q_1^t = 65$, $q_2^t = 35$, $q_3^t = 58$ and $q_4^t = 24$, for each $t \in T$. The storage costs (in €) per commodity unit stored are, respectively: $h_0 = 0.30$, $h_1 = 0.23$, $h_2 = 0.32$, $h_3 = 0.33$ and $h_4 = 0.23$.

The optimal supply policy which minimizes the total inventory cost and the total transport cost in the planning horizon T provides for two days of delivery. The solution is summarized in Tables 6.41 and 6.42.

In particular, the inventory cost shown in Table 6.41 at the four time periods is used to evaluate the effect of the choices made at the end of T. A single delivery route, $\{0, 1, 2, 4, 0\}$, is used in the first time period, whereas

Table 6.41 Inventory and transport costs (in €) in the time periods for the Split distribution problem; the time horizon is extended to the fourth time period to assess the effect of choices made at the end of the planning horizon.

Period	Inventory cost	Transport cost
1	235.48	921.00
2	237.32	206.00
3	243.06	0.00
4	250.15	–

Table 6.42 Quantities transported (in tons) from the DC to the four Split customers in the three time periods.

	Customer			
Period	1	2	3	4
1	61	35	0	24
2	69	0	116	0
3	0	0	0	0

two back-and-forth routes, {0, 1, 0} and {0, 3, 0}, are used for the next time period. On the third day, no delivery is made. Observe that the total volume transported in the first time period is equal to 120 tons (coincident with the vehicle capacity), whereas in the second time period, since the total volume transported is equal to 185 tons, two vehicles are needed (or two daily routes performed by the same vehicle).

Regarding the first customer, the quantity of product consumed in the first time period is equal to 65 tons, whereas the inventory level at the beginning of the time period is 130 tons and the quantity delivered amounts to 61 tons; as a result, at the beginning of the next time period, there are 126 tons available for the first customer. Considering that a further 69 tons are delivered in the second time period, the total availability of 195 tons is enough to satisfy the consumption rate of the first customer in the second and third time periods (note that no deliveries are scheduled in the third time period) and, in addition, the availability of 65 tons of product is ensured at the beginning of the fourth planning time period ($H + 1$).

Regarding the second customer, the consumed quantity in the first period is 35 tons, with an initial inventory level equal to 70 tons and a delivered quantity of 35 tons. Thus, at the beginning of the second period, the customer has the same inventory level as in the previous period, 70 tons, which is sufficient to handle its consumption in the subsequent time periods in which she receives no delivery.

Regarding the third customer, the supply policy, selected by the model, provides for only one delivery in the second time period. In fact, in the first time period, this customer has an initial inventory level equal to 58 tons, a consumption of about 58 tons and no deliveries. This way, at the beginning of the second time period, the inventory level is zero but a delivery of 116 tons (equal to the storage capacity of the customer) is planned. The delivered quantity can satisfy the product consumption of the subsequent time periods of the planning horizon.

The supply policy of the fourth customer is equal to that of the second customer. The initial inventory level at the beginning of the first time period is equal to 48 tons, and the product consumption is equal to 24 tons. In this situation, the model plans a delivery of 24 tons in the first time period, in order to have the same inventory level at the beginning of the second time period, enough to satisfy the customer's consumption until the end of the planning horizon.

The supplier starts from an initial inventory level equal to 510 tons and a production equal to 193 tons. In the first time period, he has to supply customers 1, 2 and 4 with a total quantity of 120 tons, so that his inventory level, at the beginning of the second period, is 583 tons. In the second time period, he supplies customers 1 and 3 by using two distinct routes, delivering a total quantity of 185 tons. Therefore, considering a constant production of 193 tons, his inventory level, at the beginning of the third time period, is equal to 591 tons, increasing to 193 tons in the third time period, thus guaranteeing an inventory level of 784 tons at the beginning of the fourth time period.

The storage cost in the planning horizon, owing to the deliveries to the customers and to the storage cost at the supplier, is equal to € 966.01.

The transport cost corresponds to the routes activated in the first two time periods of T, and is equal to € 1 127.

In order to impose an inventory level equal to zero in the fourth period for customer ($I_1^4 = 0$), the minimum quantity, sufficient for supplying the first customer, is equal to 4 tons. This policy would have yielded a total inventory cost of € 972.73. The route activated in the first would have been $\{0, 1, 0\}$, while in the second time period the routes would have been $\{0, 1, 2, 4, 0\}$ and $\{0, 3, 0\}$, with a transport cost of € 1 127.

These routes are the same as in the previous solution but are scheduled in different time periods. The total storage cost of this second solution would be greater than that of the policy selected by the model.

6.12 Case study: Air network design at Intexpress

Intexpress is a firm whose core business is express freight delivery all over North America. In order to guarantee a timely delivery to all destinations, an important part of the transport activities is performed by plane. The main services offered to the customers are of three types: (a) delivery within 24 hours; (b) delivery within 48 hours and (c) delivery within three to five days.

The Intexpress logistics system is made of *shipment centres* (SCs in the sequel of this section), a single hub, a fleet of airplanes and a fleet of trucks. The flight transport services are provided either by a company-owned fleet of cargo airplanes or by commercial airplanes, whereas ground transport is guaranteed by a fleet of trucks and vans. Loads are consolidated both in the SCs and in the hub. In particular, goods originating from the same SC are transported as a single load to the hub, whereas all the goods assigned to the same SC are sent jointly from the hub.

An initial SC and a final SC are associated to each transport service request from an origin to a destination. Freight is first transported to the origin, where it is consolidated; it is then transported to the final SC by air, by truck (*ground service*) or by using a combination of the two modes. Finally, freight is moved from the final SC to the destination by truck. Of course an SC-to-SC transfer by truck is feasible only if the distance does not exceed a given threshold.

The outgoing freight is picked up in the late afternoon from the origin and delivered in the morning of the following day. Therefore, a company-owned aircraft performs three operations every day: it leaves the hub, makes a set of deliveries and then travels empty from the last delivery point to the first pickup point, where it makes a set of pickups and finally goes back to the hub. All arrivals at the hub must take place before a pre-established arrival time (*cut-off time*, or COT), in such a way that the loads arriving in incoming airplanes can be unloaded, sorted by destination and quickly reloaded on the outgoing aircraft. For the same reasons, each SC is characterized by an 'earliest pickup time' and by a 'latest delivery time'. Since it is not economically desirable or technically feasible for airplanes to visit all SCs, a subset of SCs must be selected as aircraft loading and unloading points (AS, *air-stop*). An SC that is not an AS is connected to an AS by truck. Figure 6.51 depicts a possible freight route between an origin-destination pair.

Commercial air services, whose cost depends on the freight quantity and which have a low reliability, are used when at least one of the following conditions occurs:

- the distance between the origin of the route and the closest SC is so large that it is not possible to perform a quick ground connection;
- the company's owned airplanes have an insufficient capacity to satisfy its entire air transport demand;
- it is not possible to provide an adequate service among the origin-destination pairs by using the company-owned trucks.

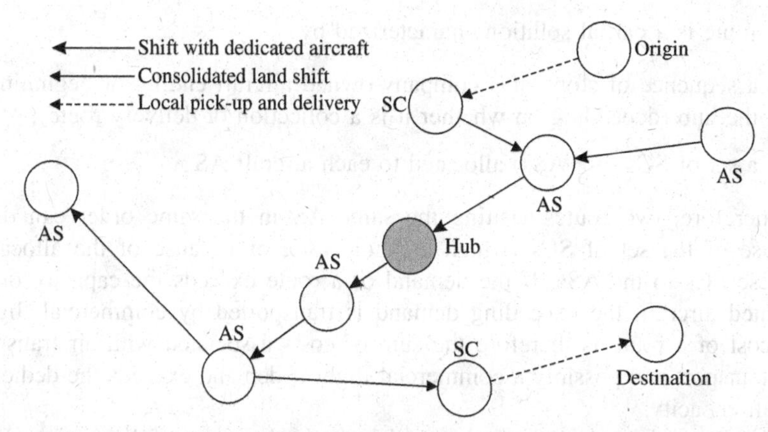

Figure 6.51 A possible freight route between an origin–destination pair at Intexpress.

Planning the Intexpress service network consists of determining:

- the set of ASs served by each aircraft of the company;
- truck routes linking SCs (which are not ASs) to ASs;
- the transport tasks performed by commercial airlines.

The objective pursued is the minimization of the operational cost subject to 'earliest pickup time' and 'latest delivery time' constraints at SCs, to the COT restriction and so on.

The solution methodology used by Intexpress is made up of two stages: in the first stage, the size of the problem is reduced, according to a qualitative analysis; in the second stage, the reduced problem is formulated and solved as an IP model. In the first stage,

- origin-destination pairs that can be serviced by truck (in such a way that all operational constraints are satisfied) are allocated to this mode and are not considered afterwards;
- origin-destination pairs that cannot be served feasibly by a dedicated aircraft or by truck are assigned to the commercial flights;
- low-priority services (deliveries within 48 hours or within three to five days) are made by truck, or by using the residual capacity of the company-owned aircraft;
- the demands of origin and destination sites are concentrated in the associated SC.

A route is a partial solution characterized by

- a sequence of stops of a company-owned aircraft ending or beginning at the hub (depending on whether it is a collection or delivery route);
- a set of SCs (not ASs) allocated to each aircraft AS.

Therefore, two routes visiting the same AS in the same order can differ because of the set of SCs (which are not ASs), or because of the allocation of these SCs to the ASs. If the demand of a route exceeds the capacity of the allocated aircraft, the exceeding demand is transported by commercial flights. The cost of a route is therefore the sum of costs associated with air transport, land transport and possibly a commercial flight if demand exceeds the dedicated aircraft capacity.

The IP model solved in the second stage is defined as follows. Let K be the set of available airplane types; U^k, $k \in K$, the set of the pickup routes that can be assigned to an airplane of type k; V^k, $k \in K$, the set of the delivery routes that can be assigned to an airplane of type k; R the set of all routes ($R = \bigcup_{k \in K} (U^k \cup V^k)$); n^k, $k \in K$, the number of company-owned airplanes k; S the set of SCs; o_i, $i \in S$, the demand originating at the ith SC; d_i, $i \in S$, the demand whose destination is the ith SC; c_r, $r \in R$, the cost of route r; q_i, $i \in S$, the cost paid if the whole demand o_i is transported by a commercial flight; s_i, $i \in S$, the cost paid if the whole demand d_i is transported by a commercial flight; α_i^r, $i \in S$, $r \in R$, a binary constant equal to 1 if route r includes picking up traffic at the ith SC, 0 otherwise; δ_i^r, $i \in S$, $r \in R$, a binary constant equal to 1 if route r includes delivering traffic to the ith SC, 0 otherwise, and γ_i^r, $i \in S$, $r \in R$, a binary constant equal to 1 if the first (last) AS of pickup (delivery) route r is the ith SC, 0 otherwise. The binary decision variables are: v_i, $i \in S$, equal to 1 if demand o_i is transported by commercial flight, 0 otherwise; w_i, $i \in S$, equal to 1 if demand d_i is transported by commercial flight, 0 otherwise, and x_r, $r \in R$, equal to 1 if (pickup or delivery) route r is selected.

The IP model is

$$\text{Minimize} \sum_{r \in R} c_r x_r + \sum_{i \in S} (q_i v_i + s_i w_i) \tag{6.111}$$

subject to

$$\sum_{k \in K} \sum_{r \in U^k} x_r \alpha_i^r + v_i = 1, \ i \in S \tag{6.112}$$

$$\sum_{k \in K} \sum_{r \in V^k} x_r \delta_i^r + w_i = 1, \ i \in S \tag{6.113}$$

$$\sum_{r \in U^k} x_r \gamma_i^r - \sum_{r \in V^k} x_r \gamma_i^r = 0, \ i \in S, \ k \in K \tag{6.114}$$

$$\sum_{r \in U^k} x_r \leq n^k, \; k \in K \tag{6.115}$$

$$x_r \in F, \; r \in R \tag{6.116}$$

$$x_r \in \{0, 1\}, \; r \in R \tag{6.117}$$

$$v_i \in \{0, 1\}, \; i \in S \tag{6.118}$$

$$w_i \in \{0, 1\}, \; i \in S. \tag{6.119}$$

The objective function (6.111) is the total transport and handling cost. Constraints (6.112) and (6.113) state that each SC is served by a dedicated route or by a commercial route; constraints (6.114) guarantee that if a delivery route of type $k \in K$ ends in SC i, $i \in S$, then there is a pickup route of the same kind beginning in i. Constraints (6.115) set upper bounds on the number of routes which can be selected for each dedicated aircraft type. Constraints (6.116) express the following further restrictions. The arrivals of the airplanes at the hub must be staggered in the period before the COT because of the available personnel and of the runway capacity.

Similarly, departures from the hub must be scheduled in order to avoid congestion on the runways. Let n_a be the number of time intervals in which the arrivals should be allocated; n_p the number of time intervals in which the departures should be allocated; f_r, $r \in R$, the demand along route r; a_t the maximum demand which can arrive to the hub in interval $t, \ldots, n_a$; TA_t the set of routes with arrival time from t on; TP_t the set of routes with departure time before t and p_t the maximum number of airplanes able to leave before t. Hence, constraints (6.116) are

$$\sum_{k \in K} \sum_{r \in U^k \cap TA_t} f_r x_r \leq a_t, \; t = 1, \ldots, n_a \tag{6.120}$$

$$\sum_{k \in K} \sum_{r \in V^k \cap TP_t} x_r \leq p_t, \; t = 1, \ldots, n_p. \tag{6.121}$$

Constraints (6.120) ensure that the total demand arriving cannot exceed the hub capacity in the time intervals from t until n_a, while constraints (6.121) impose that the total number of airplanes leaving the hub is less than or equal to the maximum number allowed by runaway capacity in each time interval t, $t = 1, \ldots, n_p$.

Other constraints may be imposed. For instance, if the goods are stored in containers one must ensure that once a container becomes empty it is brought back to the originating SC. To this end, the aircraft arriving at an SC and the aircraft leaving it must be compatible. In the Intexpress problem, there are four types of aircraft, indicated by 1, 2, 3 and 4. Aircraft of type 1 are compatible with type 1 or 2, while aircraft of type 2 are compatible with those of type 1, 2 and 3. Therefore, the following constraints hold:

$$-\sum_{r\in U^1} x_r\alpha_i^r + \sum_{r\in V^1\cup V^2} x_r\delta_i^r \geq 0, \ i \in S \tag{6.122}$$

$$\sum_{r\in U^1\cup U^2} x_r\alpha_i^r - \sum_{r\in V^1} x_r\delta_i^r \geq 0, \ i \in S \tag{6.123}$$

$$-\sum_{r\in U^2} x_r\alpha_i^r + \sum_{r\in V^1\cup V^2\cup V^3} x_r\delta_i^r \geq 0, \ i \in S \tag{6.124}$$

$$\sum_{r\in U^1\cup U^2\cup U^3} x_r\alpha_i^r - \sum_{r\in V^2} x_r\delta_i^r \geq 0, \ i \in S. \tag{6.125}$$

Moreover, some airplanes cannot land at certain SCs because of noise restrictions or insufficient runaway length. In such cases, the previous model can easily be adapted by removing the routes $r \in R$ including a stop at an incompatible SC.

The variables in the model (6.111)–(6.115), (6.117)–(6.125) are numerous even if the problem is of small size. For example, in the case of four ASs (a, b, c and d), there are 24 pickup routes ($abcd$, $acbd$, $adbc$ etc.), each of which has a different cost and arrival time at the hub. If, in addition, two SCs (e and f) are connected by truck to one of the ASs a, b, c and d, then the number of possible routes becomes $16 \times 24 = 384$ (as a matter of fact, for each AS sequence, each of the two SCs e and f can be connected independently by land to a, b, c or d). Finally, for each delivery route making its last stop at an AS d, one must consider the route making its last stop at a different SC $g \in S \setminus \{d\}$. Of course, some of the routes can be infeasible and are not considered in the model (in the case under consideration, the number of feasible routes is about 800 000).

The solution methodology is a classical branch-and-bound algorithm in which at each branching node a continuous relaxation of (6.111)–(6.115), (6.117)–(6.125) is solved. The main disadvantage of this algorithm is the large number of decision variables. Since the number of constraints is much less than the number of decision variables, only a few decision variables take a nonzero value in the optimal basic solution of the continuous relaxation. For this reason, the following modification of the method is introduced. At each iteration, in place of the continuous relaxation of (6.111)–(6.115), (6.117)–(6.125), a reduced LP problem is solved (in which there are just 45 000 'good' decision variables, chosen by means of a heuristic criterion); then, using the dual solution of the problem built in this manner, the procedure determines some or all of the decision variables with negative reduced costs, introducing the corresponding columns in the reduced problem. Various additional devices are also used to shorten the execution of the algorithm. For example, in the preliminary stages only routes with an utilization factor between 30% and 185% are considered. This criterion rests on two observations: (a) because of the reduced number of company aircraft, it is unlikely that an optimal solution will contain a route with a used capacity less than 30%, and (b) the cost structure of the air transport makes it unlikely that along a route more than 85% of the traffic is transported by commercial airlines.

The above method was used to first generate the optimal service network using the current dedicated air fleet. The cost reduction obtained was more than 7%, corresponding to a yearly saving of several million dollars. Afterwards, the procedure was used to define the optimal composition of the company's owned fleet. For this purpose, in formulation (6.111)–(6.115), (6.117)–(6.125), it was assumed that n^k was infinite for each $k \in K$. The associated solution shows that five aircraft of type 1, three of type 2 and five of type 3 should be used. This solution yielded a 35% saving (about \$ 10 million) with respect to the current solution.

6.13 Case study: Meter reader routing and scheduling at Socal

Southern California Gas Company (Socal) distributes, with its own pipe network, domestic and industrial gas in an area including all of southern California. The task of surveying user consumption is accomplished by motorized operators. Every day a meter reader makes an initial car trip from the company offices up to a parking point; afterward, she walks along several streets (or segments of streets) where she makes some surveying; finally, she reaches the parking point from which she returns to the company by car. According to the working contract, if the duration of a shift exceeds a pre-established maximum value T^{max}, a meter reader receives an additional remuneration, proportional to the overtime. For reasons of equity, Socal prefers solutions in which all service routes have a similar duration. The planning of the meter-reading activities then consists of partitioning the set of street segments into subsets yielding service routes with a duration close to T^{max}.

In 1988 Socal mandated a consulting firm, DMC, to evaluate the costs and the potential benefits resulting from the use of an optimization software for the planning of meter-reading activities. A pilot study was conducted in a sample region including the townships of Culver City, Century City, Westwood, West Hollywood and Beverly Hills. This area corresponds to about 2.5% of the entire service territory and to 242.5 working days per month. The interest of Socal in this project was prompted by the sizeable expense (about \$ 15 million per year) resulting from meter-reading activities.

DMC built a model of the problem integrating the data contained in the Socal archives with those derived from a GIS. In the case of streets having users on both sides, it was assumed that the meter reader would cover both sides separately. Therefore, the problem was represented by a multigraph $G = (V, E)$ in which the road intersections are described by vertices in V and street segments are associated to single edges or to pair of parallel edges in E. The street sides containing some consumers formed a subset $R \subseteq E$ (see Figure 6.52). With each

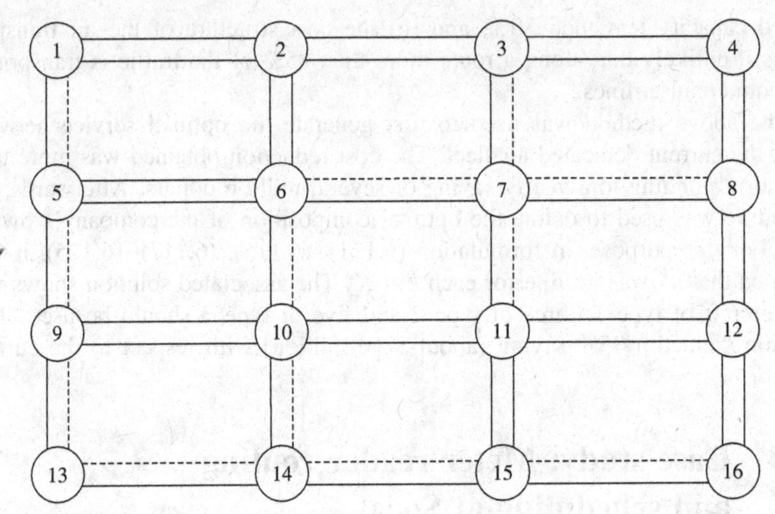

Figure 6.52 An example of multigraph used for modelling Socal problem (bold edges are required, while those with a dotted line are not required).

edge $(i, j) \in E$ was associated with a traversing time t_{ij}, and with each edge $(i, j) \in R$, was associated with a service time s_{ij} $(> t_{ij})$.

The duration $t(R')$ of a work shift corresponding to a subset of the service edges $R' \subseteq R$ has a cost equal to the sum of three terms: (a) the time spent to deal with administrative issues at the company headquarter (generally 15 minutes), (b) the travel time by car from the headquarter to a parking site and (c) the duration $RPP(R')$ of the solution of the RPP (see Section 6.8.4.2) associated with the service on foot of all edges of R'. Socal prefers work shifts associated with subsets $R' \subseteq R$ whose duration is included between $T^{min} = 7.9$ hours and $T^{max} = 8$ hours (balanced work shifts).

A partition $\{R_1, \ldots, R_n\}$ of R is feasible if each pair of parallel service arcs belongs to the same subset R_i, $i = 1, \ldots, n$ (feasible partition). The search for a feasible balanced (or at least 'almost' balanced) partition of R can be viewed as the minimization of a penalty associated to a violation of the desired length $[T^{min}, T^{max}]$ of each work shift.

Denoting by U and L the penalties associated to the two types of violation, the objective function to minimize is

$$U \sum_{i=1}^{n} \max\left\{t(R_i) - T^{max}, 0\right\} + L \sum_{i=1}^{n} \max\left\{T^{min} - t(R_i), 0\right\}.$$

DMC solved the problem by means of the following procedure:

Step 0. The workloads associated with $RPP(R')$, $R' \subseteq R$, are approximated as $W(R') = \sum_{(i,j) \in R'} t_{ij}$. This method is justified by the presence of consumers on both sides of most of the street segments. An evaluation of the average useful time $\bar{T}$ of a route (the time a meter reader will take on average to travel on foot) was calculated as the difference between T^{max} and the average time required for the administrative tasks and for driving.

Step 1. The number n of shifts was evaluated by rounding down the ratio between the total workload $W(R')$ of the service area and $\bar{T}$.

Step 2. The load of remaining workload $(W(R') - nT^{max})$ was assigned to an 'incomplete' auxiliary route, built with a heuristic criterion along the border of the service area.

Step 3. The n 'complete' shifts were defined by means of a local search procedure. The algorithm first selected n seed vertices appropriately spaced out and then progressively generated the routes by adding to each of them couples of parallel edges in order to constantly have balanced work shifts. The solution obtained in this manner was improved, where possible, with some exchanges of couples of edges (or of groups of couples of edges) among the work shifts.

The solution generated by this procedure for the sample district was characterized, compared to the one obtained manually by Socal, by a reduced number of meter readers, by a remarkable reduction of overtime and by an increase in the average useful time of an operator (see Table 6.43). Extrapolating the savings associated with these improvements, DMC estimated at about \$ 870 000 the saving expected after the use of the optimization method for the entire area served by Socal. The previous analysis convinced the Socal management to order from DMC a decision support system with a user-friendly interface, based on the optimization method just described.

Table 6.43 Comparison between the results provided by DMC procedure and the manual solution used by Socal.

	Socal solution	DMC solution	Savings (%)
Number of routes	242.5	236	2.7
Average shift duration (hours)	7.82	7.95	
Shift interval duration (in hours)	[7.03; 8.56]	[7.77; 8.11]	
Overtime (minutes)	293.4	110.4	62.3
Total deadheading time (without service)	693.4	63.1	90.9
Total travel time by car	545.5	133.2	75.6

6.14 Case study: Dynamic vehicle-dispatching problem with pickups and deliveries at eCourier

eCourier is a same-day courier company, operating in the Greater London area. It performs pickups and deliveries of parcels (documents, packages, pallets etc.) by means of a fleet of couriers, who use bicycles, motorcycles, cars or vans, depending on the commodity to be transported, as well as on the pickup and delivery locations. Each vehicle type has its own features, such as its speed under diverse traffic conditions, its capacity, or the maximum distance it can cover. Customers ask for service by calling the call center or by booking via the website. Each customer specifies the pickup and delivery locations, as well as the time windows during which he wants the service to be performed.

The traditional model of same-day courier service utilizes human controllers who assign the jobs to the couriers. In this model, because the number of requests can reach several hundreds or thousands, informational complexity grows to levels which push the limits of human analysis. In fact, when evaluating a job, a controller must take into account a number of real-time features, among which the parcel type, the vehicle type, the time windows, the travel times, the current locations of the vehicles, possible pending jobs, traffic and weather conditions. Moreover, it can be convenient to reposition idle couriers in high-density zones, to anticipate future demand. Of course, this task is not easy to perform for the controllers. Thus, managing all these aspects creates higher operational costs and lower quality of service (a higher customer inconvenience).

In order to keep the costs as low as possible, the company has decided to develop a decision support system to help the controllers automate the allocation process, thus involving them only as supervisors.

The solution implemented is based on the integration of a number of technologies: GPS and GPRS, with each courier being equipped with a GPS device embedded into a GPRS palmtop computer; GIS for the navigation system and for tracking couriers' positions; optimization techniques to route the vehicles; and forecasting techniques. The palmtop computers are also used to provide directions to couriers. Travel time forecasting is performed by a neural network that takes into account the vehicle type, as well as weather and real-time traffic data. This information is then used by the allocation system, which assigns each job to the most appropriate courier on the basis of current fleet status, time windows, possible service level agreements, road congestion as well as weather conditions.

A route is built for each courier according to the jobs allocated, satisfying all the constraints and minimizing the distance covered by the vehicle.

More specifically, at the tactical level, decisions concern shift scheduling, and the number of couriers to be allocated to each shift pattern, subject to constraints on the quality of service to be provided to customers. This problem is usually solved on a weekly or quarterly basis (the demand is usually characterized by significant yearly, weekly and daily seasonal variations) with the aim of minimizing the staffing cost. On the other hand, at the operational level, a dynamic vehicle-dispatching problem with pickups and deliveries must be solved in order to allocate each request to a vehicle, to schedule the requests assigned to each vehicle and to reposition idle vehicles.

For the sake of simplicity, we will focus on the operational problem. This problem is defined on a graph $G = (V, A)$, where V is a vertex set and A is an arc set. A fleet of m vehicles, located at a depot $i_0 \in V$ at time $t = 0$, has to service a number of pickup and delivery requests $\left\{(i_k^+; i_k^-; T_k) : k \in K\right\}$, where $i_k^+ \in V$, $i_k^- \in V$, $T_k \geq 0$ are the pickup point, the delivery point and the occurrence time of the kth request, $k \in K$, respectively. Vertices may represent individual customer locations or the zones in which the service territory is divided. Let t_{ij} be the shortest travel time from vertex $i \in V$ to vertex $j \in V$. The aim is to maximize the overall customer service level, rather than to minimize the total covered distance. Let τ_k be the delivery time of the kth request. With each customer is associated a penalty function $f_k(\tau_k)$. This definition includes the case for which $f_k(\tau_k)$ represents the customer *system time* (i.e. $f_k(\tau_k) = \tau_k - T_k,\ \tau_k \geq T_k$) or a more involved penalty function (e.g. $f_k(\tau_k) = 0,\ T_k \leq \tau_k \leq D_k$ and $f_k(\tau_k) = \tau_k - D_k,\ \tau_k \geq D_k$, where D_k is a *soft deadline* associated with the kth request).

The static version of the problem amounts to determining an ordered sequence of locations on each vehicle route such that each route starts at the depot; a pickup and its associated delivery are satisfied by the same vehicle; a pickup is always made before its associated delivery; and the total penalty incurred by the vehicles $z = \sum_{k \in K} f_k(\tau_k)$ is minimized.

In the dynamic variant, we also have to adequately distribute the waiting time along the routes, since this may affect the overall solution quality, as well as to reposition idle vehicles to anticipate future demand. The objective function to be minimized is the expected customer inconvenience over the planning horizon: $z = \sum_{k \in K} E[f_k(\tau_k)]$, where $E\left[f_k(\tau_k)\right]$ is the expected penalty associated with the delivery of the kth request, $k \in K$. Moreover, we assume that a vehicle cannot be diverted away from its current destination to service a new request in the vicinity of its current position.

The following anticipatory mechanism embedded in both an insertion and a local search procedure has been devised. Let P_k be the set of pending requests (i.e. the requests that have occurred but have not been serviced) at time T_k, when request (i_k^+, i_k^-, T_k) arrives. A reactive algorithm generates a new solution incorporating i_k^+ and i_k^- with the aim of minimizing the total inconvenience associated with the pending requests, that is $z_k = \sum_{r \in P_k} f_r(\tau_r)$.

The anticipatory algorithms aim at minimizing z_k plus the expected value (under perfect information) of the total penalty $\xi^{[t_k, t_k + \Delta t_k]}$ associated with the requests arriving in the short-term future $[t_k, t_k + \Delta t_k]$, that is $z'_k = \sum_{r \in P_k} f_r(\tau_r) +$

$E[\xi^{[t_k, t_k+\Delta t_k]}]$, where Δt_k is the short-term duration. Of course, the procedures become reactive if $\Delta t_k = 0$, $k \in K$.

The proposed procedures allowed eCourier to gain significant improvements in terms of lower operational costs. In fact, at the tactical level, the company was able to reduce its costs by approximately 10%. At the operational level, the anticipative procedures allowed the company to improve the quality of service provided to customers by about 60%, and to better distribute requests among the fleet.

6.15 Questions and problems

6.1 JKL is a national carrier operating in Australia. The transport rates for class 25 LTL palletized freight from Melbourne to Darwin are reported in Table 6.44. Determine the break weight formula (the break weight is the weight over which it is convenient to apply the rate of the subsequent range to reduce the overall transport cost). Compute the break weight for each weight range of Table 6.44.

Table 6.44 JKL transport rates for class 25 LTL palletized freight from Melbourne to Darwin.

Weight range (W, kg)	Rate (A$/kg)
$0 \leq W < 0.1$	137.13
$0.1 \leq W < 0.2$	112.46
$0.2 \leq W < 0.5$	98.23
$0.5 \leq W < 1$	72.38
$1 \leq W < 10$	60.69
$10 \leq W < 20$	54.21
$20 \leq W < 50$	47.71
$50 \leq W < 100$	46.14
$100 \leq W < 250$	41.23
$W \geq 250$	39.77

6.2 Camberra Freight is in charge of transporting auto parts for an US car manufacturer in Australia. Every week a tractor and one or two trailers move from the port of Melbourne to a warehouse located 430 km away. A tractor costs A$ 75 per day, a trailer A$ 30 and a driver A$ 7.5/h, while running costs are A$ 0.75/km. A trailer can contain 36 pallets. Derive the transport cost per pallet as a function of shipment size for the case where one or two trailers are used.

6.3 TL trucking rates from Boston, Massachusetts, to Miami, Florida (both in the United States) are usually higher than those from Miami to Boston. Why?

6.4 Freight transport costs of Class 55 are cheaper than those of Class 70 (see Table 6.1). Why?

6.5 In international freight transport, a key role is played by the free-trade zones. In such areas, freight may be entered without the intervention of the customs authorities and customs duties are due only when the goods are moved outside. Get more more information on the free-trade zones through the Internet.

6.6 Formulate the minimum-cost flow problem for the Swiss NTN (see Section 6.2.2) intermodal carrier by knowing that the demand of containers in Milan and Madrid is 70 and 25, respectively. What is the optimal solution?

6.7 Rinaldi is an Italian fast carrier located in Parma, whose core business is the transport of small-sized and high-valued refrigerated goods (such as chemical reagents used by hospitals and laboratories). Goods are picked up from manufacturers' warehouses by small vans and carried to the nearest transport terminal operated by the carrier. These goods are packed onto pallets and transported to destination terminals by means of large trucks. The merchandise is then unloaded and delivered to customers by small vans (usually the same vans employed for pickup). In order to make capital investment in equipment as low as possible, Rinaldi makes use of one-way rentals of trucks. Recently, the company has decided to enter the fast parcel transport market by opening four terminals in the cities of Bologna, Genoa, Padua and Milan. This choice made necessary a complete revision of the service network. The decision was complicated by the need to transport the refrigerated goods by special vehicles equipped with refrigerators, while parcels can be transported by any vehicle. The forecasted daily average demand of the two kinds of products in the next semester is reported in Tables 6.45 and 6.46.

Table 6.45 Forecasted transport demand of refrigerated goods (pallets per day) in the Rinaldi problem.

	Bologna	Genoa	Milan	Padua
Bologna	–	3	8	2
Genoa	–	–	1	2
Milan	4	2	–	1
Padua	3	1	1	–

Between each pair of terminals, the company can operate one or more lines. Vehicles are of two types:

- trucks with refrigerated compartments, having a capacity of 12 pallets and a cost (inclusive of all charges) of € 0.4/km;

- trucks with room temperature compartments, having a capacity of 18 pallets and a cost (inclusive of all charges) of € 0.5/km.

Table 6.46 Forecasted transport demand of goods at room temperature (pallets per day) in the Rinaldi problem.

	Bologna	Genoa	Milan	Padua
Bologna	–	3	4	2
Genoa	1	–	1	–
Milan	6	2	–	2
Padua	1	1	1	–

In addition, the company considers the possibility of transporting goods at room temperature through another carrier, by paying € 0.1/km for each pallet. A directed graph representation of the problem is given in Figure 6.53. Distances between terminals are reported in Table 6.47. Formulate the LFCND problem of finding the least-cost service network (hint: $|K| = 22$ commodities, one for each combination of an origin-destination pair with positive demand and a kind of product). Apply the drop heuristic to find a feasible solution of the problem. By using a solver, determine the optimal solution of the problem and the costs corresponding to the weak and the strong continuous relaxations. What is the quality of the feasible solution provided by the drop heuristic and the two lower bounds?

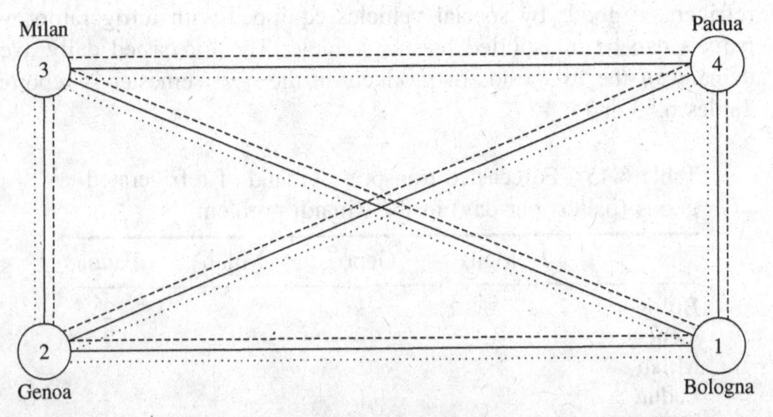

Figure 6.53 Multigraph representation of the service network design of the Rinaldi problem. In order to simplify the multigraph, a single edge for each pair of opposite arcs is represented. The dotted edges represent the connections served by truck lines of 12-pallet capacity, dashed edges represent the connections served by truck lines of 18-pallet capacity and solid lines represent connections served by the external service.

Table 6.47 Distances (in km) between terminals in the Rinaldi problem.

	Bologna	Genoa	Milan	Padua
Bologna	0	225	115	292
Genoa	225	0	226	166
Milan	115	226	0	362
Padua	292	166	362	0

6.8 Devise a local search heuristic for the service network design problem in which an individual move is to remove an existing arc or to add a new arc to the current solution.

6.9 Show that the deterministic VAP with multiple vehicle types can be modelled as an LMMCF problem on a time-expanded directed graph.

6.10 Examine how the optimal solution of the Murty problem changes whenever:
- on July 11, there is an empty vehicle in Skrikakulam;
- $d_{533} = 1$, $d_{531} = 0$;
- $d_{533} = 1$, $d_{531} = 0$, and $m_{31} = 1$.

6.11 Show that, in the DDAP, costs c_{i0}, $i \in D$, can be set equal, equivalently, to ∞.

6.12 The travel agency Gare, located in Lausanne, organizes tours by bus in several European capitals. In April there are still seven different tours to organize whose dates of departure and duration are reported in Table 6.48. The agency has to assign a driver to each tour on the basis of the driver's experience. To this end, a different value is assigned to each driver-tour pair, as reported in Table 6.49: value 3 means that driver usually makes that trip, value 1 means that driver has made that trip in the past and value 0 means that driver never made that trip in the past.

Table 6.48 Date of departure and duration (in days) of the bus tours organized by the travel agency Gare.

	City						
Tour	Madrid	Paris	Rome	London	Vienna	Budapest	Berlin
Date of departure	2	8	7	9	9	12	13
Trip duration	7	4	4	4	5	4	3

6.13 In the Fast Courier transport problem (see Section 6.6), assume that the unit cost c_H per time period of hiring a vehicle changes on the basis of the number of hired vehicles (a 10% discount is applied for any additional

Table 6.49 Ranking driver-tour value of the travel agency Gare problem.

	City						
Driver	Madrid	Paris	Rome	London	Vienna	Budapest	Berlin
1	3	3	0	0	1	3	1
2	1	3	1	0	1	3	1
3	1	3	0	0	1	3	1
4	3	1	1	0	1	3	1
5	1	0	3	0	1	3	1
6	1	0	0	3	1	3	1

hired vehicle, up to a total discount of $ 300). Determine how the original solution changes.

6.14 In which case is the problem (6.53)–(6.58) infeasible?

6.15 Estimate the travel time on a 8.5 km segment of the Dungannon road, between Cookston and Dungannon in Ireland, on the basis of the experimental measures obtained by using a small van and reported in Table 6.50.

6.16 You have an algorithm capable of solving the capacitated NRP with no fixed vehicle costs, and you would like to solve a problem where a fixed cost is attached to each vehicle. Show how such a problem can be solved using the algorithm at hand.

6.17 Show that, if there are no operational constraints, there always exists an optimal NRP solution in which a single vehicle is used (hint: least-cost path costs satisfy the triangle inequality).

6.18 Show that if the costs associated to the arcs of a complete directed graph G satisfy the triangle inequality property, then there exists in G' an ATSP optimal solution which is a Hamiltonian tour.

6.19 Show that, in the ATSP formulation, the connectivity constraints (6.61) and the subtour elimination constraints (6.62) are equivalent.

6.20 Show that the number of vertices of odd degree in a graph is even.

6.21 Let $G = (V, A)$ be a complete directed graph such that $|V| = 10$ and the costs associated to the arcs belong to A, as reported in Table 6.51:

- check if the triangle inequality property holds for each arc cost;
- formulate the corresponding ATSP;
- solve the relaxed AP;
- determine a feasible ATSP solution and evaluate its quality;
- apply a 2-exchange procedure for finding a better feasible solution.

Table 6.50 Travel time measures (in minutes) on the Dungannon road (traffic volume: 1 = low, 2 = medium, 3 = high; weather conditions: 1 = dry, 2 = light rain, 3 = heavy rain; time slot: 1 = 10:00 PM–6:00 AM, 2 = 6:00 AM–9:00 AM, 3 = 9:00 AM–3:00 PM, 4 = 3:00 PM–10:00 PM).

Traffic volume	Weather conditions	Time slot	Travel time	Traffic volume	Weather conditions	Time slot	Travel time
1	1	1	5.85	2	2	3	6.23
1	1	1	6.23	2	2	3	6.42
1	1	2	6.25	2	2	4	6.61
1	1	2	4.50	2	2	4	6.85
1	1	3	6.18	2	3	1	7.59
1	1	3	5.95	2	3	1	6.74
1	1	4	5.84	2	3	2	6.94
1	1	4	5.84	2	3	2	6.11
1	2	1	5.41	2	3	3	7.80
1	2	1	5.93	2	3	3	7.75
1	2	2	6.06	2	3	4	6.32
1	2	2	4.96	2	3	4	6.85
1	2	3	6.78	2	1	1	7.12
1	2	3	6.80	2	1	1	6.96
1	2	4	6.52	2	1	2	7.55
1	2	4	6.05	2	1	2	7.20
1	3	1	6.04	2	1	3	6.77
1	3	1	7.03	2	1	3	7.88
1	3	2	6.33	2	1	4	6.08
1	3	2	6.82	2	1	4	6.85
1	3	3	6.38	2	2	1	7.91
1	3	3	7.33	2	2	1	7.45
1	3	4	6.70	2	2	2	7.95
1	3	4	7.02	2	2	2	6.56
2	1	1	6.20	2	2	3	7.21
2	1	1	6.07	2	2	3	7.69
2	1	2	5.49	2	2	4	7.37
2	1	2	5.28	2	2	4	7.82
2	1	3	6.96	2	3	1	7.32
2	1	3	5.87	2	3	1	8.67
2	1	4	5.81	2	3	2	8.64
2	1	4	6.55	2	3	2	8.35
2	2	1	5.76	2	3	3	7.80
2	2	1	5.17	2	3	3	8.03
2	2	2	6.58	2	3	4	8.66
2	2	2	5.50	2	3	4	9.07

Table 6.51 Costs associated to the arcs of graph G' of Problem 6.21.

	0	1	2	3	4	5	6	7	8	9
0	–	28.55	12.10	43.15	66.10	38.35	79.70	17.00	77.60	55.20
1	15.80	–	23.10	28.85	52.55	53.65	66.90	4.15	71.70	61.90
2	54.95	48.90	–	52.95	88.55	54.75	77.80	42.55	91.45	76.40
3	39.80	27.95	37.85	–	79.00	41.40	79.05	19.15	44.95	35.35
4	49.80	63.95	46.15	43.50	–	71.30	41.60	35.55	67.30	78.60
5	46.50	55.70	28.10	72.30	93.65	–	55.30	54.20	93.70	94.40
6	99.55	84.25	86.15	60.40	63.70	99.85	–	72.30	88.35	81.90
7	31.00	28.80	21.35	34.55	69.55	67.65	67.95	–	77.40	68.30
8	60.35	75.35	46.85	68.15	39.05	75.05	48.90	52.60	–	80.35
9	66.00	61.10	60.75	83.50	87.30	92.50	86.40	62.00	64.00	–

6.22 Demonstrate that the optimal solution value of MSrTP is a lower bound on the optimal solution value of STSP.

6.23 Show that an optimal solution of the least-cost perfect-matching problem (6.78)–(6.79), (6.81) can be obtained by solving the following LP problem:

$$\text{Minimize} \sum_{(i,j)\in E'_M} c_{ij}x_{ij}$$

subject to

$$\sum_{i\in V'_M:(i,j)\in E'_M} x_{ij} + \sum_{i\in V'_M:(j,i)\in E'_M} x_{ji} = 1, \ j \in V'_M$$

$$\sum_{(i,j)\in E'_M, i\in W, j\notin W} x_{ij} + \sum_{(j,i)\in E'_M, i\in W, j\notin W} x_{ji} \geq 1, \ W \subset V'_M,$$

$$\left|V'_M\right| > 1, |V'_M| \text{ odd}$$

$$x_{ij} \geq 0, \ (i,j) \in E'_M.$$

(Hint: observe that, given any subset W of V'_M having odd cardinality, each perfect matching must contain at least one incident edge into the nodes of W.)

6.24 Demonstrate that the cost of the STSP solution produced by the Christofides heuristic is within $3/2$ of the optimum.

6.25 The GermanExpress transport company, based in Cologne, has to schedule a pickup service to five customers located in the cities of Bonn, Düsseldorf, Essen, Hennef and Koblenz, respectively. The distance (calculated with respect to the fastest route) between each pair of cities is reported in Table 6.52. The daily average amount of goods to pick up by each customer

Table 6.52 Distances (in km) between the depot and the five cities in the GermanExpress transport problem.

	Bonn	Düsseldorf	Essen	Hennef	Koblenz
Cologne	40	50	85	52	123
Bonn		85	114	30	65
Düsseldorf			45	76	134
Essen				107	167
Hennef					87
Koblenz					0

Table 6.53 Daily average amount of goods (in quintals) to pickup at the five customers of the GermanExpress.

Bonn	Düsseldorf	Essen	Hennef	Koblenz
4	12	7	5	8

is provided in Table 6.53. The transport vans have a capacity of 15 quintals. Determine the number of vehicles to be used and, for each of them, the daily service route starting from the depot located in Cologne.

6.26 Show that the NRPSC formulation is correct.

6.27 The Bioenergy is a wood pellet factory located in Austria; it supplies a chain of supermarkets located in Italy and Germany. The orders have to be met by minimizing the transport cost. The distances among the supermarkets, the distances between the supermarkets and the depot and the quantity ordered (in pallets) are reported in Table 6.54. Formulate the problem as a NRPCL assuming that:

- the average unloading time for a pallet is 1.5 minutes;
- the average speed is 90 km/h;
- the shift time is 8 h;
- two type of trucks can be used: those with 22-pallet capacity and those with 32-pallet capacity.

6.28 Explain why distances in Tables 6.25 and 6.26 do not necessarily satisfy the triangle inequality.

6.29 How do you solve the McNish problem (see Section 6.8.3.1) if the time windows of the sales point with ID = 5 and ID = 10 were 9:00 AM–10:00 AM and each vehicle had a capacity of 25 pallets?

6.30 Devise a local search for the capacitated ARP.

Table 6.54 Distances (in km) among the Bioenergy pellet factory and the supermarkets; in last row, the next-day orders (in pallets) are reported.

	Supermarket			
	1	2	3	4
Depot	140.0	250.0	185.0	92.0
Supermarket 1		12.0	34.0	24.0
Supermarket 2			14.5	19.3
Supermarket 3				17.0
Supermarket 4				0.0
Order	15	32	30	18

6.31 Illustrate how the Christofides and Frederickson heuristics can be adapted to the undirected general routing problem, which consists of determining a least-cost tour including a set of required vertices and edges.

6.32 Solve Problem 6.27 by imposing that a work shift has a duration of six hours.

6.33 Analyse the dynamism of an airport bus service.

6.34 Recall the GermanExpress transport problem (Problem 6.25) and assume that the depot in Cologne has a yearly facility cost of € 155 000. Furthermore, we assume that the transport service is realized 230 times in a year, according to the daily average requests reported in Table 6.53. The van travel cost is about € 1.15/km. The shipper has the possibility to rent a depot in Dortmund, with a yearly cost of € 130 000, from which the transport service can alternatively start. Establish which solution is preferable, in terms of yearly cost, by considering the distances between Dortmund and the five customers reported in Table 6.55.

Table 6.55 Distances (in km) between Dortmund and the five cities in the GermanExpress transport problem.

City	Distance
Bonn	122
Düsseldorf	70
Essen	39
Hennef	119
Koblenz	197

6.35 Modify the mathematical model formulation of IRP (6.72)–(6.85) to take into account the following constraints:

- the supplier's production is unbounded;
- the inventory level of the commodity at the supplier is $I_0^t = \infty$ for all $t \in T$;
- the initial inventory level of the commodity at the supplier $I_0^0 = 0$ and the quantity of commodity available in time period $t \in T$ is $p^t = \infty$.

In which case you can suppose that the commodity consumption rate μ is constant during a given planning horizon?

Index

Introduction to Logistics Systems Management, Second Edition. Gianpaolo Ghiani, Gilbert Laporte and Roberto Musmanno.